Women in Engineering and Science

The Springer Women in Engineering and Science series highlights women's accomplishments in these critical fields. The foundational volume in the series provides a broad overview of women's multi-faceted contributions to engineering over the last century. Each subsequent volume is dedicated to illuminating women's research and achievements in key, targeted areas of contemporary engineering and science endeavors.The goal for the series is to raise awareness of the pivotal work women are undertaking in areas of keen importance to our global community.

Jill S. Tietjen • Ilene J. Busch-Vishniac
Lauren Busch

Women Laureates 1970 – 2007

National Medal of Science and National Medal of Technology and Innovation Winners

Springer

Jill S. Tietjen
Technically Speaking Incorporated
Greenwood Village, CO, USA

Ilene J. Busch-Vishniac
Baltimore, MD, USA

Lauren Busch
Dallas, TX, USA

ISSN 2509-6427 ISSN 2509-6435 (electronic)
Women in Engineering and Science
ISBN 978-3-032-23115-4 ISBN 978-3-032-23116-1 (eBook)
https://doi.org/10.1007/978-3-032-23116-1

This Springer imprint is published by the registered company Springer Nature Switzerland AG
The registered company address is: Gewerbestrasse 11, 6330 Cham, Switzerland

To the amazing women scientific pioneers whose achievements have been recognized by these Medals and for the many other women who also trailblazed, demonstrating that women can do science.

Foreword

The current dialog in science is troubling. The integrity of science and scientists is being questioned and skepticism continues about women's ability to do science. It is obvious that the past one hundred years have seen change. Women in most countries are able to pursue careers, vote, and choose to follow a traditional path... or not. However, the path is narrow for women. It remains constricted, the women on the path are mostly unrecognized, and all too often women are less well rewarded than men.

I have chosen to provide a few examples of significant contributions to science and discovery made by women, and then, offer an example, from personal experience, of how, sometimes, an obstacle facing women can be overcome. Finally, I will share, from my own research, how a path taken by a woman is made more difficult by the refusal of the scientific community at large to accept her ideas, hypothesis, or innovation. These actions—or lack thereof—impede progress and contribute to higher human suffering than is necessary. It is only years later, when the woman's (or women's) idea, innovation, or contribution is finally accepted and incorporated into scientific acceptance. Even then, it is rarely acknowledged that the idea, innovation, or contribution was achieved by a woman, specifically because it was a woman who made the contribution. The scientific path for women is essentially a long road of embedded nails over which she must walk.

The story of one such woman is told by Dava Sobel who writes elegantly of Galileo's daughter. She labored by her father's side, did calculations, and was a partner in Galileo's discoveries. Because of the social mores of the time, however, she spent her life in a nunnery as that was the best, if not the only, way through which she could assist her father's research.

Whooping cough (pertussis) was elucidated by Pearl Kendrick who studied *Bordetella pertussis*, the pathogen that causes whooping cough, along with her laboratory assistants, Grace Eldering and Loney Clinton Gordon. Loney was an African American woman. These three women overcame medical skepticism to stop whooping cough. They saved lives, but because they were women, they remain essentially nameless.

Alice Evans (Fig. 1) overcame ridicule and excoriation because she linked raw milk to undulant fever (Brucellosis). In the late nineteenth century, Danish veterinarian Bernhard Bang had determined a bacterium was responsible for spontaneous abortions in cows. Scottish pathologist, David Bruce, isolated the bacterium that caused undulant fever (sometimes referred to as Malta fever), a disease associated with fever and muscle pain in humans.

Fig. 1. Alice Evans (courtesy of Library of Congress)

Alice Evans hypothesized that the same bacterium, later identified as *Brucella abortus*, named after David Bruce, was responsible for both the bovine and human diseases. She did a series of experiments, injecting pregnant lab animals with either the human or the bovine strain of the bacteria, proving her hypothesis that they were the same strain. But her declaration that the raw milk from cows could make you sick was unacceptable—pasteurization of milk was not yet common at the time.

After presenting her findings in 1917, Evans faced severe skepticism and criticism from both scientists and the milk industry—who said she could not have made a discovery that escaped the observation of other (i.e., male) researchers. Discredited for years, Evans persisted. Her findings were finally accepted by 1927. When Alice Evans was elected as the first woman President of the American Society for Microbiology (then known as the Society of American Bacteriologists) in 1928, she was unable to attend her own installation as President as, ironically, she was in the hospital suffering the effects of Brucellosis.

Dame Jocelyn Bell Burnell is a Northern Irish physicist who, in 1967, while conducting research as a graduate student earning her doctorate, discovered the first radio pulsars, the cosmic sources of peculiar radio pulses. That discovery resulted in the Nobel Prize in Physics in 1974, but Bell was NOT among the winners of the prize. She had earned her doctorate in radio astronomy (1969) from the University of Cambridge. Her professor, Antony Hewish, and Martin Ryle, another astronomy professor at the University, were awarded the Nobel Prize in Physics for Bell's discovery. In 2021, Bell Burnell did receive the Copley Medal—and a reward—the amount of which was greater than the Nobel Prize money.

Women scientists have to help other women scientists. Here is one of my stories and how I was able to change the lineage of the Presidency of the American Society for Microbiology from 99% men to a fair representation of women.

As I have mentioned, Alice Evans was the first woman to be elected President of this professional, scientific organization. It was fifteen years before the election of the next woman President—Rebecca Lancefield. That was in 1943. Lancefield is known for her serological classification of beta *hemolytic streptococci* into what we still know as the Lancefield Group. She was the first to show *Streptococcus pyogenes* infection was the cause of rheumatic fever. The rumor was that in 1943, all the men microbiologists were at war, hence a woman could be elected as President of ASM, at least until the men were back home.

The next woman elected as President of ASM was Helen Whiteley which occurred in 1976, more than thirty years after Lancefield. Whiteley was born in Harbin, China, of Russian parents (her father had been a pilot in the Czarist air force during World War I). She immigrated with her parents from Harbin, Manchuria, to Seattle, Washington, in 1923, following the revolution in Russia. She earned her PhD from the University of Washington and did her research there. She couldn't be hired as a Professor at the University of Washington due to the nepotism rules of the time. Her husband, Arthur Whiteley, was a professor of Zoology and thus she could not be hired as a member of the faculty. When Helen Whitely received the National Institutes of Health Research Career Award, the University of Washington was required to offer her tenure. This necessitated the university overturning its nepotism rules. That rule change has benefited many career couples since. A Professor of Microbiology, Helen Whiteley was the first to clone the gene in *Bacillus thuringiensis*, which naturally produced an insecticidal protein. She was able to introduce the gene into plants, including cotton and tobacco, to make them insect resistant. She was never elected to the National Academy of Sciences—the rumor was that she was "blackballed" (blocked from being elected) because women weren't elected to the National Academy of Sciences in the 1970s—or at least were not welcome.

I was elected President of ASM in 1985 and discovered how women were prevented from being put on the election ballot as part of the officer slate. It turned out that the ASM President serves a one-year term and then a one-year term as Past President. The Past President chairs the Nominating Committee that chooses the next President-Elect. I discovered that the nominations for voting by ASM members were determined in June but not announced until August, with the stipulation that write-in nominations had to be made by May or early June–August was too late. I covertly informed the women's committee of ASM that if they didn't want to wait 10–20 years for another woman President of ASM, they had better provide a write-in candidate before the first of June. The women's committee produced a write-in candidate, Jean Benchley, who was successful. The women's committee persisted each year thereafter and produced write-in candidates. Now just about every other year, ASM has a woman President duly nominated and overwhelmingly elected.

The major blocks to women's progress in science and most endeavors are threefold. The first block is attribution bias—women's achievements are attributed to "luck" and men's successes are attributed to "skill" or ingenuity. Second, women's mistakes are catastrophic, whereas men's mistakes are "part of the learning process." And third, women's achievements usually have to be exceptional to be recognized and then are termed "excellent." Men's achievements are "extraordinary or brilliant." These findings are from Joan Williams' studies carried out at the University of California.

Briefly, from my own research, early past work to the present, I illustrate how at least one woman scientist was able to negotiate a path forward.

Throughout my career, from my graduate student days to the present, there has been an ongoing pandemic, the agent of which is *Vibrio cholerae.* We have had to deal with cholera on a global basis. Most recently, we faced the seventh pandemic of cholera that raged simultaneously with the COVID-19 pandemic.

Cholera, the "Queen of Diarrheal Diseases," incurs 1.5 billion cases annually with nearly 2 million deaths each year, on average. Cholera is water related. It cannot be eradicated, as my team and I have shown—because *Vibrio cholerae* is native to the aquatic environment—freshwater, estuaries, and the open ocean. This is a very versatile bacterium, as we have shown.

Its host in the aquatic environment is the copepod (a small crustacean). These ideas—a bacterium being native to the aquatic environment, when and where there is NO cholera, and its host being the copepod—were ridiculed loudly at scientific meetings in the 1970s and mocked viciously from the 1960s to the 1980s. Now these findings are chapters in introductory Microbiology textbooks . . . as "givens" . . . "everybody knows that"—but this was definitely not the case in 1965–1975.

A crude model for the origin, dissemination, transmission, and disease infection was proposed in 1975. Again, ridiculed, disparaged. To provide the validity of our hypothesis, we designed a project whereby simple sari cloth filters were used to remove plankton and particulate matter by Bangladeshi village women from the water they collected from ponds and rivers for household use for their families. This experiment resulted in clean water—not bacteria-free but with numbers of bacteria below the infectious dose—and a resulting 50% reduction in cholera was achieved.

The model for transmission clearly was person to person, but the source was the aquatic ecosystem and predominantly, the copepod.

The relationship to plankton was further elucidated when NASA launched the LANDSAT mission in the 1980s. Satellite sensors measured the components of chlorophyll a, b, and c, sea surface temperature, and sea surface height. The chlorophyll served as a proxy for zooplankton (copepods)—with a 6–8 week delay from the time of the phytoplankton bloom. The correlation of cholera cases and plankton, temperature, and sea surface was dramatic and published in the *Proceedings of the National Academy of Sciences* 25 years ago.

The response was disbelief and mockery that satellites could detect cholera. However, I knew the computational model was powerful.

Today, 25 years later, the satellites we use are even more powerful with a 20-foot pond width now measurable for many more environmental parameters. A long, long way from the 20-mile quadrant off the Bay of Bengal that we used 25 years ago in that first set of experiments employing satellite monitoring. Today we are able to predict risk of cholera in Yemen and Malawi—providing the early indication of risk of a cholera outbreak in June of 2023, in sufficient time to prepare for the cholera surge that appeared as predicted in August of that year. Ongoing risk prediction maps are provided for Gaza and countries of the Middle East, like Yemen and in Ethiopia and other parts of Africa for the UN.

A timeline for the discoveries shows that from the 1960s to the present, our understanding of cholera has matured significantly. Notably, "the viable but non culturable" phenomenon has finally, 20 years later, entered the commonplace. A discovery for which I was ridiculed as late as the 1980s by the CDC as "seeing ghosts" is now understood as the viable state of the *Vibrio cholerae* commensal bacteria entering dormancy along with its host, the copepod, as the copepod spends winters in a diapause state (similar to hibernation), buried in sediment of water systems worldwide and the bacterium goes to sleep with its host.

The hierarchical model for the *Vibrio cholerae* can now be employed for all waterborne pathogens and all vibrios are native to the aquatic environment. ALL are pathogens, some pandemic like *Vibrio cholerae*, and others opportunistic like *V. cincinnatiensis*, but all are potential pathogens. Unfortunately, with climate change, the metagenomic composition, the microbial communities in aquatic ecosystems globally are being altered. Warmer waters favor *Vibrio parahaemolyticus*, *Vibrio vulnificus*, and *Vibrio cholerae* and they are increasingly infecting humans.

Along the Eastern seaboard, we are documenting a northward shift in the incidence of vibrio infections, reaching to the mid-Atlantic of the USA. The bacteria are present and becoming dominant. In fact, there has been a year-by-year increase in the number of infections from the 1960s to the present.

Florida with its warmer water temperatures reaching 100°F has seen highly dramatic increases in cases and deaths from vibriosis, namely *Vibrio vulnificus*, the so-called flesh-eating bacterium.

Heavy weather systems, including hurricanes and typhoons, create even more favorable conditions for vibrios—elevated temperatures of the water, increased nutrients from heavy winds stirring up nutrients in sediment, increased pH, and greater concentration of heavy metals in the surface water. All contribute to human tragedy.

Pathogenic vibrio infections now occur as far North as Maine on the East Coast—and Alaska on the West Coast. *Vibrio cholerae* can be detected in water samples collected on the East Coast and in the Gulf of Mexico—ominously suggestive of a return of cholera outbreaks in the USA that are homegrown.

As you read the stories of women scientists, the technologies to which they contributed, and learn about their accomplishments, understand that, despite opprobrium, ridicule, and jeers, women scientists and technologists have persevered—and humankind continues to reap the benefits of their contributions in many ways. The world is better today than it would have been without them. Society should honor and accept women—like men—who are scientists and technologists and encourage them in their education and throughout their careers.

Rita Rossi Colwell, PhD

College Park, MD, USA
Former Director, National Science Foundation
National Medal of Science, 2006
Distinguished University Professor Emerita
University of Maryland, College Park, MD, USA

Preface

I was attending my first meeting of the Society of Women Engineers' (SWE) National Board of Directors in 1988 after having been elected a Vice President. During the meeting, then President Suzanne Jenniches held up the nomination forms for the National Medal of Science and the National Medal of Technology (as it was then known) and asked for volunteers to prepare nominations. Seventeen women in that room and no one raised her hand—so I did.

I didn't know anything about the Medals. I didn't know that they were awarded by the President of the United States. I didn't know their history. And I certainly didn't know how few women had been selected to receive them. But I learned.

The first nominations I prepared were for women who were not of the right caliber. But my colleague in the Rocky Mountain Section of SWE, Alexis Swoboda, and I were actively researching historical and contemporary women in science and engineering in order to conduct a "Great Women in Engineering and Science" essay contest for sixth graders as an outreach program in Colorado and Wyoming to attract students to science, technology, engineering, and mathematics (STEM) careers. I realized that if I did a good enough job on her nomination, Admiral Grace Murray Hopper would be selected to receive the National Medal of Technology for her invention of the computer compiler.

I worked on her nomination for two years. I consulted with the Department of Commerce who oversaw the nomination process at the time. I contacted the Secretary of the Navy, the former Secretary of the Navy, the pioneers in the computer industry, the Senator from Virginia, and others. One of those computer industry pioneers, Jean Sammet, provided invaluable counsel. So did Henry Tropp, a mathematics professor and computer history scholar. In June 1991, just before I assumed the presidency of SWE, I was informed that Admiral Hopper would receive the National Medal of Technology. The ceremony would be in September 1991. Admiral Hopper was very frail, and she asked me to receive the Medal on her behalf.

In September 1991, I found myself at the White House Rose Garden to receive Admiral Hopper's National Medal of Technology. As I looked at the people around me, I kept asking myself, "Where are the women?" I know of their accomplishments; I had now been actively researching historical and contemporary women in science and engineering for four years. I learned that Admiral Hopper was the first individual woman to receive the National Medal of Technology and there had only been one woman before her—and that woman had shared the Medal. This was 1991! How was this possible?

I've since been at the White House two more times to see my nominees for the National Medal of Technology and Innovation accept their medallions—Yvonne Brill and Kristina Johnson. And now, with these volumes, my co-authors Ilene Busch-Vishniac and Lauren Busch and I get to tell the stories of all the female Medalists to date for both the National Medal of Science and the National Medal of Technology and Innovation.

It is my pleasure and honor to co-author this book and write these accomplished women into history.

Greenwood Village, CO, USA Jill S. Tietjen

Acknowledgments

In every book-writing endeavor, there are individuals who knowingly or unknowingly provide support and assistance. We appreciate them all, even if they are not explicitly acknowledged herein.

To the National Science and Technology Medals Foundation for their enthusiastic support of this project.

To Mary James and Springer for their tremendous support of the Springer Women in Engineering and Science series.

To Jill's village, who support her in all her endeavors.

To Ethan Vishniac for his encouragement of this project.

To Ruben Perlmutter for his willingness to read and comment upon draft chapters.

Competing Interests The authors have no competing interests to declare that are relevant to the content of this manuscript.

Contents

About the Authors

Jill S. Tietjen P.E., entered the University of Virginia in the Fall of 1972 (the third year that women were admitted as undergraduates after a suit was filed in court by women seeking admission), intending to be a mathematics major in the College of Arts and Sciences. But midway through her first semester, she found engineering and made all the arrangements necessary to transfer to the School of Engineering and Applied Science. In 1976, she graduated with a B.S. in Applied Mathematics (minor in Electrical Engineering; Tau Beta Pi, Virginia Alpha) and went to work in the electric utility industry.

Galvanized by the fact that no one, not even her Ph.D. engineer father, had encouraged her to pursue an engineering education and that only after her graduation did she discover that her degree was not ABET-accredited, she joined the Society of Women Engineers (SWE) and for more than 45 years has worked to encourage young women to pursue science, technology, engineering, and mathematics (STEM) careers. In 1982, she became licensed as a professional engineer in Colorado.

Tietjen started working jigsaw puzzles at age two and has always loved to solve problems. She derives tremendous satisfaction from seeing the results of her work—the electricity product that is so reliable that most Americans just take its provision for granted. Flying at night and seeing the lights below, she knows that she had a hand in this infrastructure miracle. As an expert witness, she worked to plan, license, and put into the rate base new power plants.

Her efforts to nominate women for awards began in SWE and have progressed to her acknowledgment as one of the top nominators of women in the country. Her nominees have received the National Medal of Technology and Innovation and the Kate Gleason Medal; they have been inducted into the National Women's Hall of Fame and state Halls, including Colorado, Maryland and Delaware and have received university and professional society recognition. Tietjen believes that it is imperative to nominate women for awards—for the role modeling and knowledge of women's accomplishments that it provides for the youth of our country.

Tietjen received her MBA from the University of North Carolina at Charlotte. She has been the recipient of many awards, including the Distinguished Service Award from SWE (of which she has been named a Fellow and is a National Past President), the Distinguished Alumna Award from the University of Virginia, the Distinguished Alumna Award from the University of North Carolina at Charlotte, and the Distinguished Engineering Alumni Award (special category as a non-alum) from the University of Colorado at Boulder. She has been inducted into the Colorado Women's Hall of Fame and the Colorado Authors' Hall of Fame and elected to the National Academy of Construction. Tietjen sits on the board of Georgia Transmission Corporation and served for eleven years on the board of Merrick & Company.

Her publications include the bestselling and award-winning book *Her Story: A Timeline of the Women Who Changed America* for which she received the Daughters of the American Revolution History Award Medal and *Hollywood: Her Story, An Illustrated History of Women and the Movies* which has received numerous awards. Her award-winning book *Over, Under, Around and Through: How Hall of Famers Surmount Obstacles* was released in 2022. The memoir of Dr. Antonia Novello, *Duty Calls: Lessons Learned from an Unexpected Life of Service*, which she co-wrote, was released in English and in Spanish in 2024 and has received several awards. She is the series editor of the Springer Women in Engineering and Science series, has authored or co-authored volumes in the series, served as co-volume editor for additional volumes, and written chapters for a number of volumes.

Ilene J. Busch-Vischniac entered the University of Rochester in 1972 as a Bachelor of Arts in Music student with a focus on piano. In her first term, she enrolled in a small freshman course entitled "The Physics of Music," and from that moment on she pursued a career in acoustics. The change was fortuitous as she was also learning that she had neither the necessary talent nor the drive to become a successful musician. In 1976, Busch-Vishniac earned a BS in Physics and a BA in Mathematics (both summa cum laude), with a senior thesis on the reverberation in the utility tunnels the students used to travel between buildings to avoid the snow, cold, and wind. She married fellow physics student, Ethan Vishniac, and then moved with him to Boston to pursue a PhD in Mechanical Engineering.

Busch-Vishniac studied with Professor Dick Lyon at MIT, a researcher well known for his work on quieting product noise. Dick had not had a female graduate student before Busch-Vishniac arrived. That Fall, he took on four female graduate students who shared an office and became good friends. As a team, they worked to change the graduate program to make it more collaborative and less competitive. Busch-Vishniac's work was on quieting the noise of an impact line printer, a piece of research that became obsolete shortly after her work, as the first laser printers were released. She has, since that time, advocated for acoustic output to be part of product design rather than an afterthought. As a person sensitive to noise, she hates backup beeping and loud sounds signaling locking or unlocking of a device.

After graduation, Busch-Vishniac went to Bell Laboratories in Murray Hill, NJ, as a postdoctoral fellow and then a Member, Technical Staff. She was the first female Member, Technical Staff in the Acoustic Research Department, the largest research department at Bell Laboratories. When given that title, there were a few colleagues who suggested that she was promoted because of the lack of women. Busch-Vishniac had the sense to respond that since nobody else could fill that role, she was intending to demand a salary higher than that of the person suggesting she was unqualified.

Busch-Vishniac worked with Jim West, a noted Black scientist and the co-inventor of the telephone touchpad and the microphone used in every hearing aid and telephone from 1966 until about 2000. This work produced seven US patents, one of which was awarded a patent of the year award by Bell Laboratories and has been a source of enormous satisfaction. She and Jim have worked together for over forty years in a strong partnership. They continue to challenge conventional wisdom on acoustics and signal processing issues and develop new and important technologies.

Busch-Vishniac left Bell Laboratories for a faculty appointment at The University of Texas in 1981. She continued her research on engineering sensors and drivers and broadened her acoustics work to include design of highway noise barriers. Her work showed that if these barriers were built so their height varies along their length, they are more effective at reducing community noise. A few barriers were built with this in mind, including one in Dallas that she has driven by and admired with delight.

During this period, she earned a number of research and teaching awards, including the Lindsay Award of the Acoustical Society of America, the Curtis McGraw Research Award of the American Society for Engineering Education, and the Achievement Award of the Society of Women Engineers. She took on an administrative role in her large home department out of frustration that others were doing the job so badly and it was hurting students. She also had two children: Cady Anne and Miriam Rachel Vishniac. Busch-Vishniac left the University of Texas when it became clear that the glass ceiling preventing her from moving further up in the administration was more like steel and concrete than glass. Fortunately, her talents were in great demand and she had a pool of opportunities from which to choose her next move.

In 1997, Busch-Vishniac became Dean of Engineering at Johns Hopkins University. She was the first female dean outside the School of Nursing. There she expanded the number of faculty and students in the school and introduced policies to produce a more diverse faculty. Her diversity policies were later copied by several major US universities—a source of great pride to Busch-Vishniac. During her tenure, the national ranking of every department either improved or stayed the same and research expenditures increased by 80%. She resigned as dean in 2003, after the university president suggested she "use her womanly wiles" to get what she wanted when negotiating with other deans and assumed the role of President of the Acoustical Society of America.

During her term in this role, the society celebrated their 75th anniversary and moved forward on strategic planning she led. She continued to be recognized for her leadership on diversity issues and in acoustics and in 2008, she was honored with the establishment of an annual lecture in her name at Johns Hopkins University. Now that Busch-Vishniac lives in Baltimore she is able to attend this annual lecture and thoroughly enjoys meeting the stellar women brought in to inspire students.

In 2007, Busch-Vishniac left Johns Hopkins University and became the Provost and Vice President, Academic of McMaster University in Hamilton, Ontario. There she focused on improving the quality of the academic experience for students and changing budget operations to eliminate structural deficits and distribute funds more equitably. She created a program of equity focused on Indigenous people of Canada and dramatically increased the Indigenous faculty numbers. Her work in this area, of which she is extremely proud, was noted by political leaders and she was included in important meetings to bring a reckoning for past treatment of Indigenous children by the government. She was elected Chair of the Ontario Council of Academic Vice Presidents by her peers. Busch-Vishniac thoroughly enjoyed working with President Peter George and learned that in Canada it was possible to get in trouble for not taking enough vacation time.

In 2012, Busch-Vishniac moved to The University of Saskatchewan as the President and Vice Chancellor. There she focused on three issues: improving the quality of the medical school (which was twice put on probation by the accrediting agency), eliminating the structural deficit, and providing greater service to the Indigenous community which was roughly 17% of the province's population. She made strong progress in all these areas but was removed without cause from office after only two years, largely as a result of pressure on the board from the Premier and the former university president. Busch-Vishniac sued the university and the premier and ultimately reached a settlement agreement in 2015. Getting her current dog, Nipin (the Plains Cree word for summer) is the best thing that happened to her in Saskatchewan.

After working as a consultant on higher education issues, Busch-Vishniac joined Sonavi Labs, a startup company spun out of research at Johns Hopkins University in 2017, as Chief Innovation Officer. Sonavi Labs produces a smart stethoscope able to diagnose serious lung problems based on lung sounds. It reflects the culmination of a career in acoustics, as it truly uses sound to save lives. In her four years with Sonavi Labs, Busch-Vishniac was able to raise significant funding for the company, partly through two NIH grants for which she was principal investigator.

She retired from Sonavi Labs in 2022 and is now working on passion projects, such as this book.

Lauren K. Busch is an engineer with degrees from Washington University in St. Louis (WUSTL) and George Washington University. Her career as an operations research analyst has included positions in industry and government contracting.

Busch entered WUSTL with the intention of majoring in mathematics at the School of Arts and Sciences. During her first year, she became aware of a class in operations research, a discipline she had previously never heard of, in the School of Engineering & Applied Science. After taking this course, Busch switched her focus to applied mathematics and engineering. Being in the School of Engineering and Applied Science meant fewer humanities requirements and more courses that appealed to Busch's love of solving equations. It also resulted in fewer courses that required participating in class discussions—a bonus for the introverted Busch.

Upon completing her BS in Systems Science and Mathematics at WUSTL in 1983, Busch began her professional career at Johns Hopkins University's Applied Physics Laboratory. Here, she worked on projects for the United States Navy, creating Monte Carlo simulations of naval weapons systems. During this time, she also pursued further education, attending graduate school at George Washington University in the evenings and earning an MS in Operations Research in 1986.

A few years later, Busch moved to Science Applications International Corporation (SAIC), creating economic models for the Department of Energy's long-term forecasting of natural gas prices.

Busch's career took an unexpected turn when she met her husband, a New York attorney, and they decided to settle in Philadelphia. There, she worked as an operations research analyst for Rohm and Haas, a chemical manufacturer. However, her time in Philadelphia was cut short when the family, now with a young son, relocated to Mexico City for her husband's job. Busch was able to continue working part-time for Rohm and Haas in an era when telecommuting was not the norm. After four years in Mexico City, the family, which by then had added a daughter, returned to the USA to Busch's hometown of Dallas, Texas. Busch continued telecommuting for work for Rohm and Haas but eventually left the workforce to focus on raising her two children.

She then spent countless hours volunteering in the community. Later, she was a search engine optimization consultant. In recent years, Busch returned to SAIC (now Leidos, Inc.) to assist with updates to some mathematical models for the Department of Energy.

Having now officially retired from Leidos, Inc., Busch was excited to become involved in writing this book to highlight the amazing women who have won the National Medal of Science and the National Medal of Technology and Innovation.

Chapter 1
Introduction

Keywords National Medal of Science · National Medal of Technology · National Medal of Technology and Innovation · Nobel Prize · National Inventors Hall of Fame · National Women's Hall of Fame · U.S. postage stamp · U.S. quarter

When the President of the United States awards the Presidential Medal of Freedom or the Oscars are presented, a photo from the event is usually above the fold on the front page of newspapers around the country. However, when the National Medal of Science and National Medal of Technology and Innovation are awarded by the President, there might be no press coverage at all or a very small paragraph somewhere in the back of one of the newspaper sections. Since science and technology drive our country's economy, this press coverage seems backward. The overwhelming majority of the more than seven hundred recipients of both Medals have been men. However, over the decades that the Medals have been awarded through 2024, 77 women have received either the National Medal of Science or the National Medal of Technology and Innovation. We are writing this book, and its companion book, to tell the stories of the amazing women who have received either Medal.

1.1 National Medal of Science

The National Medal of Science was established in 1959 by the U.S. Congress and "is the highest recognition the nation can bestow on scientists and engineers. The presidential award is given to individuals deserving of special recognition by reason of their outstanding contributions to knowledge in the physical, biological, mathematical, engineering, or social and behavioral sciences, in service to the Nation." Since the medal was first presented in 1962, there have been 529 total awardees, of whom 60 have been women (U.S. National Science Foundation 2025). The first woman to win the National Medal of Science was Barbara McClintock, who received the Medal in 1970 for her work in genetics. Table 1.1 shows the female recipients by category and year.

J. S. Tietjen et al., *Women Laureates 1970 – 2007*, Women in Engineering and Science, https://doi.org/10.1007/978-3-032-23116-1_1

Table 1.1 National Medal of Science female recipients by year and category (NSTMF 2025)

Name	Year	Accomplishment for which recognized
Behavioral and social sciences		
Anne Anastasi	1987	Differential psychology
Eleanor J. Gibson	1992	Perceptual learning
Shelley Taylor	2022	Mental health and human connection
Anne Treisman	2011	Attention limits—Human mind/brain
Biological sciences		
Huda Akil	2022	Brain biology of emotions
Nancy C. Andreasen	2000	Integrated study of mind, brain, and behavior
Mary Ellen Avery	1991	Premature infant respiratory distress
Bonnie L. Bassler	2024	Bacteria communication systems
May Berenbaum	2012	Chemical coevolution
Helen M. Blau	2024	Cells can reprogram and rejuvenate
Sallie W. Chisholm	2011	Ocean organisms
Mildred Cohn	1982	Mechanisms of enzymatic catalysis
Rita R. Colwell	2006	Marine microbes
Elizabeth C. Crosby	1979	Neuroanatomy
Katherine Esau	1999	Plant structure and development
Nina V. Fedoroff	2006	Plant molecular biology
Elaine Fuchs	2008	Biology of the skin
Ann M. Graybiel	2001	Anatomy and physiology of the brain
Mary-Claire King	2014	BRCA1 breast cancer gene
Susan Lee Lindquist	2009	Protein folding
Eve Marder	2022	Neural circuits
Lynn Margulis	1999	Development, structure, and evolution of living things
Barbara McClintock	1970	Plant genes, chromosomes, jumping genes
Rita Levi-Montalcini	1987	Nerve growth factor
Elizabeth F. Neufeld	1994	Lysosomal storage diseases
Ruth Patrick	1996	Diatoms, biodiversity, water pollution
Janet D. Rowley	1998	Chromosomal translocations
Berta Scharrer	1983	Neurosecretion
Lucy Shapiro	2011	Integrated genetic circuits in bacteria
Maxine F. Singer	1992	Molecular biology, genetic code
Joan A. Steitz	1986	Cellular machinery
Salome G. Waelsch	1993	Mammalian genetics
Evelyn M. Witkin	2002	DNA mutagenesis
Teresa Kaye Woodruff	2024	Blending oncology and female health
Rosalyn S. Yalow	1988	Radioimmunoassay
Chemistry		
Jacqueline Barton	2010	Long-range electron transfer
Gertrude B. Elion	1991	Drug-receptor interactions
Joanna S. Fowler	2008	Medical imaging compounds

Table 1.1 (continued)

Name	Year	Accomplishment for which recognized
Marye Anne Fox	2009	Excited state and charge-transfer processes
Darleane C. Hoffman	1997	Primordial plutonium
Isabella L. Karle	1995	Molecular structure through X-ray analysis
Judith P. Klinman	2012	Enzyme catalysis
Geraldine Richmond	2013	Molecular characteristics of water surfaces
Susan Solomon	1999	Cause of Antarctic ozone hole
JoAnne Stubbe	2008	Mechanisms of ribonucleotide reductases
Engineering		
Angela Marie Belcher	2024	New materials using nanoscience and materials science
Mildred S. Dresselhaus	1990	Electronic properties of metals and semimetals
Mathematics and computer science		
Ingrid Daubechies	2024	Wavelet discoveries
Cynthia Dwork	2024	Privacy, algorithms, statistical validity
Cathleen Synge Morawetz	1998	Partial differential equations and wave propagation
Karen K. Uhlenbeck	2000	Global geometry
Physical sciences		
Fay Ajzenberg-Selove	2007	Nuclear physics
E. Margaret Burbidge	1983	Observational astronomy—Stars and quasi-stellar objects
Esther M. Conwell	2009	Electron and hole transfer in semiconductors
Sandra M. Faber	2011	Extra-galactic astronomy
Wendy L. Freedman	2024	Hubble constant, astrophysics
Shirley Ann Jackson	2014	Condensed matter and particle physics
Vera Rubin	1993	Observational cosmology, dark matter
Myriam Sarachik	2022	Experiments in solid-state physics
Chien-Shiung Wu	1975	Experiments in nuclear decay

1.2 National Medal of Technology and Innovation

The National Medal of Technology and Innovation (NMTI) (originally known as the National Medal of Technology) was established in 1980 by the U.S. Congress and first awarded in 1985. It is the highest honor for technological achievement, bestowed by the President. "The Medal is awarded to individuals, teams (up to four individuals), companies or divisions of companies for their outstanding contributions to America's economic, environmental and social well-being. The purpose of the NMTI is to recognize those who have made lasting contributions to America's competitiveness, standard of living, and quality of life through technological innovation, and to recognize those who have made substantial contributions to strengthening the nation's technological workforce." To date, there have been 240 awardees, of whom 17 have been women (US Patent and Trademark Office 2025).

Table 1.2 National Medal of Technology and Innovation female recipients

Name	Year	Accomplishment for which recognized
Helen Edwards	1989	Particle accelerator
Admiral Grace Murray Hopper	1991	Computer compiler and languages
Stephanie Kwolek	1996	Aramid fibers
Dace Viceps Madore	2005	Pneumonia vaccine
Maya Koster	2005	Pneumonia vaccine
Esther Sans Takeuchi	2008	Silver vanadium oxide battery
Helen M. Free	2009	Dip and read techniques
Yvonne C. Brill	2010	Rocket propulsion systems
Frances Arnold	2011	Biofuels and chemicals
Edith M. Flanigen	2012	Chemistry and materials
Cherry A. Murray	2012	Telecommunications devices
Mary Shaw	2012	Computer science curriculum
Nancy Ho	2013	Technology to develop biofuels
Mary-Dell Chilton	2022	Genetically modified organisms
Kristina M. Johnson	2024	Optoelectronic devices
Paula T. Hammond	2024	Nanoscale engineering
Jennifer A. Doudna	2024	CRISPR—Gene editing technology

The first woman to receive the National Medal of Technology was Helen Edwards, who shared the 1989 Medal with her colleagues for their work on the TEVATRON particle accelerator. The first individual woman to receive the Medal was Admiral Grace Murray Hopper in 1991 for her invention of the computer compiler. Table 1.2 provides information on each of the 17 women to have received the NMTI.

1.3 Nobel Laureates

Six Nobel Laureates can be found among these female NMS and NMTI laureates (The Nobel Prize 2025):

- Rosalyn Yalow (NMS), 1977 Nobel Prize in Physiology or Medicine.
- Barbara McClintock (NMS), 1983 Nobel Prize in Physiology or Medicine.
- Rita Levi-Montalcini (NMS), 1986 Nobel Prize in Physiology or Medicine.
- Gertrude B. Elion (NMS), 1988 Nobel Prize in Physiology or Medicine.
- Frances H. Arnold (NMTI), 2018 Nobel Prize in Chemistry.
- Jennifer Doudna (NMTI), 2020 Nobel Prize in Chemistry.

In addition, Susan Solomon (NMS) was a member of the Intergovernmental Panel on Climate Change, which shared the 2007 Nobel Peace Prize with Al Gore.

The work that two of these Medal Laureates were partners in or heavily involved in did lead to Nobel Prizes for the men involved—but not for the women. Chien-Shiung Wu proved the theory of conservation of parity. The two men who proposed the theory won the Nobel Prize in Physics in 1957. Isabella Karle solved the problem of x-ray diffraction in practice. For developing the associated theory, her husband Jerome Karle and Herbert Hauptman shared the 1985 Nobel Prize in Chemistry.

1.4 Other Honors and Awards

Members of this distinguished group have received many honors and awards in addition to those above. Here is just a sampling—there are many others.

Barbara McClintock and Chien-Shiung Wu have each been featured on a U.S. postage stamp (Smithsonian National Postal Museum a, b). Vera Rubin was featured on a U.S. quarter (U.S. Mint 2025).

Inductees into the National Inventors Hall of Fame include Frances Arnold, Yvonne Brill, Mary-Dell Chilton, Jennifer Doudna, Mildred Dresselhaus, Gertrude Elion, Edith Flanigen, Helen Murray Free, Kristina Johnson, Stephanie Kwolek, and Esther Sans Takeuchi (NIHF 2025).

Inductees into the National Women's Hall of Fame include Mildred Cohn, Rita Colwell, Gertrude Elion, Helen Murray Free, Admiral Grace Murray Hopper, Shirley Ann Jackson, Stephanie Kwolek, Barbara McClintock, Ruth Patrick, Janet Rowley, Susan Solomon, Chien-Shiung Wu, and Rosalyn Yalow (NWHF 2025).

Places have been named for the women—on Earth, on planets, and on planetary bodies. Asteroids have been named for Margaret Burbidge, Vera Rubin, and Chien-Shing Wu. The Vera Rubin Ridge is a feature on Mars. The Solomon Glacier and the Solomon Saddle—both in Antarctica—were named in honor of Susan Solomon. And an orchid has been named for Rita Levi-Montalcini: *Ophrys* × *montalciniae*.

Other ways of honoring the women include the USS Hopper—an Arleigh Burke class destroyer—named for Admiral Grace Murray Hopper. The Vera Rubin Observatory is in Chile. Nvidia named one of its new chips Vera Rubin, and its new GPU Architecture is named for Admiral Hopper.

A number of the laureates have been inducted into state women's halls of fame: Rita Colwell (Maryland), Elizabeth Crosby (Alabama and Michigan), Kristina Johnson (Colorado), Isabella Karle (Michigan), Stephanie Kwolek (Delaware), Barbara McClintock (Connecticut), and Susan Solomon (Colorado).

1.5 Conclusion

The National Medal of Science and the National Medal of Technology and Innovation, awarded by the President, have been awarded to 77 women over the decades that they have been in existence. Profiles of each of the 77 women who have received one of the Medals through 2024 are provided in the pages of this book and its companion volume.

References

National Inventors Hall of Fame (NIHF). National Inventors Hall of Fame Inductees have changed the world. https://www.invent.org/inductees. Accessed July 3, 2025.

National Science and Technology Medals Foundation (NSTMF). Laureates. https://nationalmedals.org/timeline/. Accessed June 25, 2025.

National Women's Hall of Fame (NWHF). Discover the Women of the Hall. https://www.womenofthehall.org/women-of-the-hall/. Accessed July 3, 2025.

The Nobel Prize. Nobel Prize—awarded women. https://www.nobelprize.org/prizes/lists/nobel-prize-awarded-women/. Accessed June 24, 2025.

Smithsonian National Postal Museum a. Forever Chien-Shiun Wu single. https://postalmuseum.si.edu/object/npm_2021.2010.77. Accessed July 3, 2025.

Smithsonian National Postal Museum b. Women on Stamps, Part 2: Barbara McClintock. https://postalmuseum.si.edu/exhibition/women-on-stamps-part-2-science-improving-lives/barbara-mcclintock. Accessed July 3, 2025.

U.S. Mint. United States Mint Begins Shipping 2025 American Women Quarters Honoring Dr. Vera Rubin on June 2. https://www.usmint.gov/news/press-releases/united-states-mint-begins-shipping-2025-american-women-quarters-honoring-dr-vera-rubin-on-june-2. Accessed July 3, 2025.

U.S. National Science Foundation, "National Medal of Science," https://www.nsf.gov/honorary-awards/national-medal-science, accessed February 27, 2025.

U.S. Patent and Trademark Office. "National Medal of Technology and Innovation (NMTI)." https://www.uspto.gov/learning-and-resources/ip-programs-and-awards/national-medal-technology-and-innovation-nmti. Accessed June 24, 2025.

Chapter 2
Barbara McClintock

For establishing the relations between inherited characters in plants and the detailed shapes of their chromosomes, and for showing that some genes are controlled by other genes within chromosomes

Keywords Jumping genes · Transposons · Chromosomes · Transposable elements · Genes · Maize · Genetics · Cytology · Mendel · Plant breeding · Botany · Corn

The first woman to receive the National Medal of Science (1970), Nobel Laureate Barbara McClintock (Fig. 2.1) dedicated her life to science, persisting despite decades of significant skepticism about her results from her scientific colleagues, and proving that genes can change places on the chromosome, the so-called jumping genes. In 1983, she was the first woman to win an unshared Nobel Prize in Physiology or Medicine and the third woman to have an unshared win of any of the Nobel Prizes in a science category.

J. S. Tietjen et al., *Women Laureates 1970 – 2007*, Women in Engineering and Science, https://doi.org/10.1007/978-3-032-23116-1_2

Fig. 2.1 Barbara McClintock. (Courtesy of the Library of Congress)

2.1 Life (1902–1992)

Barbara McClintock was born in 1902 in Hartford, Connecticut, and named "Eleanor." However, soon she was called Barbara, because her parents thought Eleanor was a delicate name and she wasn't delicate! She changed her name legally to Barbara as an adult. Her father was a doctor, and her mother was a professional pianist, a writer, and a painter. Barbara was a solitary child even as a baby. She liked being alone, playing with a toy by herself. This trait would serve her in good stead in her career as a plant scientist. When she was two and a half, her parents sent her to live with a paternal aunt and her husband in Massachusetts. Barbara loved living with them. She moved back with her family when she was six—to go to school—and the family moved to Brooklyn (Heiligman 1994; Ware 2004).

McClintock attended Erasmus High School, where she discovered her love of math and science. She wanted to go to college, but her mother was against the idea. Although McClintock hadn't completed the paperwork to enroll at Cornell, she traveled to Ithaca, New York, stood in line to register, and through some events that she never understood, was allowed to sign up for classes in the fall of 1919 at Cornell University in the agricultural college. McClintock enjoyed college, was elected president of the freshman class, and played the banjo in a jazz band. By the time she was a junior, however, almost all of her attention was focused on her studies. She became fascinated with genetics, receiving her BS degree in 1923. Corn was the focus of genetics at Cornell and corn became McClintock's focus. She began her graduate studies immediately, but the plant breeding department did not accept women. Thus, she joined the botany department. She took classes in cytology (the study of cells) in the botany department and genetics courses from the plant-breeding department. The corn she studied was maize (Indian corn) (Heiligman 1994; McGrayne 1993; The Nobel Prize 1983; Shiels 1985; Ware 2004).

As a graduate student, McClintock changed the direction of maize genetics by identifying the individual members of the set of chromosomes within each maize cell. People previously had been able to see the chromosomes under a microscope and count them, but no one until McClintock had been able to tell them apart. She was able to do so by using a new staining technique. Her research was incorporated in Lester Sharp's textbook *An Introduction to Cytology*, so she received recognition in the field very early in her career. McClintock received her MA in 1925 and her PhD in 1927, both from Cornell (McGrayne 1993; The Nobel Prize 1983; Shiels 1985).

Dr. McClintock stayed on at Cornell as a botany instructor after she completed her PhD and published nine papers on maize chromosomes in scientific journals between 1929 and 1931 that are each considered a major scientific contribution. The paper she published with Harriet Creighton in 1931 documents a discovery that is still considered a milestone in the history of genetics. They proved that the exchange of genetic information that occurs during cell division for reproduction (meiosis) is accompanied by an exchange of chromosomal material. Drs. McClintock and Creighton also proved that genes for physical traits are carried on the chromosomes (McGrayne 1993; Heiligman 1994; Shiels 1985).

The paper published with Dr. Creighton made Dr. McClintock's reputation. In the book *Great Experiments in Biology*, the authors declared, "Beyond any question, this is one of the truly great experiments in modern biology." The editor of *Classic Papers in Genetics* opined, "This paper has been called a landmark in experimental genetics. It is more than that—it is a cornerstone." Despite this success, the faculty was opposed to giving permanent faculty positions to women, and Dr. McClintock knew she would have to leave Cornell (McGrayne 1993).

Dr. McClintock left Cornell in 1931 and accepted a fellowship from the National Research Council. For the two years of the fellowship, she divided her time between the University of Missouri and the California Institute of Technology. She was the first woman postdoc at Caltech—Caltech would not hire its first woman professor until 1971. In 1933, she received a Guggenheim Fellowship to go to Germany. However, the timing was bad because of the rise of Hitler in Germany, and she arrived back at Cornell in the middle of the Great Depression. She received a grant from the Rockefeller Foundation, which enabled her to stay and work at Cornell. She stayed at Cornell for two years. For a man, such fellowships would have been considered significant stepping-stones to a professorship. But for a woman, they were stopgap financial measures and did not lead to professorships. Almost all of her fellowships and positions resulted from the efforts of friends and colleagues to enable her to continue her research and provide enough funds for her to live on (McGrayne 1993; Shiels 1985).

Knowing that she didn't want to teach at a women's college—or teach at all—limited her employment opportunities as did the fact that industry, government, and most colleges and universities did not hire women (McGrayne 1993). Dr. McClintock accepted a job offer as an assistant professor at the University of Missouri in 1936. She stayed at the University of Missouri until 1941, during which time she published a significant series of articles, discovered ring chromosomes (a chromosome whose two ends have broken and then joined to form a ring (Wikipedia 2025)), and served as Vice President of the Genetics Society of America. At that point, realizing she was not comfortable at a university and that she just wanted to do her research, she joined the Carnegie Institution of Washington's Genetics Laboratory at Cold Spring Harbor. She started in a one-year position, but the position soon became permanent. Dr. McClintock remained there until her death and became well known for working 12 hours a day, six days a week (Heiligman 1994; McGrayne 1993; NSTMF 2025; Shiels 1985; Ware 2004).

In 1944, Dr. McClintock was elected President of the Genetics Society of America and was elected a member of the National Academy of Sciences—only the third woman in that organization's history. In 1951, she presented the results of her research that transposition was occurring in her ears of corn. Transposition is the movement of a DNA sequence from one location on the genome to another location. This discovery, for which she would receive the Nobel Prize in 1983, was met with reactions ranging from "perplexed to hostile" and she stopped publishing her results—although she wrote them up for publication and filed them, publishing only concise summaries of her work in the Carnegie Institution annual *Year Book* (Fedoroff 1996). Although she received the Achievement Award from the American Association of University Women in 1947, it was not until the 1970s that her work received widespread acknowledgement and understanding (Shiels 1985). In 1970, Dr. McClintock became the first woman to receive the National Medal of Science.

The many awards she received in the 1980s included the John D. and Catherine T. MacArthur Foundation Fellow Laureate, the Albert Lasker Basic Medical Research Award, and the Wolf Prize in Medicine. In 1983, she received the Nobel Prize in Physiology or Medicine, the first woman to win that prize unshared and only the third woman at the time to receive any Nobel Prize in the sciences unshared (Shiels 1985). The citation was "for her discovery of transposable genetic elements" (Ware 2004). She was inducted into the National Women's Hall of Fame in 1986 (NWHF 2025). In 2005, the United States Postal Service issued a postage stamp honoring her accomplishments (Smithsonian 2025).

As stated by cytogeneticist Marcus Rhoades (Fedoroff 1996),

> One of the remarkable things about Barbara McClintock's surpassingly beautiful investigations is that they came solely from her own labors. Without technical help of any kind, she has by virtue of her boundless energy, her complete devotion to science, her originality and ingenuity, and her quick and high intelligence made a series of significant discoveries unparalleled in the history of cytogenetics. A skilled experimentalist, a master at interpreting cytological detail, a brilliant theoretician, she has had an illuminating and pervasive role in the development of cytology and genetics.

2.2 Contributions

In the 1940s, scientists thought genes lined up on a chromosome like pearls or beads on a string and did not move. But Dr. McClintock discovered that genes did move. Through six years of research, she was able to prove that genes were not stable pearls, but instead they can move around and turn on and off during the development of a cell. Sometimes they jumped around on one chromosome. Sometimes, they jumped from one chromosome to another. When they jumped, very unusual things happened (Heiligman 1994; McGrayne 1993).

Dr. McClintock discovered that there were two types of jumping genes, which she called "transposable elements" or "transposons." One type is a controlling element—like a switch—that can turn genes that determine physical characteristics, such as color or size, on and off. The second is an activator. It can make that controlling element—the switch—jump around from one part of a chromosome to another. The process that Dr. McClintock discovered is today called genetic transposition (Heiligman 1994; McGrayne 1993; NSTMF 2025).

Dr. McClintock thus determined that genes are unstable and so are their mutation effects. She showed that environmental conditions could cause mutations to reverse and turn genes back on. This meant that genetics was much more fluid than previously thought. Dr. McClintock understood that transposons explained the incredible diversity of organisms. Unfortunately, she was decades ahead of her time (McGrayne 1993).

In the late 1960s, scientists discovered transposable elements in bacteria. Molecular biologists were now willing to admit that mobile genetic elements were reasonably common—and even found in humans. By the 1980s, scientists began to understand that jumping genes play an important role in evolution, in how bacteria become resistant to antibiotics, and in how normal cells become cancerous. The Nobel Prize Committee said her work was "one of the two great discoveries of our time in genetics" (the other being the structure of DNA). Today, entire scientific meetings are devoted to the subject of jumping genes. Her work has enabled the mapping of the corn genome (Heiligman 1994; McGrayne 1993; NSTMF 2025).

2.3 Conclusion

The first woman to receive the National Medal of Science and the first female Nobel Laureate to win an unshared Nobel Prize in Physiology or Medicine, Barbara McClintock loved her corn plants and persisted despite decades of skepticism from her colleagues. Her findings over the course of her career in many ways dictated the direction of the field of genetics.

References

Fedoroff, Nina V. 1996. Two Women Geneticists. The American Scholar. Volume 65, Number 4, Autumn. Pp. 587-592.

Heiligman, Deborah. *Barbara McClintock: Alone in Her Field.* New York, New York: W.H. Freeman and Company. 1994.

McGrayne, Sharon Bertsch. 1993. Nobel Prize Women in Science: Their Lives, Struggles, and Momentous Discoveries. New York, New York: Birch Lane Press.

National Science & Technology Medals Foundation. "Barbara McClintock." https://nationalmedals.org/laureate/barbara-mcclintock/. Accessed June 10, 2025.

National Women's Hall of Fame. Barbara McClintock. https://www.womenofthehall.org/inductee/barbara-mcclintock/. Accessed June 21, 2025.

The Nobel Prize. 1983. Barbara McClintock—Biographical. https://www.nobelprize.org/prizes/medicine/1983/mcclintock/biographical/. Accessed June 21, 2025.

Shiels, Barbara. 1985. Winners: Women and the Nobel Prize. Minneapolis, Minnesota: Dillon Press, Inc.

Smithsonian—National Postal Museum. Barbara McClintock. https://postalmuseum.si.edu/exhibition/women-on-stamps-part-2-science-improving-lives/barbara-mcclintock. Accessed June 21, 2025.

Ware, Susan, Editor. 2004. Notable American Women: A Biographical Dictionary—Completing the Twentieth Century. Cambridge, Massachusetts: The Belknap Press of Harvard University Press.

Wikipedia. Ring Chromosomes. https://en.wikipedia.org/wiki/Ring_chromosome. Accessed December 17, 2025.

Chapter 3
Chien-Shiung Wu

For her ingenious experiments that led to new and surprising understanding of the decay of the radioactive nucleus

Keywords Beta decay · Law of conservation of parity · Nuclear physics · Isotopes · Manhattan Project · Neutrino · Proton · Electron · K-meson · Neutron

One of the world's foremost nuclear physicists, Chien-Shiung Wu (Fig. 3.1) is best known for a classic experiment in beta decay that confirmed a theory put forward by two physicists, Tsung-Dao Lee and Chen Ning Yang, for which Lee and Yang received the 1957 Nobel Prize in Physics. The theory was that parity, the symmetry between left and right, was not always conserved in nature. For beta particles, the prediction was that the result of beta decay would not be symmetrical because of certain weak forces. Dr. Wu's experiments showed clear evidence that violation of parity conservation was observed, and the results were announced in 1957. The second female National Medal of Science laureate, Dr. Wu received the award in 1975.

J. S. Tietjen et al., *Women Laureates 1970 – 2007*, Women in Engineering and Science, https://doi.org/10.1007/978-3-032-23116-1_3

Fig. 3.1 Chien-Shiung Wu—Courtesy of the Library of Congress

3.1 Life (1912–1997)

Chien-Shiung Wu was born in Liuhe (also Liu Ho), a small town near Shanghai, China. In Chinese, Chien-Shiung means "Courageous Hero," which is what her father hoped she would become. Her father, Wu Zong-Yee, a principal at a school for girls that he had established in Liuhe, surrounded Wu and her two brothers with books, encouraging them to read. He said, "I want every girl to have a school to go to." Her father recognized that life was changing in China and wanted to prepare his children for those changes as China modernized, as well as ground them in respect and knowledge of the old China. He was also a strong advocate of equal rights for women and told his daughter, "Ignore the obstacles. Just put your head down and keep walking forward." Her mother, Fan Fuhua, visited local families to encourage them to give their daughters an education and to stop binding their feet. Wu said of her childhood, "It was a wonderfully happy life. I had a fortunate and happy childhood" (McGrayne 1993; Yost 1959).

When Wu was nine years old, she completed the education that was available in her community and was sent to Soochow Girls School, where she stayed except for winter and summer vacations until she had completed her high school work, graduating in 1930 when she was seventeen years old. She had enrolled in the teacher training program because it was free but realized that she wanted to learn more science like the students in the academic school. Since all the students lived together, she worked out a solution: she borrowed the science books from the girls enrolled in the academic school after they had finished their homework and taught herself mathematics, physics, and chemistry. She liked mathematics and science, but when she discovered her affinity for physics, "Somehow I soon knew it was what I wanted to go on with" (McGrayne 1993; Yost 1959).

During the summer of 1930, Wu received a letter informing her that she had been selected to attend National Central University, an elite Chinese university, in Nanking. She wanted to major in physics but wasn't sure her background was good enough. She expressed this to her father, who said, "There's ample time to prepare yourself," and he brought home math, chemistry, and physics books for her to use. She studied all summer, enrolled as a mathematics student, and then transferred to physics. Years later, she said, "If it hadn't been for my father's encouragement, I would be teaching grade school somewhere in China now." She had a good college experience and graduated with her BS in 1934. She was the university's top student (McGrayne 1993; Yost 1959).

After her college graduation, Wu taught for a year and then did research for a year at the National Academy of Sciences in Shanghai. There, a woman colleague strongly suggested that Wu pursue graduate education in the USA. In fact, the woman suggested her alma mater—the University of Michigan. Funded by her uncle, Wu left China in 1936 and sailed to San Francisco, intending to attend the University of Michigan. She would never see her family again (McGrayne 1993).

Instead of going to Michigan, while she was in San Francisco, through family friends, she met Chia-Liu Yuan (later Luke Yuan), a physics graduate student at the University of California at Berkeley. He persuaded her to come to Berkeley and would later become her husband (1942). She had learned that Berkeley and its physics department were very famous, that the University of Michigan didn't allow women to use its student union building, and that there were already 600 Chinese students at Michigan. She didn't need too much persuading (McGrayne 1993; Shearer and Shearer 1997).

At Berkeley, she worked in the laboratory of Ernest Orlando Lawrence, inventor of the atom-smashing cyclotron, who received the Nobel Prize in Physics for that work in 1939, while she was there. Robert Oppenheimer was teaching there before he was called to the Manhattan Project. Wu received her PhD in 1940. She had focused on two areas. The first was X-ray radiation from beta decay in which she developed new methods of separating two types of rays during the disintegration and getting good agreement between the theory and the experiment. In the second, she worked with future Nobel Laureate in Physics (1959) Emilio Segré with whom she established "two complete chains of radioactive decay with the half-lives, radiations and isotope numbers completely identified." The latter work was classified but shared with the Manhattan Project investigators at Los Alamos National Laboratory. Segré treated her like his own daughter—lecturing her about working too hard and complimenting her incredible work ethic. She said, "I have always felt that in physics, and probably in other endeavors, too, you must have total commitment. It is not just a job. It is a way of life" (McGrayne 1993; Yost 1959). In 1937, Yuan accepted an offer at Caltech and moved to Southern California. Over the next 45 years, Dr. Wu and Yuan would be married but a commuting couple (McGrayne 1993).

After earning her PhD, Wu stayed at Berkeley as a research assistant in Dr. Lawrence's Radiation Laboratory for two years. She became regarded as a fission expert. However, Berkeley would not hire her—not one of the nation's top twenty physics institutions had hired a woman physics professor (McGrayne 1993). Instead, in 1942, when Yuan was working in Princeton, New Jersey, she moved to Smith College, where she taught physics. They met in New York City on weekends (McGrayne 1993). With a shortage of physicists during World War II, Ernest Lawrence recommended Dr. Wu for university positions. She got offers from eight universities and accepted the offer from Princeton. She and Yuan could be together. Now, at age thirty-one, Dr. Wu began teaching nuclear physics to Princeton's all-male student body. She was the first woman instructor ever hired at Princeton (McGrayne 1993).

Her stay at Princeton was short. She was called to an interview by the Division of War Research at Columbia University in New York City, where she would be working on the Manhattan Project. In March 1994, she interviewed with two physicists in their offices who did not reveal what they were working on but questioned her about physics for an entire day. At the end of the interview, they asked if she knew what they were doing, and she answered, "I'm sorry, but if you wanted me not to know what you're doing, you should have cleaned the blackboards." She started work the next day (McGrayne 1993; Yost 1959).

After the war, Dr. Wu was asked to stay on at Columbia University and returned to her studies of beta decay. Her son Vincent was born in 1947 (McGrayne 199). Dr. Wu was immersed in beta decay from 1946 to 1952. She was promoted to associate professor in 1952 with tenure and was looking for a new field to focus on (Yost 1959). In 1954, she and Yuan became naturalized US citizens (McGrayne 1993).

When Tsung Dao Lee came into her office in 1956 to ask for advice, she found a new area to explore—the laws of parity and symmetry. The experiment on the conservation of parity that she conducted solidified her worldwide reputation. Lee and Chen Ning Yang, the two men who developed the theory, won the Nobel Prize. Her experiment took about nine months during 1956 and 1957 (McGrayne 1993). Columbia University called a press conference to announce the results and Dr. Wu became a worldwide sensation. From the *New York Post*,

> This small modest woman was powerful enough to do what armies can never accomplish: she helped destroy a law of nature. And laws of nature, by their very definition, should be constant, continuous, immutable, indestructible (McGrayne 1993).

Dr. Wu eventually became a full professor at Columbia (1958), then held an endowed professorship (1972), and retired from Columbia in 1981 (Ogilvie and Harvey 2000; Shearer and Shearer 1997). She and Steven Moszkowski wrote the classic textbook *Beta Decay,* which was published in 1966. It became the standard reference on the topic and is still in use today (Wikipedia 2025).

Although she did not receive a Nobel Prize for her efforts, Dr. Wu was elected to the National Academy of Sciences in 1958 (the seventh woman). That same year, she received the first honorary doctorate of science awarded to a woman by Princeton University. Princeton said that she had "richly earned the right to be called the world's foremost female experimental physicist." She would receive almost a dozen more honorary degrees in the USA and abroad, including from Yale and Harvard. She was also elected to membership of the Academia Sinica (Academy of Sciences in China) (McGrayne 1993; Ogilvie and Harvey 2000; Yost 1959).

Dr. Wu received the first Wolf Prize awarded by Israel (1978) and the first Research Corporation Prize given to a woman (1959). She was the first woman to win the Comstock Award of the National Academy of Sciences (1954). In 1975, she won the National Medal of Science. She was the first living scientist with an asteroid named after her (1990) (McGrayne 1993). She became the first woman president of the American Physical Society (McGrayne 1993). Dr. Wu was named a Fellow of the American Academy of Arts and Sciences, the American Association for the Advancement of Science, and the American Physical Society. She was an honorary Fellow of the Royal Society of Edinburgh (Ogilvie and Harvey 2000). Posthumously, Dr. Wu was inducted into the National Women's Hall of Fame (NWHF 2025). A U.S. postage stamp featuring Dr. Wu was released in 2021 (Smithsonian National Postal Museum 2021; USPS 2021).

A strong advocate for women in science, Dr. Wu said, "Men have always dominated the fields of science and technology. Look what an environmental mess we are in." She also said,

> Bringing a womanly point of view may be advantageous in some areas of education and social science, but not in physical and mathematical sciences, where we strive always for objectivity. I wonder whether the tiny atoms and nuclei, or the mathematical symbols, or the DNA molecules have any preferences for either masculine or feminine treatment.
>
> I sincerely doubt that any open-minded person really believes in the faulty notion that women have no intellectual capacity for science and technology. Nor do I believe that social and economic factors are the actual obstacles that prevent women's participation in the scientific and technical field. The main stumbling block in the way of progress is and always has been unimpeachable tradition.
>
> It is shameful that there are so few women in science. This is the fault of men. In Chinese society, a woman is valued for what she is, and men encourage her to accomplishments yet she remains eternally feminine (McGrayne 1993; NSTMF 2025).

3.2 Beta Decay and the Law of Parity

Beta decay occurs when the nucleus of an atom ejects a very fast electron and a neutrino. In the process, it becomes a different element. Generally, electrons are not in the nucleus. However, in beta decay, a neutron within the nucleus breaks into three different parts—a proton, an electron, and a neutrino. The electron and the neutrino come out of the nucleus at high speed leaving only the proton which is now inside a new, stable nucleus (McGrayne 1993). Physicist Enrico Fermi had postulated a theory in 1933 that predicted how the nucleus would behave during beta decay. Previous experiments conducted around the world did not prove this theory. Dr. Wu's experiments resulted in exactly the speeds predicted by Fermi (McGrayne 1993).

During the spring of 1956, Tsung Dao Lee, a physics professor at Columbia University, came to Dr. Wu's office to ask for advice. He was working with Chen Ning Yan of the Institute for Advanced Study at Princeton, and they had a problem. A new particle, the K-meson, did not decay in the manner that had been expected. Sometimes when it decayed, two particles resulted. But sometimes, three particles resulted. The physics world assumed that the K-meson was actually two different particles. However, Lee and Yang postulated that the K-meson was one particle, but it was decaying in two different manners (McGrayne 1993; Yost 1959).

But there was an issue. If Lee and Yang's theory was correct, then a basic law of physics would be found to be violated. At that time, the law of parity and the law of symmetry said that molecules, atoms, and nuclei behave symmetrically. This meant that it would look the same whether observed directly or in a mirror. But Lee and Yang thought that the molecules in the nucleus might be either right-handed or left-handed (McGrayne 1993).

Dr. Wu's advice to Lee was to check the literature to see if anyone had ever proven through experiments that parity was valid inside the nucleus. Lee and Yang discovered that no one had. They wrote a paper that outlined possible experiments that could be conducted but admitted that such experiments were too difficult (McGrayne 1993).

Dr. Wu wanted to test it. "This was a golden opportunity for a beta decay physicist to perform a crucial test, and how can I let it pass?" Dr. Wu and Yuan had booked a cruise to celebrate the 20th anniversary of their departure from China in the Spring of 1956. Yuan went by himself. Dr. Wu had work to do, and she realized that the experiment had to be initiated immediately (McGrayne 1993).

Months of preparation were required to design and test equipment. She collaborated with the National Bureau of Standards Low Temperature Physics Group in Washington, DC, which had a laboratory that could cool materials almost to absolute zero. Her team needed to make cerium magnesium nitrate crystals. After being told it would take months, she and her team consulted a German chemistry book and made the crystals themselves. Her graduate student made the first crystals in a beaker in her kitchen while she was cooking dinner! Other materials used included a dentist's drill, Duco cement, dish soap (Palmolive was good, but Ivory was better), and nylon string. The team worked around the clock. And the experiment showed that more electrons come out of one end of the nucleus than out of the other. The particles were not symmetrical. The bottle of champagne to celebrate their success was opened on January 9, 1957. Very quickly, in order to avoid being scooped, they wrote the article to publish the results (McGrayne 1993; Proffitt 1999).

Dr. Wu said,

> The overthrow of the parity law drives home again the idea that science is not static, but ever-growing and dynamic. It is the courage to doubt what has long been established, the incessant search for its verification and proof that pushed the wheel of science forward (Yost 1959).

She also said,

> These were moments of exhilaration and ecstasy! A glimpse of this wonder can be the reward of a lifetime. Could it be that excitement and ennobling feelings like these have kept us scientists marching forward forever? (Shearer and Shearer 1997).

Later, Dr. Wu conducted another experiment using beta decay, this time to confirm, instead of to refute, a different law of physics. This law was that the vector current, electric charge per time per area, would be conserved during beta decay. This would mean that vector current is analogous to the electromagnetic force. This law of physics had been hypothesized by physicists Richard Feynman and Murray Gell-Mann of Caltech. In 1959, Gell-Mann asked Dr. Wu to conduct the needed experiment. Finally, in 1963, she performed the experiment that confirmed the law (McGrayne 1993).

3.3 Conclusion

Nuclear physicist Chien-Shiung Wu conducted the experiment that proved the law of conservation of parity was not observed in nature and for which the two men who developed the theory received the Nobel Prize in Physics. The first woman to receive many scientific prizes and other awards, Dr. Wu was a strong advocate for women in science.

References

McGrayne, Sharon Bertsch, *Nobel Prize Women in Science: Their Lives, Struggles, and Momentous Discoveries*, New York, New York: Carol Publishing Group, 1993.

National Science and Technology Medals Foundation (NSTMF). Chien-Shiung Wu. https://nationalmedals.org/laureate/chien-shiung-wu/. Accessed July 11, 2025.

National Women's Hall of Fame (MWHF). Chien-Shiung Wu. https://www.womenofthehall.org/inductee/chienshiung-wu/. Accessed July 11, 2025.

Ogilvie, Marilyn and Joy Harvey, Editors. 2000. The Biographical Dictionary of Women in Science: Pioneering Lives from Ancient Times to the Mid-20th Century. New York, New York: Routledge.

Proffitt, Pamela, Editor, *Notable Women Scientists*, Detroit, Michigan: The Gale Group, 1999.

Shearer, Benjamin F. and Barbara S. Shearer, Editors. 1997. Notable Women in the Physical Sciences. Westport, Connecticut: Greenwood Press.

Smithsonian National Postal Museum. 2021. Forever Chien-Shiung Wu single. https://postalmuseum.si.edu/object/npm_2021.2010.77. Accessed July 11, 2025.

United States Postal Service (USPS). 2021. Nuclear Physicist Chien-Shiun Wu to be Honored on a U.S. Postal Service Commemorative Forever Stamp. https://about.usps.com/newsroom/national-releases/2021/0201ma-nuclear-physicist-chien-shiung-wu-to-be-honored-on-forever-stamp.htm. Accessed July 11, 2025.

Wikipedia.org. Chien-Shiung Wu. https://en.wikipedia.org/wiki/Chien-Shiung_Wu. Accessed July 13, 2025.

Yost, Edna. 1959, Women of Modern Science. New York, New York: Dodd, Mead & Company.

Chapter 4
Elizabeth C. Crosby

For outstanding contributions to comparative and human neuroanatomy and for synthesis and transmission of knowledge of the entire nervous system of the vertebrate phylum

Keywords Brain · Anatomy · Anatomic structure · Neuroanatomy · Forebrain · Midbrain · Diencephalon · Telencephalon · Brain structure · Vertebrates · Vertebrate brain · Optic tectum · Nervous system

Dr. Elizabeth C. Crosby, shown in Fig. 4.1, studied the vertebrate nervous system with an emphasis on the detailed electrical connections of the brain. By examining brain structures in reptiles, birds, and mammals, she was able to determine the impact of evolution. Dr. Elizabeth C. Crosby received the National Medal of Science in 1980.

J. S. Tietjen et al., *Women Laureates 1970 – 2007*, Women in Engineering and Science, https://doi.org/10.1007/978-3-032-23116-1_4

Fig. 4.1 Dr. Elizabeth Crosby. (Smithsonian Institution)

4.1 Life (1888–1983)

Elizabeth Caroline Crosby was born in Petersburg, Michigan, on October 25, 1888, to parents Lewis Frederick Crosby and Frances Kreps Crosby. At birth, she was frail, and her parents would occasionally put her in the warm oven to bring color to her cheeks. She grew up as an only child and was very curious from a young age. Elizabeth's mother read to her but often got pulled away by household duties. Impatient with the breaks in the stories, Elizabeth taught herself how to read. By the time she reached elementary school, she had already read books ranging from *Jane Eyre* to *The History of the French Revolution*. Throughout grade school, Elizabeth was a model student, displaying a high level of curiosity and ability (Woodburne (a) 1959).

Upon graduating from high school, Elizabeth's father promised to support her through four years of college. She attended Adrian College in Adrian, Michigan, majoring in mathematics. Although it was a four-year program, Elizabeth completed her BS in 1910, after only three years. While attending Adrian College, Elizabeth was mentored by Dr. Elmer Jones, her teacher in science classes, including elementary zoology. He was sufficiently impressed with Elizabeth's abilities that he suggested she go to university to pursue a graduate degree. He specifically recommended that she study neuroanatomy under Dr. C. Judson Herrick at the University of Chicago. With her father's promise of financial support for one more year at school, Elizabeth opted to go to the University of Chicago as recommended (Woodburne (a) 1959; Saleh et al. 2022).

When Elizabeth showed up at the University of Chicago, she met with Dr. Herrick and explained her interest in working with him. Although Dr. Herrick knew Dr. Jones and respected his opinion of Elizabeth, he noted that Elizabeth had only attended a single biology course. At the University of Chicago, all the anatomy courses required at least as much general science as is demanded for medical school admission, so Dr. Herrick suggested that she enroll in the Department of Zoology and devote a year to making up the deficiencies in her background (physics, chemistry, and biology) before taking classes in the Department of Anatomy. In response, Elizabeth is reported to have said, "No, I came to study with Professor Herrick. Let me try (Herrick 1959)."

Impressed by the determination of Elizabeth, Dr. Herrick offered her the chance to join medical students in the dissecting room and to let her register for the class in gross anatomy. Elizabeth immediately accepted these offers. Dr. Herrick also let her attend his course in neuroanatomy, but he suggested she not take the final exam as she would be so ill prepared. Elizabeth did attend the class and spent the first few weeks furiously trying to learn the background material required to understand the lectures. After those first few weeks, she was performing as well as the medical students in the class, so she did take and pass the final exam (Herrick 1959).

As she was pursuing her Master's degree, Elizabeth spent so much time in the anatomy laboratory that faculty grew concerned about her health. They confiscated her keys to prevent her from entering during the evenings and weekends. A few days later, Dr. Herrick found her walking home with books under one arm and a microscope and slides under the other. He concluded that it simply wasn't possible to keep her from working, and so he returned the keys to the lab to her (Woodburne (b) 1959). She completed her Master's degree in 1912 and was granted a fellowship in the anatomy department to pursue a PhD.

From 1912 to 1915, Elizabeth worked on her PhD dissertation on the structure of the forebrain (front brain organs) of the American alligator, graduating magna cum laude. As stated by Dr. Herrick, "Her dissertation on the microscopic structure of the forebrain of the alligator is a model of exact and concise presentation of intricate detail illustrated by beautifully clear drawings (Herrick 1959)." The dissertation was widely read by neuroanatomists and established her as an important researcher in the field.

Upon completing her PhD, Dr. Crosby returned home to Petersburg to help care for her parents, who were becoming frail. She initially taught Latin, mathematics, and zoology at the local high school but was elevated to principal of the school in 1916. She also served as the coach of the boys' basketball team. In 1918, Dr. Crosby became the Superintendent of Schools in Petersburg, a role she served for two years. It was punctuated by the death of her mother in 1918 (Woodburne (a) 1959).

In 1920, Dr. Crosby sought to return to neuroanatomy research by moving to the University of Michigan. Dr. G. Carl Huber, Director of the Anatomical Labs there, appointed her as a junior instructor, marking the beginning of her long career at the University of Michigan Medical School's Anatomy Department and her long collaboration with Dr. Huber. Dr. Crosby and her father moved to Ann Arbor, Michigan, and lived together until her father's death in 1923 (Woodburne (b) 1959).

Throughout much of her career, the aim of Dr. Crosby's work was to study the brain from a phylogenetic perspective—to understand differentiation of the brain structures as one climbed the evolutionary ladder. She and her collaborators studied reptiles, birds, and mammals (including humans). With Dr. Huber, Dr. Crosby became an expert in the forebrain of vertebrates, detailing the fiber pathways connecting various parts of the brain and allowing information exchange. They also studied the optic tectum, the part of the brain that processes visual information and controls eye and head movements.

In 1926, Dr. Crosby went abroad to study at the University of London and at the Institute for Brain Research in Amsterdam. Additionally, her appointment at the University of Michigan was upgraded to Assistant Professor in the Department of Anatomy. She was promoted to Associate Professor in 1929 (Woodburne (b) 1959).

Drs. Crosby and Huber were approached by Dr. C. U. Ariëns Kappers at the Netherlands Central Institute for Brain Research to help with updating a textbook on the nervous system of vertebrates. Largely through Dr. Crosby's work, the book was greatly revised and expanded. The two-volume set, entitled *The Comparative Anatomy of the Nervous System of Vertebrates, Including Man*, comprised 1900 pages with 700 images (Kappers et al. 1936). It was published in 1936 and immediately became the textbook used in training medical students and neuroanatomists. Unfortunately, Dr. Huber had been diagnosed with leukemia in 1934 and died only a few months after his diagnosis. He did not live to see the publication of the work that had occupied Dr. Crosby and him for years (Woodburne (a) 1959).

After Dr. Huber died, Dr. Crosby took over his teaching and research in neuroanatomy. She was promoted to professor in 1936, becoming the first woman to earn that title at the University of Michigan Medical School. She gained a reputation as a stellar teacher, which she attributed to having had to teach herself the basics when she arrived at the University of Chicago. Her philosophy of teaching was that a course should be pitched to challenge the best mind, but that ample opportunities for helping students understand should be built in (Wikipedia; Woodburne (b) 1959).

The transition to a senior role in the Department of Anatomy was not without its difficulties. Dr. Crosby wrote a resignation letter sent to Dr. Albert Carl Furstenberg, Dean of the Medical School, dated February 21, 1935. She cited her gratitude for the promotions she had been given, but then went on to say,

> It would seem that a woman in my position, as second ranking member in the Anatomy Department, might, under all the circumstances, prove a source of embarrassment to you as Dean and to Executive Faculty of the Medical School. This has been apparent to me since Dr. Huber's death (Ziehn 2022).

Dr. Crosby proposed that she be demoted but kept on until she could find another position. Dean Furstenberg refused to accept her resignation.

At the time, the Chair of the Department of Anatomy was Dr. Bradley Patten. Dr. Patten forced Dr. Crosby to use a large model of the human nervous system for her classes. It was nicknamed "Patten's Folly." He also dismantled the research library that Dr. Crosby had set up for Dr. Huber. In October 1937, Dr. Crosby again wrote a resignation letter, this time addressing it to Dr. Patten and Dr. Furstenberg. Dean Furstenberg wrote back to Dr. Crosby saying that "she had the profound admiration and respect of the President of the university, myself, the Director of your Department, and every member of the Medical Faculty as well as the student body (Ziehn 2022)." She was once again asked to remain and chose to stay.

In 1939 and 1940, Dr. Crosby visited the Marischal College of the University of Aberdeen (Scotland) and organized a curriculum for them in microscopical anatomy and neuroanatomy. The University of Aberdeen offered her a permanent appointment, which she considered seriously before deciding to return to the University of Michigan. When she did return, she brought with her Kathleen Rosena Robb, a girl she had met in Aberdeen and whom she later adopted. It is unclear what circumstances led to this adoption, but the Robb family were Dr. Crosby's neighbors in Aberdeen, and the adoption was with their permission and likely at their request. A few years after returning to Michigan, Dr. Crosby also took on the care and education of Suzanne McCotter, a native of Detroit, supporting her until she landed a teaching job (Woodburne (a) 1959).

In her research, Dr. Crosby engaged in long programs of study of regions of the brain. She began with the forebrain, then progressed to the midbrain (the upper part of the brain stem) in the 1940s. Although Dr. Crosby regularly rounded with neurosurgical staff and was frequently consulted to discuss underlying neurological deficits patients exhibited, her research through the 1940s was solely focused on brain structure and not on function. That changed in the 1950s, when Dr. Crosby became more interested in the link between brain structure and function. She studied patient cases and even expanded lab work to include studies of the impact of damage to various brain areas.

In 1958, Dr. Crosby retired from the University of Michigan. Her resignation was insisted upon by Dr. Patten, who wanted Dr. Crosby to retire before him so that she wouldn't overshadow him at retirement, and he could be honored. As of her retirement, Dr. Crosby had supervised 39 PhD students. Of these, over a quarter were women, an unusually large fraction for that time. She had also taught 8500 medical students (Saleh et al. 2022; Ziehn 2022).

In 1963, Dr. Crosby joined the University of Alabama at Birmingham as an Emeritus Professor. At the University of Alabama at Birmingham, Dr. Crosby organized a group of researchers to study comparative neurobiology. She was involved in the research of 16 PhD students there. She maintained a pace of teaching and research not much different from her pre-retirement years (Saleh et al. 2022).

Dr. Crosby published two books in addition to *The Comparative Anatomy of the Nervous System of Vertebrates, Including Man.* She co-authored *Correlative Neurosurgery* with Dr. Edgar A. Kahn in 1969 (Kahn 1969). A second edition was issued in 1969 and an expanded third edition in 1982. In 1962, Dr. Crosby published *Correlative Anatomy of the Nervous System*, with co-authors Drs. Tryphena Humphrey and Edward W. Lauer (Crosby et al. 1962).

Dr. Crosby amassed an impressive number of awards throughout her career. In 1946, she was the first woman named as the Henry Russell Lecturer at the University of Michigan. The Henry Russell Lectureship is considered the highest honor for a senior member of the active faculty and recognizes exceptional achievement in research or teaching. In 1950, Dr. Crosby received the Achievement Award of the American Association of University Women, an award given to an outstanding scholar whose work spans at least 20 years. In 1956, she was awarded the Distinguished Faculty Achievement Award of the University of Michigan. This award recognizes outstanding achievement in research, teaching, or mentoring. In 1969, Dr. Crosby received the Karl Spencer Lashley Award of the American Philosophical Society in recognition of her work to understand the neuroscience of behavior. In 1972, she received the Henry Gray Scientific Achievement Award of the American Association of Anatomists, given to recognize unique and exceptional work in the science of anatomy. In 1980, in addition to receiving the National Medal of Science, Dr. Crosby was named a Distinguished Faculty Lecturer at the University of Alabama at Birmingham. Additionally, Dr. Crosby was inducted into several international societies and received ten honorary doctorates from various national and international universities.

Dr. Crosby's legacy is preserved through her published papers and books. In 1957, the University of Michigan Medical School created the Elizabeth C. Crosby Award for outstanding teaching in the basic sciences. A professorship in her name has also been created at the University of Michigan in cell and developmental biology. Additionally, there is a campus award in her name given to young female academics in science and engineering with funding awards up to $20,000 (Saleh et al. 2022).

In her later years, Dr. Crosby shared a home off and on with Tryphena Humphrey, a former graduate student, and they often traveled together. Dr. Crosby died in 1983. She was remembered as the foremost authority in the world on human and comparative neuroanatomy, possessing an unusually detailed and yet broad knowledge of the entire nervous system of the entire vertebrate phylum. She was also remembered fondly as a nice person.

> Despite the demands of her profession, Dr. Crosby was a warm and compassionate person and extremely generous… In a world hungry for power, wealth, and status, Elizabeth Caroline Crosby was an anachronism of modesty and self-effacement" (Alabama Women's Hall of Fame).

4.2 The Structure of the Brain

The brain is the most complex organ in the body, tasked with regulating bodily functions, interpreting senses, establishing intelligence, controlling emotions and behavior, and initiating movement. In humans, the brain weighs about three pounds and is comprised of gray matter on the outside and a white matter interior. The gray matter is made up of neuron somas, cell bodies that are responsible for processing information. The white matter of the brain is primarily composed of axons wrapped in myelin, which are responsible for transmitting information (Johns Hopkins Medicine). This is shown schematically in Fig. 4.2.

The brain structure is divided into three regions based upon geographic location in the skull: the forebrain or prosencephalon, midbrain or mesencephalon, and hindbrain or rhombencephalon. This is shown in Fig. 4.3. The work of Dr. Crosby concentrated on understanding the forebrain structure, although she did study the midbrain in the latter part of her career. She did not carry out a long study of the hindbrain (Britannica).

The forebrain is further subdivided into the diencephalon and the telencephalon. The diencephalon is a fluid-filled sac containing the thalamus, hypothalamus, epithalamus, and subthalamus, as shown in Fig. 4.4. This region of the brain is responsible for relaying sensor and motor signals to other parts of the brain, regulating the sleep-wake cycle, signaling the endocrine system to release hormones, regulating body temperature, controlling appetite, helping establish the sense of smell, regulating consciousness, and controlling the autonomic nervous system. These functions require connections between various brain organs. It is these fiber paths that Dr. Crosby described in her research (Britannica; Johns Hopkins Medicine).

The telencephalon is the cerebrum that covers the diencephalon, shown in darker color in Fig. 4.4. This part of the brain initiates and coordinates movement and

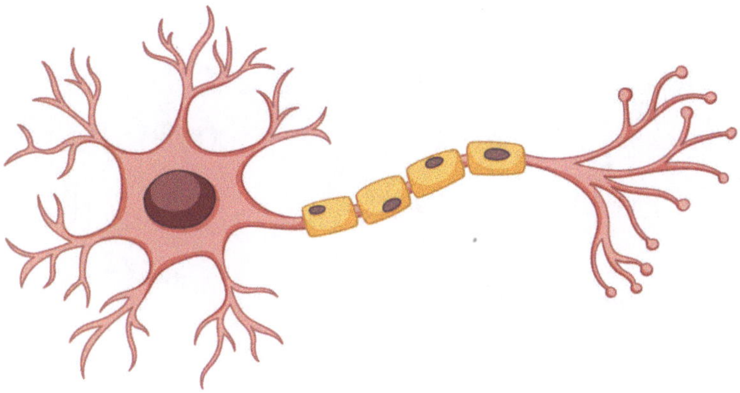

Fig. 4.2 The gray and white matter of the brain. (Modified from Freepik)

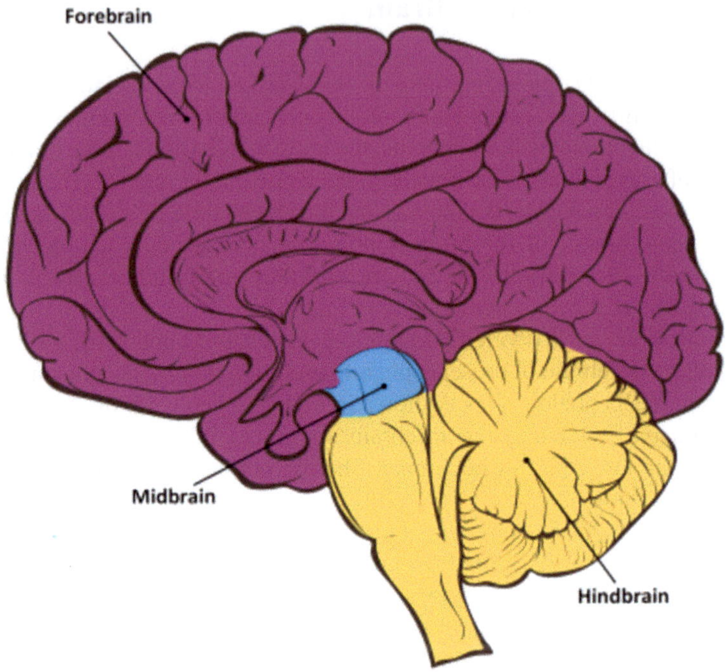

Fig. 4.3 Major divisions of the brain. (National Institute of Neurological Disorders and Stroke, public domain)

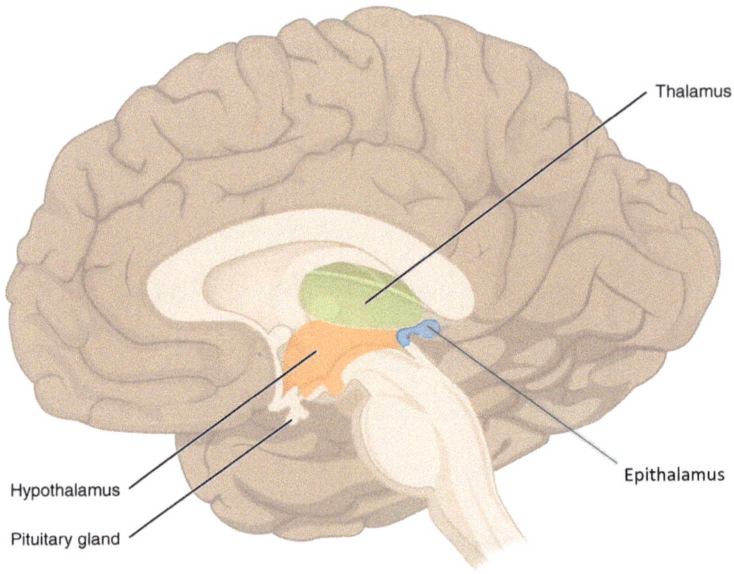

Fig. 4.4 The diencephalon. (Lumen anatomy Section 6)

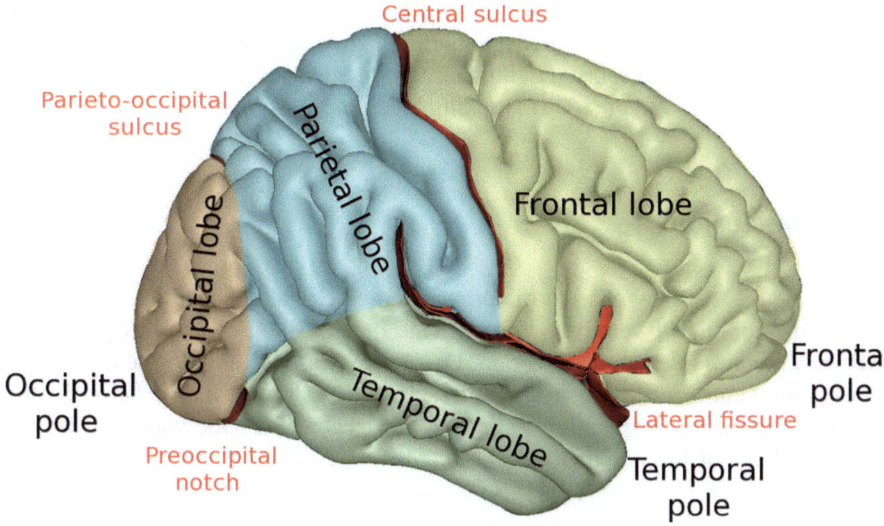

Fig. 4.5 Lobes of the cerebrum. (From Wikipedia CC-BY-SA 4)

enables speech, judgment, thinking and reasoning, emotions, and learning. It also plays a role in sensing. The cerebrum is divided into two halves with a large fissure running between them from the front to the back of the brain. These halves are connected by a large white matter organ called the corpus callosum, which allows them to communicate with each other. The right half of the cerebrum controls the left side functions of the body, while the left side of the cerebrum controls the right side. The cerebrum covering is gray matter that is called the cerebral cortex. It is this covering which exhibits intricate folds, thus creating a large surface area (Johns Hopkins Medicine).

The cerebrum is divided into four lobes, as shown in Fig. 4.5: occipital, parietal, frontal, and temporal. The frontal lobe is associated with personality traits, decision-making, movement, recognition of smells, and speech generation. The parietal lobe deals with identifying objects, understanding spatial relationships, interpreting pain and touch, and comprehending speech. The occipital lobe is associated with vision. The temporal lobe controls short-term memory, musical rhythm, and speech and smell recognition (Johns Hopkins Medicine).

The midbrain connects the cerebrum with the spinal cord. It facilitates hearing, movement, calculating responses, and responding to changes in the environment. Contained in the midbrain is the optic tectum, an organ Dr. Crosby studied extensively. In vertebrates such as fish, reptiles, and birds (but not humans), the optic tectum is the main processor of vision. Dr. Crosby studied the optic tectum as one means of showing evolutionary changes in the brain. The midbrain also includes the medulla, which regulates involuntary functions such as heart rhythm, breathing, blood flow, and oxygen levels. The medulla additionally produces reflexive activities such as sneezing, vomiting, coughing, and swallowing (Johns Hopkins Medicine).

The hindbrain consists of the cerebellum. This part of the brain is responsible for voluntary muscle movements and maintains posture and balance. Although it is small compared to the cerebrum in humans, it contains nearly a third of all the neurons of the nervous system (Johns Hopkins Medicine).

4.3 Conclusion

Dr. Elizabeth C. Crosby charted the organs and electrical connections in the forebrain and midbrain of vertebrates in intricate detail. She used the information to study how evolution changed the brain. Dr. Crosby was also able to use her knowledge of the central nervous system to teach neurosurgeons and to help identify structural deficits corresponding to neurological behaviors. She was awarded the National Medal of Science in 1980.

References

Alabama Women's Hall of Fame, Elizabeth Caroline Crosby (1888-1983), https://www.awhf.org/crosby.html. Accessed May 27, 2025.

Britannica, Brain, https://www.britannica.com/science/brain. Accessed June 30, 2025.

Crosby, Elizabeth, Humphrey, Triphena, and Lauer, Edward. 1962. *Correlative Anatomy of the Nervous System*. Macmillan.

Herrick, Charles. 1959. Tribute, *J Comparative Neurology* 112(1).

Huber, G. and Crosby, Elizabeth. 1929. The Nuclei and Fiber Paths of the Avian Diencephalon, with Consideration of Telencephalic and Certain Mesencephalonic Centers and Connections, *J. Comparative Neurology* 48(1).

Johns Hopkins Medicine, Brain Anatomy and How the Brain Works, https://www.hopkinsmedicine.org/health/conditions-and-diseases/anatomy-of-the-brain. Accessed June 30, 2025.

Kahn, Edgar, Crosby, Elizabeth, et al. 1969. Correlative Neurosurgery, Charles C. Thomas.

Kappers, C, Huber, G, and Crosby, Elizabeth. 1936. The Comparative Anatomy of the Nervous System of Vertebrates, Including Man (2 Volumes). Macmillan.

Lumen Anatomy Section 6:The Central Nervous System, The Diencephalon: Thalamus, Hypothalamus, Epithalamus, https://courses.lumenlearning.com/suny-dutchess-anatomy-physiology/chapter/the-diencephalon-thalamus-hypothalamus-epithalamus/#:~:text=The%20diencephalon%20is%20composed%20primarily,make%20contact%20in%20the%20middle. Accessed May 27, 2025.

National Institute of Neurological Disorders and Stroke, Brain Basics: Know Your Brain, https://www.ninds.nih.gov/health-information/public-education/brain-basics/brain-basics-know-your-brain. Accessed May 27, 2025.

Saleh, Sara, Holste, Katherine, et al. 2022. Elizabeth Crosby: Titan of Neuroanatomy and Friend of Neurosurgery, *Neurosurgery* 90:214–242.

Wikipedia, Elizabeth C. Crosby, https://en.wikipedia.org/wiki/Elizabeth_C._Crosby. Accessed June 30, 2025.

Woodburne, Russel (a). 1959. Elizabeth C. Crosby. A biographical sketch, Wiley Online Library, https://onlinelibrary.wiley.com/doi/pdf/10.1002/cne.901120105. Accessed May 27, 2025.

Woodburne, Russell (b) 1959. Tribute, *J. Comparative Neurology* 112(1).

Ziehn, Lara. 2022. No Resignation, *Michigan Medicine* Feb 7, https://bentley.umich.edu/news-events/magazine/no-resignation/. Accessed May 27, 2025.

Chapter 5
Mildred Cohn

For pioneering the use of stable isotropic tracers and nuclear magnetic resonance spectroscopy in the study of the mechanisms of enzymatic catalysis

Keywords Nuclear magnetic resonance · NMR · Biophysics · Stable isotopes · Electron spin · Molecular biology · Metabolites · Isotropic tracers · Spectroscopy · Enzymatic catalysis · Metabolic processes · Biochemistry · MRI · Magnetic resonance imaging · Contrast agents · Electron paramagnetic resonance · EPR

Mildred Cohn (Fig. 5.1), born and raised in New York City by Russian Jewish immigrants, was encouraged to pursue education from an early age. Although she experienced discrimination throughout her education and career due to her gender and only became a faculty member twenty years after she completed her PhD, Dr. Cohn was among the first scientists to work with nuclear magnetic resonance in investigating metabolism. A leader in using physical chemistry to examine biological questions, Dr. Cohn pioneered in understanding chemical reactions at the atomic level. She used the new and emerging equipment of the time—often building it herself—mass spectrometer, nuclear magnetic resonance, and electron paramagnetic resonance—to determine organic chemical reactions. Her areas of research included studies of metabolism, electronic spin, enzyme catalysis (how enzymes behave during chemical reactions in the body), and ATP (adenosine triphosphate). She received the National Medal of Science in 1982 (American Chemical Society; American Philosophical Society; National Science & Technology Medals Foundation; Zagorski a 2009).

J. S. Tietjen et al., *Women Laureates 1970 – 2007*, Women in Engineering and Science, https://doi.org/10.1007/978-3-032-23116-1_5

Fig. 5.1 Mildred Cohn. (Courtesy of the American Philosophical Society)

5.1 Life (1913–2009)

Mildred Cohn was born in the Bronx (New York City) on July 12, 1913, the second of two children of Russian Jewish emigrants (American Philosophical Society). Her father, Isidore Cohn, told her that she could achieve great success, if she put her mind to it. He had trained as a rabbi in Russia and worked in the garment and printing industries once he was in the USA (Barrer). Although not very mechanically inclined, he had invented a machine for cutting trousers (Gortler 1987&1988). Her mother, Bertha Klein Cohn, also valued learning and education. Thus, both she and her brother Albert learned the value of education and were encouraged to excel at an early age. When Mildred was 13 years old, the Cohn family moved to a Jewish collective that emphasized learning and social justice (Kaiser).

Cohn graduated from high school at age 14, having skipped grades along the way. She became interested in chemistry through the father of one of her friends from school. He owned beauty salons and manufactured beauty products and told her that chemistry was a wonderful career for women (Gortler 1987&1988). Although the dean at her high school expected Cohn would enter the workplace after her high school graduation, Cohn knew she was going to college. Her family was aware that Hunter College was free and in New York City, so at age 15, off to Hunter College she went (American Chemical Society; Gortler 1987&1988).

She studied physics and chemistry at Hunter, but there was no physics major there (Gortler 1987&1988). "It wasn't offered as a major probably because it was assumed that girls shouldn't or didn't want to be physicists" (Zagorski b). She majored in chemistry. A professor of physical chemistry was "highly competent. And that's probably the reason I decided to become a physical chemist." She graduated from Hunter College in 1931 with a BA in chemistry cum laude during the Great Depression and its paucity of jobs (American Chemical Society; Gortler 1987&1988).

Cohn was the only student from her graduating class at Hunter College who went to graduate school. She chose to attend Columbia University because she had saved money from her summer and weekend jobs to pay tuition and could live at home—so there was no cost for room and board (Gortler 1987&1988). She had tried to get scholarships but was unsuccessful. She earned her MS in chemistry from Columbia in 1932 (Gortler 1987&1988; Kresge et al. 2009).

After completing her master's degree, Cohn needed to earn money. She had taken the civil service examination for chemists while she was a senior in college, and near the end of the year that it took her to get her master's, she got a job offer. The National Advisory Committee for Aeronautics (NACA), the predecessor organization to NASA, offered her a job at Langley Field in Hampton, Virginia, performing computations in aerodynamic research as a junior scientific aide, not as a junior chemist. She accepted the offer but did not enjoy the work. She instead found a job elsewhere at NACA. She worked in the engine laboratory as a chemist for a couple of years, publishing two papers and devising the equipment that she needed (Gortler 1987&1988). She was the only woman along with 70 men. The workplace was not promising for women scientists, and pay equity was out of the question. She took a leave of absence and returned to Columbia to work on her PhD (Gortler 1987&1988).

The work she had done in the engine laboratory at NACA had interested Cohn, and she thought she would get a PhD in chemical engineering at Columbia, especially with her papers. However, the chair of chemical engineering at Columbia told her they hadn't had any women students in their department, and they didn't plan to start with her (Gortler 1987&1988).

Back to the chemistry department she went. Cohn says, "I had long since decided [when she was working on her master's if she were to pursue a PhD] that I wanted to work with [Harold] Urey and then he was awarded the Nobel Prize that year (1934). So I went to see him." Urey told Cohn that she didn't want to be his graduate student—that he ignored them. Nevertheless, she convinced him to take her on (Gortler 1987&1988). Urey was the first of a number of Nobel Laureates she would be affiliated with over the course of her education and career (Bernard Becker Medical Library).

Under Urey, she studied ways to separate different isotopes of carbon but experienced equipment problems. The experiments associated with the theoretical calculations she performed needed to be conducted on a mass spectrometer, but the one at Columbia was not yet operational—it was still being built. She drove her samples to Princeton and used the mass spectrometer there, but her results were not good. Thus, instead of carbon, she turned to the behavior of isotopes of oxygen. With a now operational mass spectrometer at Columbia, she was able to complete the needed experiments. Her PhD in physical chemistry was awarded in 1938, the same year she married physicist Henry Primakoff (American Chemical Society). Her dissertation title was *Oxygen Exchange Reactions of Organic Compounds and Water* (Proffitt 1999). Her thesis is considered a foundation paper in organic chemistry. It not only had equations but also mechanisms and was still being cited fifty years later (Gortler 1987&1988).

Finding a job after she completed her PhD was difficult for Dr. Cohn. In fact, discrimination against women would affect her during her entire career. She was ineligible for teaching assistant positions at Columbia while pursuing her graduate studies because such positions were awarded to men only (American Chemical Society). In addition, at the time she was completing her PhD, corporations posted notices to upcoming graduates at Columbia that read "Interviewing all Ph.D. candidates, male and Christian" (Barrer; Kresge et al. 2009). Therefore, she wasn't qualified for and couldn't interview for those job postings—being female and Jewish.

Urey even looked for jobs for her but said no one wanted to hire a woman chemist. With this awareness of the rampant discrimination of the time, he told Dr. Cohn that she could continue in his lab as a postdoc, which she agreed to. Almost immediately, Dr. Cohn was contacted by personnel at Columbia's medical school who told her that they had a postdoc position for her with Vincent du Vigneaud in biochemistry who was using deuterium as a tracer, i.e., as a substance inserted into specimens and followed (Gortler 1987&1988). Dr. Cohn said, "every member in the chemistry faculty personally congratulated me when I got the position with du Vigneaud, because obtaining an academic position, even as a lowly postdoc, was such a rare achievement, particularly for a woman" (Zagorski a 2009).

du Vigneaud was at George Washington University (GWU) Medical School in Washington, DC, and initially he did not want to hire Dr. Cohn because she was a woman. He relented when Urey told him that Dr. Cohn was the only person in the country who was qualified. She knew how to handle isotopes—stable or otherwise. Urey strongly urged Dr. Cohn to accept the postdoc, and she did (Gortler 1987&1988; Proffitt 1999). She now began her association with a Nobel Laureate-to-be (Gortler 1987&1988). Dr. Cohn worked with du Vigneaud at GWU, without Henry (who had stayed in New York City, where he was teaching physics at the Polytechnic Institute of Brooklyn), until du Vigneaud moved—and she moved—to New York City to work at the Department of Biochemistry at Cornell Medical College (American Institute of Physics). From this point forward, Henry and Dr. Cohn would live and work together in the same location (Gortler 1987&1988). Dr. Cohn worked with du Vigneaud as a postdoc through 1946 using isotopic tracers to better understand mechanisms of chemical reactions in animals, specifically metabolism of sulfur-containing amino acids (American Chemical Society; Proffitt 1999; Science History Institute).

With du Vigneaud, Dr. Cohn worked with isotopes—particularly on amino acid metabolism with a primary interest in sulfuric amino acids—methionine and cysteine. Methionine is an essential amino acid. Without the essential amino acids, animals cannot grow. Essential amino acids constitute about half of the twenty amino acids. While she worked with him, he discovered transmethylation in animals. Transmethylation, a process crucial for a variety of biological functions, refers to the chemical process in which a methyl group (CH_3) is transferred from one molecule to another. Dr. Cohn spent most of her time with du Vigneaud working on transmethylation, thus gaining an understanding of the mechanics of chemical reactions in animals (American Chemical Society; Gortler 1987&1988).

In addition, she assembled the mass spectrometer required to do the analysis. The first two of Dr. Cohn's three children were born during the years she worked in New York City with du Vigneaud (Gortler 1987&1988). du Vigneaud would receive the 1955 Nobel Prize in Chemistry for synthesizing the first polypeptide hormone (Gortler 1987&1988).

In 1946, Dr. Cohn's husband received a faculty appointment at Washington University in St. Louis, Missouri. du Vigneaud arranged a lunch with himself, Carl Cori (who had a biochemistry lab at Washington University with his wife Gerty Cori), and Dr. Cohn in New York City. Dr. Cohn was fortunate in that Washington University only had partial nepotism rules, and she was hired by the Coris to join their lab. Carl and Gerty Cori won the 1947 Nobel Prize in Physiology or Medicine for the discovery of the catalytic conversion of glycogen (Gortler 1987&1988, Nobel Prize; Perelman School of Medicine; Zagorski a 2009).

Dr. Cohn began her time at the Coris' lab as a research associate. Dr. Cohn said of Gerty Cori, "She sort of adopted me as soon as I got to the lab" (Zagorski b). In order to conduct her research, Dr. Cohn built a mass spectrometer identical to what she used at George Washington University. She set up a radioactive isotope laboratory for the Coris. (Gortler 1987&1988). While Dr. Cohn worked in the Coris' laboratory, she studied reactions to acceleration of chemical changes in enzymes (enzyme catalysis). She used an oxygen isotope to gain insight into organic phosphates. During that time, Henry was invited to MIT for a year, so Dr. Cohn spent 1950–1951 at the Harvard Medical School (Gortler 1987&1988). In 1952, Dr. Cohn became an Established Investigator for the American Heart Association—the first woman with that title. She retained that award until 1958 (Gortler 1987&1988; Unpublished 2006).

Nuclear magnetic resonance (NMR) and electron paramagnetic resonance (EPR) have a significant advantage over mass spectroscopy in that they don't destroy the molecules being investigated (DeBakcsy). Dr. Cohn said she started thinking about using NMR when she was in England during Henry's sabbatical in 1955. "I found out about NMR and EPR because of the fact that my husband was in a department at Washington University where research in magnetic resonance was very actively pursued" (Gortler 1987&1988). They went to Oxford where she worked in Hans A. Krebs' laboratory. Krebs won the Nobel Prize in Physiology or Medicine in 1953. And she started thinking about NMR with regard to ATP (adenosine triphosphate) and ADP (adenosine diphosphate)—both crucial compounds for metabolism (Gortler 1987&1988). Her work was enhanced by the development of commercial NMR spectrometers to which she had access—first at facilities other than Washington University and later at the university itself (Gortler 1987&1988).

In 1958, Dr. Cohn finally obtained a faculty position (Associate Professor at Washington University). She published work on paramagnetic (weakly magnetic) ions, indicating that she thought they would be promising. They were. They are used as contrast agents for medical imaging—an important element of magnetic resonance imaging (MRI), which is NMR used for medical purposes (Gortler 1987&1988, Science Guys 2003).

In 1960, Henry was appointed professor in the physics department at the University of Pennsylvania. Dr. Cohn joined the Johnson Research Foundation/Department of Biophysics at the University of Pennsylvania as well. She was faculty now—associate professor. Although she asked for a joint appointment with the Department of Biochemistry that did not happen. Years later (1979), the departments merged, and her appointment was in the Department of Biochemistry and Biophysics. After a year, she was promoted to Full Professor (1961). From 1964 until she turned 65 in 1978, Dr. Cohn was a Career Investigator with the American Heart Association. Cohn spent the rest of her career at the University of Pennsylvania, being named the Benjamin Rush Professor of Physiological Chemistry in 1978. By the time she retired in 1982, Dr. Cohn had published over 160 scientific papers (Adler 2022; Gortler 1987&1988).

At Penn, Dr. Cohn used NMR to look at chemical reactions with ATP. ATP is the primary source of energy for our body. She defined each phosphate group present on ATP. She also determined how ATP was converted to ADP—which produces energy—and how ADP was converted back to ATP to store energy. This was the identification of the metabolic pathway (Kaiser). She also used NMR to examine RNA. Her work with NMR is considered her most important contribution to science and medicine (Proffitt 1999).

After her retirement, Dr. Cohn stayed active in scientific pursuits. From 1982 to 1985, she was a Senior Member, Institute for Cancer Research, Fox Chase, Philadelphia, Pennsylvania (Gortler 1987&1988). She remained involved in university life. For her 90th birthday, Dr. Cohn went hang gliding (Science History Institute).

A significant pioneer, Dr. Cohn achieved the following firsts as a woman: (1) to do independent research at NACA (1933); (2) PhD hired by du Vigneaud (1937); (3) Member of the editorial board of the *Journal of Biological Chemistry* (1958–1963; 1968–1973); (4) Lifetime career investigator of the American Heart Association (1964); and (5) President, American Society of Biological Chemists (1978) (today the American Society of Biochemistry and Molecular Biology—ASBMB) (American Philosophical Society; Bernard Becker Medical Library).

Dr. Cohn was active in professional organizations and received many honors and awards, including nine honorary doctorates, Fellow of the American Academy of Arts and Sciences, and the 1982 National Medal of Science (American Philosophical Society; Debakcsy; Gortler 1987&1988; Kričaj 2018; University of Cincinnati 2003; Unpublished 2006):

- 1963, American Chemical Society, Garvan Medal
- 1971, National Academy of Sciences
- 1972, American Philosophical Society
- 1975, The Franklin Institute/Cresson Medal

- 1979, NMR and Biochemistry: A Symposium Honoring Mildred Cohn, 1979 International Organization of Women Biochemists Award
- 1986, Columbia University, Chandler Medal
- 1994, The Magnetic Resonance in Biology Founders Medal
- 1997, Protein Society Stein-Moore Award
- 2000, Ralph and Helen Oesper Award, University of Cincinnati
- 2009, National Women's Hall of Fame.

The ASBMB awards the Mildred Cohn Young Investigator Award every other year. In 2023, ASBMB combined the Mildred Cohn Award in Biological Chemistry and the ASBMB Young Investigator Award to create the ASBMB Mildred Cohn Young Investigator Award, which "recognizes outstanding research contributions to biochemistry and molecular biology" (ASBMB).

Dr. Cohn and her husband Henry Primakoff had two daughters and one son. Primakoff died in 1983 (American Philosophical Society). She said her husband was "cerebral and very European in the regard of housework" (Kričaj 2018). "My greatest piece of luck was marrying Henry Primakoff, an excellent scientist who treated me as an intellectual equal and always assumed that I should pursue a scientific career and behaved accordingly" (American Chemical Society).

Of her career and the limitations she faced as a woman, Dr. Cohn said,

> My career has been affected at every stage by the fact that I am a woman, beginning with my undergraduate education, which was very inferior in chemistry, and physics was not even offered [as a major] at Hunter College, unlike the excellent scientific education that my male counterparts received at City College. In my day, I experienced discrimination in academia, government, and industry. (Kresge et al. 2009)

Her advice to women wanting to pursue a scientific career was,

> A woman student should always go into a <u>new</u> area and become expert in that new area. Then when she's ready for a job, there are really so few people who are capable of handling the job, that a woman will be chosen whether they want a woman or not. (Gortler 1987&1988)

5.2 Nuclear Magnetic Resonance

A spectrometer (Fig. 5.2) used for NMR (which goes by magnetic resonance imaging (MRI) when used for medical purposes) consists of a large magnet, a radio wave generator, and a console to collect and analyze data. It complements X-ray technology. X-rays show bone well. NMR shows soft tissue well due to the presence of water (H_2O) with two atoms of hydrogen in every molecule—hydrogen is the element with the strongest magnetic resonance (Science Guys 2003).

NMR uses strong magnetic fields. The nuclei of atoms, which are spinning, have magnetic properties and will align themselves in a magnetic field. When exposed to radio waves, the nuclei will absorb the energy associated with certain radio waves—what is known as "resonance." The energy thus absorbed causes the spinning characteristics of the nuclei to change. This is observed through analysis

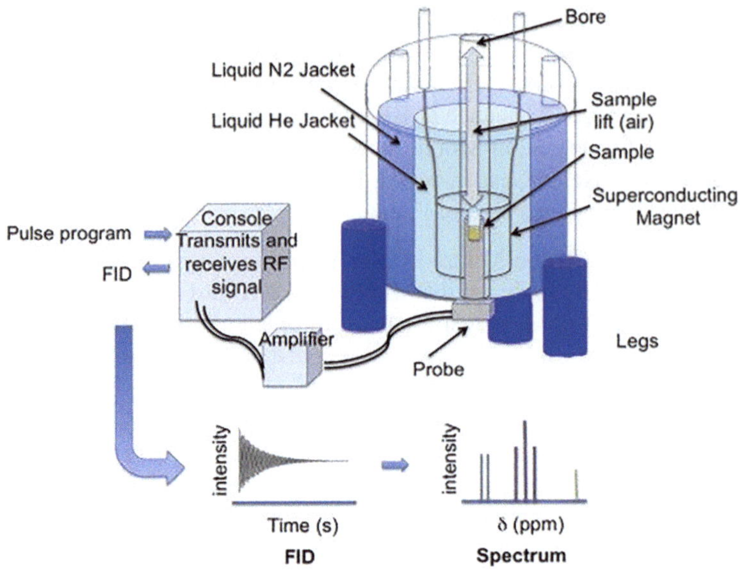

Fig. 5.2 Simplified diagram of a nuclear magnetic resonance spectrometer. (Rankin et al. 2014)

of the NMR spectrum. The operator of the machine controls what type of atom is absorbing the energy from the radio waves by controlling the frequency of the radio waves (Millipore Sigma; Science Guys 2003). Absorption is determined by monitoring the radio waves sent and received—a drop in reception indicates absorption.

Because the chemical environment surrounding the nuclei also affects how the nuclei absorb the radio waves, the researcher can determine the molecular structure of chemicals. This is accomplished by studying the NMR "spectrum of absorbed waves." NMR is considered the world's most useful and versatile tool for determining and understanding the structures of chemicals. NMR can also be used to understand the rates of chemical processes (Gortler 1987&1988; Science Guys).

5.3 Conclusion

A pioneer in understanding chemical reactions at the atomic level, Dr. Mildred Cohn used mass spectrometry, nuclear magnetic resonance, and electron paramagnetic resonance to determine organic chemical reactions. Overcoming significant prejudice due to her race and religion, Dr. Cohn served as a trailblazer for women while contributing significantly to science and medicine.

References

Adler, Richard. 2022. Mildred Cohn. EBSCO Information Services. https://www.ebsco.com/research-starters/biography/mildred-cohn. Accessed July 30, 2025.

American Chemical Society. Mildred Cohn (1913-2009). https://www.acs.org/education/whatischemistry/women-scientists/mildren-cohn.html. Accessed July 30, 2025.

American Institute of Physics. H. Primakoff. https://history.aip.org/phn/11608020.html. Physics History Network. Accessed August 24, 2025.

American Philosophical Society Manuscript Collections Search. Mildred Cohn Papers. Collection—Box:1-53. Identified Mss. Ms. Coll. 17. https://as.amphilsoc.org/repositories/2/resources/2617. Accessed July 30, 2025.

ASBMB. ASBMB Mildred Cohn Young Investigator Award. https://www.asbmb.org/career-resources/awards-grants/young-investigator. Accessed July 30, 2025.

Barrer, Betty. Mildred Cohn. Jewish Women's Archive. https://jwa.org/encyclopedia/article/cohn-mildred. Accessed July 30, 2025.

Bernard Becker Medical Library. Washington University in St. Louis. Mildred Cohn (b. 1913). https://beckerexhibits.wustl.edu/legacy-exhibits/mowihsp/bios/cohn.htm. Accessed July 30, 2025.

DeBakcsy, Dale. From ATP to MRI: Mildred Cohn's Pioneering Work in Nuclear Magnetic Resonance. The Women in Science Archive. https://www.wisarchive.com/post/from-atp-to-mri-mildred-cohn-s-pioneering-work-in-nuclear-magnetic-resonance. Accessed July 30, 2025.

Gortler, Leon. Interview of Mildred Cohn. At the University of Pennsylvania, Philadelphia Pennsylvania. December 15, 1987 and January 6, 1988. Philadelphia: Chemical Heritage Foundation, Oral History Transcript #0080.

Kaiser, John, et.al. Mildred Cohn. The University of Kansas Libraries. https://opentext.ku.edu/deichemistry/chapter/mildred-cohn/. Accessed July 30, 2025.

Kresge, Nicole; Simoni, Robert D.; Robert L. Hill. Succeeding in Science Despite the Odds; Studying Metabolism with NMR by Mildred Cohn. Journal of Biological Chemistry. https://www.jbc.org/article/S0021-9258(18)66379-3/fulltext. 2009. November 6. 284 (45), e12-e13.

Kričaj, Neža. 2018. Mildred Cohn. She Thought It. https://shethoughtit.ilcml.com/biography/mildred-cohn/. February 21. Accessed July 30, 2025.

Millipore Sigma. Nuclear Magnetic Resonance (NMR). https://www.sigmaaldrich.com/US/en/applications/analytical-chemistry/nuclear-magnetic-resonance. Accessed August 22, 2025.

National Science & Technology Medals Foundation. Mildred Cohn. https://nationalmedals.org/laureate/mildred-cohn/. Accessed July 30, 2025.

The Nobel Prize. Gerty Cori. https://www.nobelprize.org/stories/women-who-changed-science/gerty-cori/. Nobelprize.org. Accessed August 21, 2025.

Perelman School of Medicine, University of Pennsylvania. Dr. Cohn, Biochemistry and Biophysics. In Memoriam, 2009. https://www.med.upenn.edu/asef/dr-cohn.html. Accessed July 30, 2025.

Proffitt, Pamela. Notable Women Scientists. Detroit, Michigan: Gale Group. 1999.

Rankin, Naomi & Preiss, David & Welsh, Paul & Burgess, Karl & Nelson, Scott & Lawlor, Debbie & Sattar, Naveed. (2014). The emergence of proton nuclear magnetic resonance metabolomics in the cardiovascular arena as viewed from a clinical perspective. Atherosclerosis. 237. 287–300. https://doi.org/10.1016/j.atherosclerosis.2014.09.024.

Science Guys. 2003. What is nuclear magnetic resonance (NMR) and how does it work? Union University. https://www.uu.edu/dept/physics/scienceguys/2003Feb.cfm. February. Accessed August 21, 2025.

Science History Institute. Mildred Cohn. https://www.sciencehistory.org/education/scientific-biographies/mildred-cohn/. Accessed July 30, 2025.

University of Cincinnati. 2003. The 2000 Ralph and Helen Oesper Awardee—Honoring Mildred Cohn. https://www.artsci.uc.edu/content/dam/refresh/artsandsciences-62/docs/cohn.pdf. Accessed July 30, 2025.

Unpublished nomination of Mildred Cohn to the National Women's Hall of Fame. In the files of Jill S. Tietjen. January 25, 2006.
Zagorski, Nick.a. 2009. Mildred Cohn: Isotopic and Spectroscopic Trailblazer. American Society for Biochemistry and Molecular Biology. ASBMB today. https://www.asbmb.org/Asbmb.Web/media/files/atoday/ASBMBToday-2009-09.pdf. September. Accessed August 23, 2025.
Zagorski, Nick.b. Mildred Cohn: Trailblazing Scientist 1913-2009. Jewish Women's Archive. https://jwa.org/weremember/cohn-mildred. Accessed July 30, 2025.

Chapter 6
Berta V. Scharrer

For her pioneering contributions to establishing the concept of neurosecretion and demonstration of the central role of neurosecretion and neuropeptides in the integration of animal function and development

Keywords Neurons · Neurosecretion · Neuroendocrinology · Brain · Chemical pathways · Ernst Scharrer · Invertebrates · Anatomy · Scharrerae · Neuroimmunology · Synapses · Neurotransmitters

Dr. Berta V. Scharrer, shown in Fig. 6.1, and her husband Dr. Ernst Scharrer established that nerve cells not only establish electrical connections in the brain but are also able to secrete chemicals to establish chemical pathways of communication. Dr. Berta Scharrer focused on invertebrates and was able to show that all of those she studied had nerve cells capable of secreting chemicals. She was awarded a National Medal of Science in 1983 by President Reagan.

J. S. Tietjen et al., *Women Laureates 1970 – 2007*, Women in Engineering and Science, https://doi.org/10.1007/978-3-032-23116-1_6

Fig. 6.1 Dr. Berta V. Scharrer. (Courtesy of Springer)

6.1 Life (1906–1995)

Berta Vogel was born in December 1906 in Munich, Germany, to parents Johanna Weiss Vogel and Karl Phillip Vogel. Her father was a judge and Vice President of the federal court of Bavaria. The family ultimately included four children, and Berta had a happy childhood (National Academies of Sciences, Engineering, and Medicine 1998).

Berta developed an interest in biology while attending gymnasium (public school) and decided she would like to be a research scientist. However, she recognized this aspiration as unlikely to unfold for a woman.

Berta attended the University of Munich, earning first a BS degree in biology and then a PhD in 1930. During her studies, she worked in the lab of Dr. Karl von Frisch, a bee behavioralist who received the Nobel Prize in Physiology or Medicine in 1973. Berta researched the relative tastes and nutritional values of various sugars to the honeybee. While pursuing her PhD, Berta met Ernst Scharrer, another of Dr. Frisch's students, a man who became her spouse and her research collaborator. In 1928, while still a student, Ernst Scharrer found that some neurons (nerve cells) in the hypothalamus of fish secreted droplets. This was a very controversial discovery, as it was thought that neurons could only transmit electrical activity and that only endocrine cells could secrete droplets. As a result, Ernst Scharrer's results were greeted with great skepticism (National Academies of Sciences, Engineering, and Medicine 1998).

The Scharrers moved to the Research Institute of Psychiatry once they both completed their graduate work. There, Dr. Ernst Scharrer pursued a medical degree, and Dr. Berta Scharrer studied spirochetes and brain infections in birds and amphibians. Recognizing that it might be impossible to forge a lasting career as a research scientist, she also pursued a teaching certificate. Later, commenting on this aspect of her life, Dr. Scharrer said,

> An academic career at that time did not look promising at all for a woman. I must say that I could not have done what I have if I had not been married to a biologist who gave me a chance to work. (Millen 1989)

In 1934, the Scharrers moved to the University of Frankfurt, where Dr. Ernst Scharrer was appointed the director of the Edlinger Institute for Brain Research. Nepotism rules at that time prevented Dr. Berta Scharrer from holding a paid position at the university, so she worked without title or salary as a research scientist. She and her husband decided to pursue further documentation of the secretion of chemicals by neurons in what they called "nerve-gland" cells. They agreed that Dr. Ernst Scharrer would study vertebrate organisms, while Dr. Berta Scharrer would study invertebrates. Dr. Berta Scharrer's work led to her first papers on neuron secretion in 1936 and 1937, describing the phenomenon in aplysia (a snail) and nereis (a worm) (National Academies of Sciences, Engineering, and Medicine 1998).

By the mid-1930s, Germany was becoming uncomfortable for the Scharrers. A law passed in 1933 required the dismissal of Jews from all aspects of civil service. While this did not directly affect the Scharrers, they disagreed with the persecution of the Jews and were resistant to pressure to join Nazi organizations. They determined they needed to leave Germany but knew this would be a difficult task given the importance assigned by the government to Dr. Ernst Scharrer's role as a doctor (National Academies of Sciences, Engineering, and Medicine 1998).

Starting in 1937, the Scharrers made a series of moves to land them with a permanent home in the USA. First, they left for Dr. Ernst Scharrer to have a one-year fellowship at the University of Chicago, telling the government they would return after that year abroad. Although they only traveled with two suitcases and $4 each, they did not intend to return to Germany.

In 1938, the Scharrers moved to New York City for Dr. Ernst Scharrer to be a visiting investigator at the Rockefeller Institute for Medical Research. They then moved to Western Reserve University in Cleveland in 1940, where Dr. Ernst Scharrer was appointed as an assistant professor of anatomy. In 1946, the pair moved to the University of Colorado Medical School in Denver, where Dr. Ernst Scharrer was appointed as an associate professor. Finally, in 1955, they moved to the Albert Einstein College of Medicine at Yeshiva University in New York City, where Dr. Ernst Scharrer was made the founding chairman of the Department of Anatomy (National Academies of Sciences, Engineering, and Medicine 1998).

Throughout these moves, Dr. Berta Scharrer was given lab space for her work but no salary and generally no appropriate title. At Western Reserve University, she was an instructor and fellow, teaching histology to medical students, but remained unpaid. She was not permitted to attend seminars until she agreed to prepare the tea for the faculty attending them (National Academies of Sciences, Engineering, and Medicine 1998). At the University of Colorado, Dr. Berta Scharrer was given a Guggenheim Fellowship and then a special fellowship from the US Public Health Service. In 1950, Dr. Berta Scharrer was given her first academic title—"assistant professor (research) unsalaried." Finally, in 1955, Yeshiva University made her a full professor. The dean of the Albert Einstein Medical College said,

> I know about the nepotism rules, but we are an entirely new school. We can do something a little progressive (Purpura 1998).

Even with that appointment, Dr. Berta Scharrer worked full-time but was paid half-time.

Throughout these moves, Dr. Berta Scharrer continued to work on understanding the secretion of chemicals by neurons. In Chicago, she worked on cockroaches as these were available (nearly unavoidable) free of charge in the building. Again, she found that these invertebrates had neurosecretory cells. In 1940, Dr. Scharrer found woodroaches in a shipment of monkeys from South America and switched to using them as her research subject, as they were larger and slower than their North American counterparts. Through her work, she was seeking to understand the function of neurosecretion and various neuroglandular bodies in the brain. For instance, she studied the corpus allatum and corpus cardiacum of the woodroaches, showing that removal of the corpus allatum caused nymphs to metamorphosize prematurely. Removal of the corpus cardiacum caused no similar effect, from which Dr. Scharrer concluded that the corpus cardiacum was simply there to store secretions (National Academies of Sciences, Engineering, and Medicine 1998).

By the early 1950s, the idea of chemical secretion by neurons was accepted, given the large body of evidence the Scharrers had collected. Their findings were documented in a book they jointly wrote, entitled *Neuroendocrinology,* which became an invaluable text for anatomists and medical students (Scharrer and Scharrer 1963).

In 1965, the Scharrers went to Miami, Florida, to attend a conference. Dr. Ernst Scharrer chose to swim in the ocean but was pulled under and drowned. Recalling this event's impact on her career, Dr. Scharrer has said,

> There was a very important turning point after my husband's death. I had to show that I could go it on my own (Purpura 1998).

Indeed, it was at this point that Dr. Scharrer finally began to receive recognition for her work. She was elected to the National Academy of Sciences in 1967, as well as the American Academy of Arts and Sciences.

Returning to Yeshiva University, Dr. Scharrer continued her work but now with full salary. The electron microscope was newly invented, and Dr. Scharrer made good use of it to study the structure of neurosecretory cells in greater detail. She showed that sometimes the targets of the secretions were nearby neurons and sometimes the secretions were simply released into the bloodstream to be taken up by targets far away. In the 1980s, she and her colleagues also showed that the neurosecretions help to regulate the immune system in invertebrates. Her work in this area is credited with giving rise to the field of neuroimmunology (National Academies of Sciences, Engineering, and Medicine 1998).

Dr. Scharrer served two years as the acting chair of the Department of Anatomy and was elected President of the American Association of Anatomists for the 1978–1979 year. She served as an associate editor of two prestigious journals in her field and on the National Academy of Sciences committee on human rights. In 1978, she received the Kraepelin Gold Medal, awarded for outstanding work in the field of psychiatry. In 1980, Dr. Scharrer received the Fred C. Koch award of the Endocrine Society and in 1982, the Henry Gray Award of the American Society of Anatomists. She received the Schleiden Medal of the Leopoldina, the German Academy of Natural Science, in 1983, and the National Medal of Science from President Ronald Reagan. In 1983, Dr. Scharrer also was nominated for a Nobel Prize for her pioneering work on brain chemicals. In 1994, Dr. Scharrer received the Order of Merit from Bavaria. She was given honorary degrees from eleven universities. But perhaps the most important honor paid to Dr. Scharrer was that a species of cockroach was named in her honor, scharrerae (National Academies of Sciences, Engineering, and Medicine 1998).

In the winter of 1994, Dr. Scharrer closed her lab, desiring to write papers and devise some lab experiments for others to carry out. She sent her cockroaches to George Stefano at SUNY Old Westbury, on Long Island, New York. Commenting upon their departure, she said, "The great exodus. I felt nostalgic. I said goodbye to them" (Siebert 1995).

Dr. Scharrer died in July 1995, having been laid low with heart failure. She was survived by a brother and a sister.

6.2 How Neurons Work

The nervous system of the human includes four key components: the brain, spinal cord, ganglia, and nerves. Ganglia are structures of multiple nerve cell bodies that act as relay stations for nerve impulses. The basic unit of the nervous system is the neuron, which is a cell that transmits information throughout the body. Nerves are made of many bundled neurons. It is estimated that the human brain contains about 86 billion neurons, each connected to about 1000 others (Newman 2023).

Figure 6.2 shows a schematic of a neuron. The fibers leading to the cell body, the dendrites, provide information to the cell. The axon, leading from the cell body to neighboring neurons, outputs the information to the structures to which it is attached. The attachments are called synapses and are of two types: electrical and chemical.

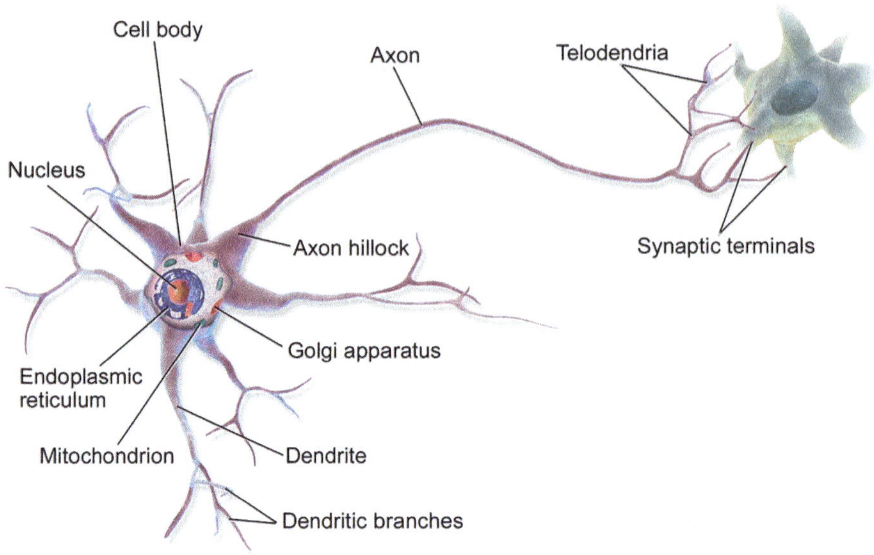

Fig. 6.2 Schematic of a human neuron. (Blausen_0657_MultipolarNeuron, Wikipedia)

In electrical synapses, there is a small gap between the end of the axon and the structure closest to it. Current that travels down the axon from the cell body passes through this tiny gap. This type of synapse is less common and faster than chemical synapses (Newman 2023).

In chemical synapses, such as shown in Fig. 6.3, a signal reaching the axon end causes release of a chemical called a neurotransmitter. It was these chemicals that the Drs. Scharrer observed in vertebrates and invertebrates. These neurotransmitters interact with receptors on a nearby neuron and trigger a response in them (Newman 2023).

A key characteristic of each neuron is that it has a set threshold above which it responds by firing. If input signals from the dendrites to the cell body are below that threshold, the axon will not send a signal along. If the input signals combine to a level above the threshold, then the axon will send an electrical impulse, called the action potential. The action potential is a fixed size for a given neuron, so doubling the excitation above the threshold from the dendrites will not change it. It is an electrical voltage created by the motion of ions across the axon membrane through ion channels that open when the threshold is exceeded. The ions are normally potassium (K^+) and sodium (Na^+). As the action potential moves down the ion, the region that just exhibited ionic flow with opening and closing of an ion channel relaxes, meaning that the action potential can only travel in one direction—namely, further down the axon (Newman 2023).

The frequency of neuron excitation is a measure of the intensity of the signal it is transmitting. For instance, loud sounds result in more frequent firing of the neurons associated with hearing them than soft sounds do. The neuron body

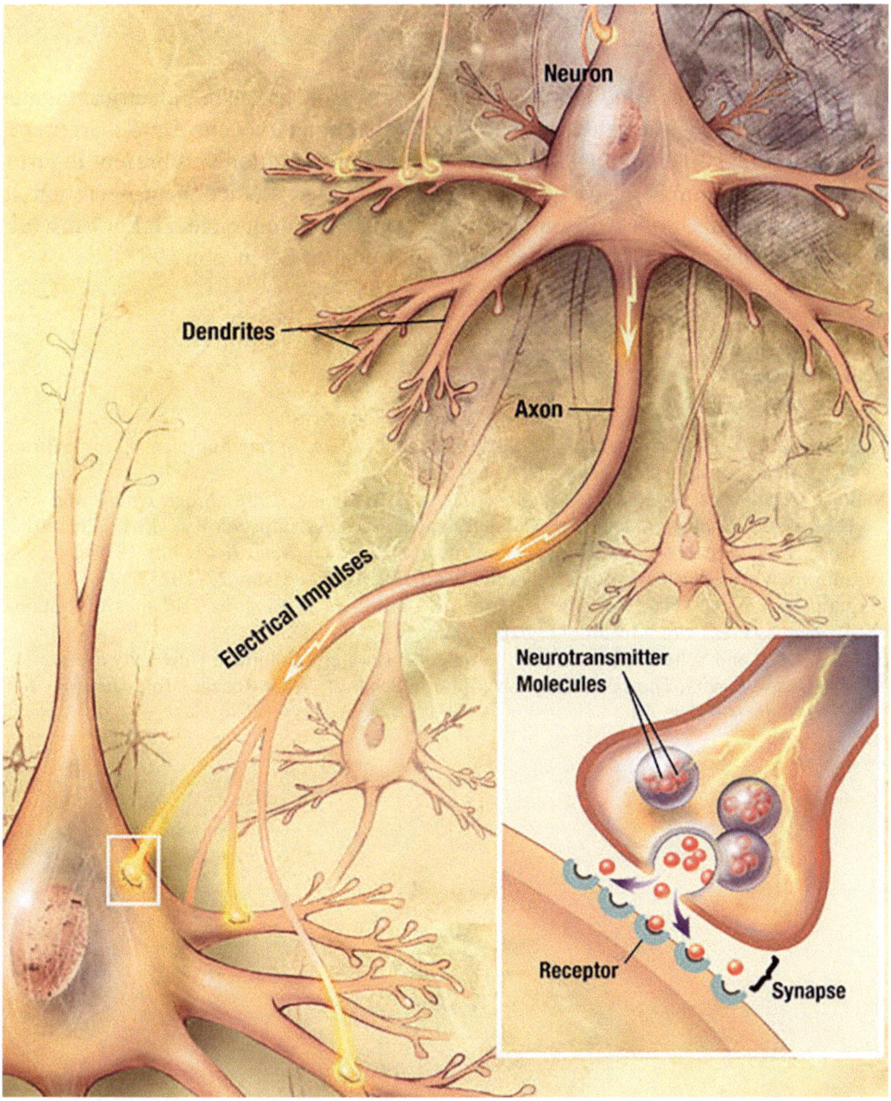

Fig. 6.3 Connected neurons and an enlarged view of a chemical synapse. (Public domain, Wikipedia)

releases neuropeptides that can modulate the neurotransmitters if excitation continues for a long period of time. Neuropeptides are small-chain proteins that act to increase or decrease the release of neurotransmitters. For instance, when a continuous action causes pain, the neuropeptides act to reduce the neurotransmitters, which we sense as the pain decreasing. This is a form of habituation (Newman 2023).

6.3 Conclusion

Dr. Berta Scharrer showed that nerve cells can not only transmit electrical signals but also secrete chemicals that signal various functions to occur. Dr. Scharrer specifically studied invertebrates and showed that neurosecretion was present in virtually all these animals. Her husband, Ernst Scharrer studied neurosecretion in vertebrate animals, prior to his untimely death in a drowning incident. For her work, Dr. Berta Scharrer was awarded the National Medal of Science in 1983.

References

Millen, S (ed). 1989. On Journeys Well Traveled, in *Einstein*, Spring/Summer, pp. 3-6, Bronx, N.Y.: Albert Einstein College of Medicine.

National Academies of Sciences, Engineering, and Medicine. 1998. *Biographical Memoirs: Volume 74*. Washington, DC: The National Academies Press. https://doi.org/10.17226/6201.

Newman, T. 2023. All you need to know about neurons. *Medical News Today* July 18. Available at https://www.medicalnewstoday.com/articles/320289. Accessed May 27, 2025.

Purpura, Dominick. 1998. Berta V. Scharrer. In *Biographical Memoirs: Volume 74*. National Academies Press. doi:https://doi.org/10.17226/6201

Scharrer, Ernst and Scharrer, Berta. 1963. *Neuroendocrinology*. Columbia University Press.

Siebert, Charles. 1995. The Lives They Lived: Berta Scharrer; What Roaches Told Her. *New York Time Magazine*. Dec., 31.

Chapter 7
E. Margaret Burbidge

For leadership in observational astronomy. Her spectroscopic investigations have provided crucial information about the chemical composition of stars and the nature of quasi-stellar objects

Keywords Astronomy · Stellar nucleosynthesis · Nucleosynthesis · Quasars · Star composition · Spectroscopy · Observational astronomy · Black holes · Quasi-stellar objects · Hubble Space Telescope

Dr. E. Margaret Burbidge, shown in Fig. 7.1, was an observational astronomer best known for her work to determine the chemical composition of stars and for her observations of quasars. She was also involved in designing instrumentation for the Hubble Space Telescope and in fighting against discrimination of women in science. Dr. Burbidge received the National Medal of Science in 1983.

J. S. Tietjen et al., *Women Laureates 1970 – 2007*, Women in Engineering and Science, https://doi.org/10.1007/978-3-032-23116-1_7

Fig. 7.1 Dr. E. Margaret Burbidge. (Wikipedia CC4.0)

7.1 Life (1919–2020)

Eleanor Margaret Peachey was born in Davenport (England) on August 12, 1919, exactly nine months after the World War I armistice, to parents Marjorie Stott Peachey and Stanley John Peachey. Her father was a lecturer in chemistry at Manchester School of Technology, near to their home. Her mother was seventeen years younger than her father and had been one of his students. The marriage had been opposed by Marjorie's father, who had also opposed her decision to attend university. The family was completed when Margaret's younger sister, Audrey, was born (Sargent and Longair 2021).

A few years after Margaret's birth, Stanley Peachey obtained a patent on a process to speed the vulcanization of rubber. This patent provided enough income for the family that they chose to move to London, where Margaret's father set up his own industrial chemistry lab. There, Margaret attended Heysham, a small private school for girls, for early education. She was notoriously shy, choosing to immerse herself in books. Given her interest in math and natural sciences, her parents gave Margaret a microscope and a chemistry set. Her mother's father gave her a science book by Sir James Jeans, a distant relative (DeVorkin 1978; Burbidge 1994).

Margaret attended high school at the Frances Holland Church of England School, another girls' school, where her interest in math was encouraged despite the school's emphasis on the humanities. Having completed her course requirements early and taken the entrance exam for university at fifteen, Margaret spent her last year of high school attending classes she chose. At that point, her father was in ill health and largely bedridden. He died when Margaret was seventeen (DeVorkin 1978).

Margaret's mother chose where she would go to university and wanted her to stay in London, so Margaret enrolled at University College London in 1936, when she was seventeen years old. It was her first experience with a coeducational school. She discovered astronomy courses in her first year and chose that as her major. Margaret graduated with first-class honors and a bachelor's degree in 1939 (DeVorkin 1978). By this time, Europe was again engaged in war, but bombing of London had not begun.

Immediately after graduation, Margaret went to work at the company computing the mathematical tables for the Nautical Almanac. On her first day of work, her boss asked her to join him for lunch. His interests in her seemed unrelated to her work and quite inappropriate, so she quit the job immediately (Burbidge 1994; DeVorkin, 1978).

Rather than seeking a different job, Margaret enrolled as a graduate student at the University of London, the parent school of her undergraduate college. She was paid to work at the University of London Observatory and gladly collaborated with Dr. Gregory, her former teacher and boss at the observatory. She was presented with a choice of three topics to pursue for her PhD work: astrometry (focused on distance to stars and planets), spectroscopy (used to determine the properties of stars), and photometry (to rank stars by their brightness). She chose to work on spectroscopy (Burbidge 1994).

The London Blitz, beginning in Fall 1940, changed Margaret's activities to some extent. In addition to her work maintaining telescopic equipment, she was engaged by day in building wartime equipment, such as rangefinders. Further, her duties at the observatory were expanded to include caretaking and making repairs from bombing damage as necessary. At this point, Margaret moved outside of London to an apartment at the home of Dr. Gregory and his family to avoid the worst of the bombing. She planted a vegetable garden on the observatory grounds and contributed the harvest to the Gregory household (Burbidge 1994; DeVorkin 1978).

Despite the war's interruption of classes, Margaret managed to continue her nighttime observing. She found the London blackout requirements helpful, as they decreased the light pollution she observed. However, she was twice interrupted by nearby V-1 flying bombs going off while she was observing. She also once saw a bomb strike the path she had just been on while bicycling from the observatory to the Gregory home. She completed her PhD in 1943 and continued her work at the University of London Observatory (Sargent and Longair 2021; Wikipedia, Margaret Burbidge 2025).

Margaret applied for a postdoctoral position from the Carnegie Observatories, but she was turned down in 1945. She received a note from the director, I. S. Bowen, that essentially said: "You should have known better, you should have known that this was not open to women" (DeVorkin 1978). This was the first time Margaret had encountered blatant discrimination on account of her gender. The problem, as described by Dr. Bowen, was that the fellowship would require observing at Mount Wilson Observatory, and that facility did not have bathrooms appropriate for a woman. Margaret opted to remain in London and at the University of London Observatory.

As World War II wound down toward the end of 1945, the lenses for the telescopes of the University of London Observatory that had been removed to prevent damage from bombs were reinstalled. Classes were restarted, and Margaret chose to attend a couple of these. It was there that she met Dr. Geoffrey Burbidge, who was attending some of the same classes. He was a theoretical physicist working on synchrotron radiation at the time, but as the two became friends, Margaret's enthusiasm for astronomy convinced him to consider that field as a career as well. They established a strong research collaboration and married in 1948 (DeVorkin 1978).

The decade of the 1950s saw the Drs. Burbidge make a series of moves. In 1951, they left London. Dr. Geoffrey Burbidge had an Agassiz Fellowship to work at Harvard, while Dr. Margaret Burbidge had an International Astronomical Union grant to work at Yerkes Observatory in Williams Bay, Wisconsin. Yerkes Observatory is officially part of the University of Chicago, and her appointment was with that university. At this point, Dr. Margaret Burbidge was studying Be stars. Stars are often classified by their temperature. The ranking of star temperatures follows the rule OBAFGKM, where O are the hottest stars and M the coolest. B stars are blue-white and range in temperature from 10,000 to 30,000 K. They are a very small fraction of the stars that burn hydrogen (known as main sequence stars). B stars are much larger than our sun and tend to be short-lived (5,000,000–10,000,000 years). Be stars are B stars that show very prominent hydrogen emission lines (DeVorkin 1978).

In 1952, Dr. Geoffrey Burbidge joined his wife at Yerkes, and the Drs. Burbidge also spent time at McDonald Observatory in West Texas. To get there from Wisconsin, they drove a 1939 Ford they had purchased for $100. The car was unable to go faster than 35 miles per hour and seemingly consumed a quart of oil every 80 miles, but it got them to their destination (DeVorkin 1978).

In 1953, the Drs. Burbidge returned to England. Dr. Geoffrey Burbidge had an appointment at Cambridge University. Dr. Margaret Burbidge did not have a position but was a voluntary research associate and continued to work on material she had gathered at Yerkes. There, they met Dr. Fred Hoyle, who had hypothesized that the chemical elements in the universe could be produced in stars through nucleosynthesis (fusion of atoms). Dr. William Fowler, an American astronomer visiting for a term, was conducting experiments in Hoyle's lab on nuclear reactions that might support Hoyle's theory. Drs. Hoyle and Fowler invited the Drs. Burbidge to gather astronomical data to support or destroy the theory. Dr. Fowler encouraged the pair to emigrate to the USA permanently to work with him (DeVorkin 1978).

In 1955, the Drs. Burbidge returned to the USA. Dr. Margaret Burbidge had once again applied for a fellowship at Mount Wilson Observatory and once again been turned down based on her gender. In exasperation, Dr. Geoffrey Burbidge applied for the Mount Wilson Observatory Fellowship and Dr. Margaret Burbidge the Kellogg Fellowship at the California Institute of Technology. Both were awarded, although Dr. Margaret Burbidge's fellowship paid her part-time despite her working full-time. Of course, the Mount Wilson Observatory Fellowship was for observing, and Dr. Geoffrey Burbidge was a theorist rather than an observer, so whenever observations were needed, Dr. Margaret Burbidge would accompany her husband as his supposed assistant. The reality was that she would do the telescope work while her husband was in the darkroom. This situation was generally known at the telescope facility and handled by having the Drs. Burbidge live away from the other observers. This also suited the couple, as Margaret was concealing her pregnancy to avoid staff at the observatory getting fussy about her working. Daughter Sarah was born in 1956 (DeVorkin 1978).

The observations by Dr. Margaret Burbidge feeding into calculations of elemental abundances in stars led to a series of papers and culminated with a 100-page article appearing in 1957 (Burbidge et al. 1957). This article became known as B2FH, referring to the first letters of the last names of the authors in the order in which they appeared. Dr. Margaret Burbidge was the lead author. This article lays down Hoyle's theory of nucleosynthesis and reviews all the data that were available at the time to support it, including the astronomical observations and calculations of the Drs. Burbidge. It is considered one of the fundamental papers of astrophysics to this day. In 1983, Dr. Fowler was a Nobel Prize recipient in Physics for his work on nucleosynthesis. He expressed surprise that Dr. Margaret Burbidge was not included in the prize (Wikipedia, Margaret Burbidge 2025).

In 1957, the University of California at San Diego (UCSD) was just being established. Both Drs. Burbidge were offered jobs there, but they chose instead to return to the University of Chicago and Yerkes, as there were better telescopes available to them. At the University of Chicago, Dr. Geoffrey Burbidge was an Assistant Professor. Dr. Margaret Burbidge was the holder of the Shirley Farrar Fellowship. Observing often involved traveling to McDonald Observatory, and once Dr. Margaret Burbidge fell about ten feet off the platform there. This is an unfortunately common problem at observatories, and luckily, Dr. Burbidge was bruised but not seriously injured (DeVorkin 1978).

In 1962, Dr. Margaret Burbidge and Dr. Geoffrey Burbidge made their final career move to positions at the University of California at San Diego (UCSD). Because of nepotism rules, they could not be in the same department, so Dr. Margaret Burbidge was officially in the chemistry department while Dr. Geoffrey Burbidge was in physics. At this point, Dr. Margaret Burbidge turned her attention to measuring the masses, compositions, and rotations of galaxies. She also performed early spectrographic studies of quasars and discovered QSO B1442 + 101 in 1974. This was the most distant object seen at that time and remained the record until 1982. Dr. Burbidge's work on quasars helped establish that supermassive black holes power them (Wikipedia, Margaret Burbidge 2025; DeVorkin 1978).

In 1972, Dr. Margaret Burbidge made news by refusing to accept the Annie J. Cannon Award of the American Astronomical Society, an award specifically given to women in the field. When she refused the award, she said, "It is high time that discrimination in favor of, as well as against, women in professional life be removed" (Humphreys 2018). Her action led directly to the establishment of a committee on the status of women in that society.

Also in 1972, Dr. Margaret Burbidge was offered the position of director of the Royal Greenwich Observatory in England. She asked for time to consider this opportunity, but was told news of her appointment had leaked to the British press and that it would be bad for the science community in England if she now turned the offer down. Reluctantly, Dr. Burbidge accepted the offer, although she took a leave of absence from UCSD rather than resigning. Notably, the position of director of the Royal Greenwich Observatory had always also carried the title of Astronomer Royal, but the two were separated with her appointment, and the Astronomer Royal title given to a man (Burbidge 1998).

As Greenwich director, Dr. Burbidge tried to arrange for the Isaac Newton Telescope to be moved as it was useless at its current location due to light pollution. However, this caused a political hailstorm, and after only three months, Dr. Geoffrey Burbidge and Sarah left to return to the USA. Dr. Margaret Burbidge lasted 16 months and then resigned. In looking back on this period, she has said:

> Despite my foray into the US political scene as a spearhead for abolition of gender discrimination, I knew then, and have always known since, that I do not have a temperament to direct a major scientific establishment. (Burbidge 1994)

From 1976 to 1978, Dr. Burbidge was the first woman to be President of the American Astronomical Society. During that time, she convinced the membership that they should ban meetings in states that had not ratified the Equal Rights Amendment. She also took the "American" in the title seriously and became a US citizen. In 1983, Dr. Burbidge was elected the President of the American Association for the Advancement of Science, again serving as the first woman in that role. She was also at that time serving as the founding director of the Center for Astronomy and Space Sciences at UCSD, a position she held until 1988 (Wikipedia, Margaret Burbidge 2025).

Toward the end of her career in the 1980s and 1990s, Dr. Burbidge contributed to the design of the Faint Object Spectrograph, an instrument for use on the Hubble Space Telescope. She retired in 1988, becoming a professor emeritus, and continuing to be active in research (Wikipedia, Margaret Burbidge 2025).

Dr. Burbidge was the recipient of numerous awards and recognitions. In 1959, she and her husband were awarded the Helen B. Warner Prize of Astronomy for the B2FH paper. In 1964, she was made a Fellow of the Royal Society. In 1969, Dr. Burbidge was made a Fellow of the American Academy of Arts and Sciences. In 1978, she was elected to the National Academy of Sciences. In 1982, she received the Catherine Wolfe Bruce Medal of the Astronomical Society of the Pacific. The National Medal of Science was awarded to Dr. Burbidge in 1983 by President Ronald Reagan. In 1984, Dr. Burbidge received the Henry Norris Russell Lectureship, the highest honor of the American Astronomical Society. In 1988, she received the Albert Einstein World Award of Science. In 2005, she and her husband were awarded the Gold Medal of the Royal Astronomical Society. In 2020, shortly before her death, Dr. Burbidge was made the inaugural fellow of the American Astronomical Society.

There also were a few honors afforded Dr. Burbidge through naming opportunities. There is an asteroid named in her honor: Asteroid 5490 Burbidge. Additionally, there is a group of galaxies in Cetus named Burbidge's Chain. The American Physical Society has established an award in her name. There is a visiting professorship at UCSD named for her as well (Wikipedia, Margaret Burbidge 2025; University of California Academic Senate).

Dr. Margaret Burbidge died in April 2020. Her husband had predeceased her by ten years. She was the author of more than 370 papers in her lifetime.

7.2 Stellar Nucleosynthesis and Quasars

The Big Bang theory of the origin of the universe postulates there was a single enormous explosion that created the universe. A few minutes after that bang, the first atomic nuclei were formed, and shortly after that, nuclear reactions took place. Twenty minutes after the bang, the universe had expanded and cooled to the extent that nuclear reactions could not occur. Thus, the universe at this point was about three quarters hydrogen and a quarter helium, i.e., made of the lightest chemical elements with the simplest nuclei (Wikipedia, Nucleosynthesis 2025). The question this leaves us with is, how did we get heavier elements in the universe? Stellar nucleosynthesis is the answer to that question.

In the cores of stars, the temperatures and pressures are high enough to cause fusion of atoms. Light elemental atoms are pushed together under such high pressures that they can form a single atomic unit, thus creating a new element. This allows for the buildup of the heavier elements step by step. For instance, four hydrogens combined yield a helium. Three helium atoms fused produce a carbon, and four helium atoms fused produce oxygen. Based on what we know of the temperatures and pressures in stars, this process can produce elements as heavy as iron and nickel. That still leaves many heavier chemical elements unaccounted for. These, so the theory goes, can be produced through neutron capture. Energetic collision of a neutron with an atomic nucleus crams the neutron into the nucleus. That makes the atom unstable, so that the neutron decays to a proton, and an electron is kicked off the atom, thus creating a new stable element. This process allows for the heaviest elements we find in the universe to be created. Taken together, these processes of fusion and neutron capture are called stellar nucleosynthesis (Wikipedia, Nucleosynthesis 2025).

The B2FH paper used data on the age and chemical composition of stars to test this theory. In particular, the theory predicts that older stars should have fewer heavy chemical elements because the heavier elements wouldn't have formed yet when the stars were made as matter coalesced. The abundance of elements in stars is found from spectroscopic data—i.e., determining the content from the frequency of the emitted light. The star age can be found from Hertzsprung-Russell diagrams, which relate star age to its mass, brightness, and temperature. The Hertzsprung-Russell diagrams are based on a model of the life cycle of a star based on a large amount of observational data. The data from Dr. Margaret Burbidge's observations supported the stellar nucleosynthesis theory for creation of chemical elements in the universe (Wikipedia, B^2FH 2025).

Ironically, although Dr. Burbidge's observations supported the Big Bang theory for the origin of the universe, she was not a proponent of this theory. Instead, she agreed with Dr. Hoyle, who suggested that the universe had no beginning nor end and just had always been. In this theory, the universe continuously creates matter (Fox 2020). This is sometimes called the steady-state universe theory and is generally not accepted by astrophysicists today.

In addition to her work on stellar nucleosynthesis, Dr. Burbidge was cited in her National Medal of Science for her work on quasi-stellar objects, i.e., quasars.

Fig. 7.2 Artistic rendering of a quasar showing a black hole in the center and surrounding gases. (Young 2021)

Quasars are highly luminous objects that are far away (outside our galaxy), suggesting that they must have an enormous energy source. They appear as bright point light sources moving away from us at high speed. The light they emit is in a broad energy spectrum rather than peaked at frequencies indicative of particular chemical elements, indicating that their luminosity is not produced by a star. Quasars are the brightest objects in the known universe (Wikipedia, Quasars 2025).

Today, we know that the light source comes from the presence of a massive black hole, a fact that Dr. Burbidge was instrumental in establishing. A black hole is a region of space having such strong gravitational pull that no matter or radiation can escape the region. As matter is captured into the region, it relinquishes its energy in the form of light. It is this light from the material surrounding the black hole that Dr. Burbidge and other astronomers observed. Images of black holes thus look like Fig. 7.2, with dark centers ringed by bright light. The black holes at the center of quasars range from 1 to 10,000 million solar masses (i.e., multiples of the mass of our sun) (Wikipedia, Quasars 2025).

To date, about a million quasars have been identified. The nearest is about 600 million light-years away, meaning it would take 600 million years for light to travel from it to us (Wikipedia, Quasars 2025).

7.3 Conclusion

Dr. Margaret Burbidge was an observational astronomer famous for work showing that all the elements in the universe could be explained by stellar nucleosynthesis and for her identification of quasars, including one that held the record for the most distant object known for many years. Dr. Burbidge was also a strong advocate for equal treatment of women scientists, having encountered gender barriers to her work multiple times. She was awarded the National Medal of Science in 1983.

References

Burbidge, E. 1994. Watcher of the Skies. *Ann. Rev. of Astronomy and Astrophys.* 32:1–37.

Burbidge, E, Burbidge, Geoff, Fowler, William, and Hoyle, Fred. 1957. Synthesis of the Elements in Stars. *Rev. Mod. Phys.* 29:547.

DeVorkin, David. 1978. E. Margaret Burbidge, AIP, https://repository.aip.org/burbidge-margaret-1978-july-13. Accessed Jun 1, 2025.

Fox, Margalit. April 6, 2020. E. Margaret Burbidge, Astronomer Who Blazed Trails on Earth, Dies at 100, *New York Times*.

Humphreys, Roberta. 2018. The award that shook astronomy. *Physics Today*. https://pubs.aip.org/physicstoday/online/30999/The-award-rejection-that-shook-astronomy. Accessed Jun 1, 2025.

Sargent, Anneila and Longair, Malcolm. 25 Aug 2021 Eleanor Margaret Burbidge 12 August 1919 - 5 April 2020. *Biographical Memoirs of Fellows of the Royal Society*. Available at https://royalsocietypublishing.org/doi/10.1098/rsbm.2021.0017. Accessed 30 June 2025.

University of California Academic Senate, In Memorium: E Margaret Burbidge, https://senate.universityofcalifornia.edu/in-memoriam/files/margaret-burbidge.html. Accessed 30 June, 2025.

Wikipedia, B^2FH Paper. 19 June 2025. https://en.wikipedia.org/wiki/B2FH_paper. Accessed 30 June 2025.

Wikipedia, Margaret Burbidge. 24 June 2025. https://en.wikipedia.org/wiki/Margaret_Burbidge. Accessed 30 June 2025.

Wikipedia, Nucleosynthesis. 18 May, 2025. https://en.wikipedia.org/wiki/Nucleosynthesis. Accessed Jun 1, 2025.

Wikipedia, Quasars. 28 May, 2025. https://en.wikipedia.org/wiki/Quasar. Accessed Jun 1, 2025.

Young, Monica. September 28, 2021. What is a Quasar? *Sky and Telescope*. https://skyandtelescope.org/astronomy-resources/what-is-a-quasar/. Accessed Jun 1, 2025.

Chapter 8
Joan A. Steitz

For her major contributions to the basic molecular biology of bacterial and mammalian cells. Her discovery of at least six new components of the cellular machinery is an accomplishment of great distinction in both basic molecular biology and in the clinical treatment of autoimmune disease

Keywords RNA · mRNA · messenger RNA · James Watson · snRNPs · snRNAs · Ribosome · Cell biology · Molecular biology · Tom Steitz · Protein · Nucleotides · Splicing

Dr. Joan A. Steitz, shown in Fig. 8.1, is a leading cell biologist who has focused her entire career on the function of RNA. She is credited with having discovered that ribosomes, organelles in human cells that are critical in the production of proteins, start the process of protein production by binding to messenger RNA, RNA that is floating in the cell cytoplasm and that provides the recipe to build a protein by chaining segments of nucleotides together. She also discovered small ribonucleic proteins, snRNPs, that function to identify which segments of messenger RNA are required for coding and which are not. Dr. Steitz was awarded the National Medal of Science in 1986.

J. S. Tietjen et al., *Women Laureates 1970 – 2007*, Women in Engineering and Science, https://doi.org/10.1007/978-3-032-23116-1_8

Fig. 8.1 Dr. Joan A. Steitz. (Courtesy of Yale School of Medicine)

8.1 Life (1941–)

Joan Argetsinger was born in Minneapolis, Minnesota, in 1941. Her mother was a speech therapist and her father a guidance counselor. They encouraged Joan's interest in science, which began with Joan's rock and butterfly collecting as a child. Joan attended Northrop Collegiate High School, a small girls' high school which offered only three science classes, so she took them all (Leslie; Wikipedia 2025).

Upon graduating from high school, Joan attended Antioch College, her father's alma mater in Ohio, and earned a BS in chemistry in 1963. Antioch encouraged their students to participate in work programs to enhance their learning. On one such internship, Joan was a technician in the MIT laboratory of molecular biologist Dr. Alexander Rich and learned about ribonucleic acid, RNA, which, at the time, was a fairly recently described molecule. This exposure piqued Joan's curiosity and led to her lifelong interest in RNA despite the experiment she was pursuing at MIT failing. She has described this moment:

> And I just remember thinking, 'Wow!' There's a molecular basis for this genetic phenomenon that I've learned about and been fascinated by (CSH Oral History Collection).

Toward the end of her undergraduate days, Joan determined she would go to medical school, figuring that it was the only option open to a woman interested in a career related to chemistry. She applied and was accepted to several medical schools and decided to attend Harvard. Having been away from home for most of her undergraduate days, Joan opted to spend the summer prior to starting medical school with her family. She managed to get a position working for cell biologist Dr. Joseph Gall at the University of Minnesota. Unlike Joan's prior work experiences, Dr. Gall let her pursue her own project. This excited Joan so spectacularly that she decided she wanted to do research. She managed to switch her Harvard graduate work from the medical school to their new biochemistry and molecular biology graduate program (Leslie).

Joan's introduction to Harvard's biochemistry and molecular biology program did not go well. She visited her first choice of a graduate advisor, but he turned her away, saying, "But you are a woman. What will you do when you get married and have a family?" (The Jackson Lab 2017). According to Joan, she was "able to make it out of his office before I burst into tears" (The Jackson Lab 2017). Joan approached another Harvard faculty member, James Watson, and convinced him to accept her as his student. Dr. Watson had won a Nobel Prize in Physiology or Medicine just the year before, 1962, with Dr. Francis Crick, for determining the structure of DNA. Joan became his first female graduate student, a fact she did not know at the time (Heath 2022; CSH Oral History Collection).

Joan earned her PhD from Harvard in 1967. Her PhD project was a study of a bacteria-attacking virus, a bacteriophage, whose genetic identity was coded not on a double-stranded chain of nucleotides, DNA, but on single-stranded RNA (Leslie). Along the way, Joan also met and married another student in the molecular biology area—Tom Steitz.

Upon completing their PhDs, the Drs. Steitz moved to Cambridge, England, for postdoctoral studies at the Medical Research Council Lab for Molecular Biology (MRC's LMB). They made this choice predicated on Dr. Tom Steitz's work as a crystallographer (Heath 2022). The MRC was the leading place for crystallography work at the time. LMB was headed by Dr. Francis Crick. Although Dr. Joan Steitz had been told there would be a postdoctoral position for her, she arrived in Cambridge to find that no lab or office space had been dedicated to her. Further, Dr. Crick suggested that she do a literature project in the library, clearly something that would not have made use of her meticulous laboratory skills. Instead, Dr. Joan Steitz spoke to MRC researchers to find an appropriate lab-oriented project, ultimately opting to pursue a project that other postdocs had rejected as too challenging and thus unlikely to help them find permanent employment on a faculty after their postdoctoral fellowship (Heath 2022; Leslie). Dr. Steitz figured that since there were no jobs for women on the science faculty of US universities anyway, she had less at risk by working on a very difficult problem. The topic was to elucidate the interaction between messenger RNA (mRNA) and ribosomes, the organelles in cells where protein production takes place, in the process of protein production. Specifically, the question was how a ribosome chooses where to latch onto an mRNA molecule (Leslie; The Jackson Lab 2017).

During her stay at MRC, Dr. Steitz managed to sequence a piece of mRNA and received a lot of recognition for this work. Despite this accomplishment, Dr. Steitz thought that return to the USA would mean that she would be relegated to a research associate position in some man's lab. She had, after all, never seen a female science professor or lab leader and assumed that she could not therefore be one. Dr. Steitz was quite surprised when she was offered faculty positions as she neared completion of her postdoctoral term (Heath 2022). She has commented,

> Women of my day had six kids and a station wagon. At that time there were no women professors in the natural sciences at any major university. Consequently, I never envisioned myself being where I am today. I never thought I would teach; I never thought I would mentor graduate students; I never thought I would be on the faculty of a prominent university. I really thought I would be a research associate in someone's lab – a man of course (Connecticut Women's Hall of Fame; The Jackson Lab 2017).

Dr. Tom Steitz had accepted a junior faculty position at the University of California at Berkeley prior to attending his Cambridge postdoc, so the Drs. Steitz went to Berkeley upon leaving England. Dr. Tom Steitz reported to the chair of his department and showed him letters of offer that both he and his wife had received for faculty positions. He asked what the university was prepared to do for his wife. In response, he was told, "But our wives are research associates in our labs, and they love it" (Heath 2022). The Drs. Steitz returned to Cambridge for a few months rather than staying in California and left only after they had negotiated faculty positions at Yale University in 1970.

The early days at Yale University were idyllic for Dr. Joan Steitz. She was part of a department overhaul in which new faculty were being hired, and everyone supported one another. This helped her to gain confidence in her work, which was continuing to focus on mRNA and ribosomes (Heath 2022). At that time, there were only about three undergraduates per class, enabling her to craft a research project for each student. Dr. Steitz remained at Yale for the rest of her career, making inroads on understanding different types of RNA and their role in protein production. She served for a time as the Chair of the Department and retired in 2021 (Connecticut Women's Hall of Fame).

Dr. Steitz has received numerous awards for her discoveries. A partial list includes the Passano Foundation Young Scientist Award in 1975, the National Academy of Sciences Award in Molecular Biology in 1982, the National Medal of Science in 1986, the Dickson Prize for Science in 1988, the Warren Triennial Prize in 1989, the Golden Plate Award of the American Academy of Achievement in

1992, the L'Oreal-UNESCO Award for Women in Science in 2001, the RNA Society Lifetime Achievement Award in 2004, the EB Wilson Medal of the American Society for Cell Biology in 2005, the Gairdner Foundation International Award and the Rosalind Franklin Award for Women in Science of the National Cancer Institute in 2006, the Albany Medical Center Prize in Medicine and Biomedical Research in 2008, the Lasker-Koshland Special Achievement Award in Medical Science in 2018, and the Wolf Prize in Medicine and the Warren Alpert Foundation Prize in 2021. Dr. Steitz has also been elected to the American Academy of Arts and Sciences (1982), National Academy of Sciences (1983), the American Philosophical Society (1992), and the Royal Society of London (2014). She enjoyed a fellowship from the Howard Hughes Medical Institution for 35 years. She has also received 11 honorary degrees in recognition of her work. The Lab for Molecular Biology in Cambridge, England, in 1984 established a postdoctoral prize in her honor. The award is given for outstanding research by postdoctoral fellows with less than six years in their post.

In addition to her work on mRNA, Dr. Steitz is well known to be a strong supporter of women in science, saying, "We cannot expect to capture the interest and talents of girls and women for the scientific enterprise unless they can view their own participation as possible" (The Jackson Lab 2017). She served on the National Academy of Sciences committee focused on identifying the root causes and solutions for the underrepresentation of women in science. In 2006, Dr. Steitz was one of the founders of the Rosalind Franklin Society, a nonprofit organization that "recognizes, fosters, and advances the important contributions of women in the life sciences and affiliated disciplines" (Panek 2019). Dr. Steitz was also one of the authors of a paper titled, "Increasing Gender Diversity in the STEM Research Workforce," published in 2019 (Greider et al. 2019).

Dr. Steitz has also served as a scientific director of the Jane Coffin Childs Memorial Fund (1991–2002) and on the editorial board of *Genes and Development* (Wikipedia 2025). She was also elected to the board of The Jackson Laboratory in 2014 (The Jackson Lab 2017). In her spare time, she has enjoyed antique hunting and hiking in the mountains (Heath 2022; Naughton 2021).

The Drs. Steitz had a son, Jon, in 1980. He earned an undergraduate degree from their department at Yale and initially went on to be a major league baseball player. He subsequently earned a law degree and has been employed by McKinsey (Kolata 2018).

Dr. Joan Stietz's husband received the Nobel Prize in Chemistry in 2009. He died in 2018 of pancreatic cancer. Speaking of her husband at the time of his passing, Dr. Stietz said, "I was very, very lucky to have married Tom. He really believed I should have as equal an opportunity to succeed as he" (Kolata 2018).

8.2 Messenger RNA and Ribosomes

The so-called central dogma of biology is that DNA makes RNA, and RNA makes protein (Khan Academy). It is the second part of this statement that has been the focus of Dr. Steitz's research, namely the creation of proteins from RNA. At the time that Dr. Steitz started a postdoctoral fellowship at LMB, it was known that ribosomes were responsible for the production of proteins. It was also known that the ribosomes interact with mRNA in the process, and that it is the mRNA that contains the instructions for precisely which base elements, which nucleotides, to put together in a chain to form a particular protein. The question Dr. Steitz addressed was how do the mRNA and ribosomes interact. This question continued to occupy her work as she started her career at Yale University. Her work showed that ribosomes, being themselves mostly RNA, bind to specific sites on the mRNA in the manner of complementary base pair binding seen in DNA. To achieve this discovery, Dr. Steitz mixed mRNA and ribosomes, then after ribosomes attached to the mRNA, she added chemicals to slice up the unbound, exposed mRNA. She then sequenced the mRNA that had been protected by the attachment to a ribosome and found that the ribosomes homed in on a particular sequence of nucleotides identified by base pairing. Upon making this discovery, Dr. Steitz commented, "One of the greatest joys in science is realizing that you know something that nobody else knows" (CHS Oral History Collection). Her research result was reported in a landmark paper in 1975 (Steitz and Jakes 1975).

By the late 1970s, biomolecular researchers had determined that eukaryote cells (those with nuclei) produce a rough draft of mRNA, labeled pre-mRNA, rather than mature mRNA. The pre-mRNA contains sections that code for proteins (exons) but also sections that do not code for proteins (introns). The cells were known to cut out the introns and splice exons together to form the mature mRNA. The question that Dr. Steitz and other researchers investigated was how the cells pared those introns from the pre-mRNA. Dr. Steitz had discovered a small particle and suspected these were involved in the splicing process. The particles are now labeled snRNPs, small nuclear ribonucleoprotein particles called "snurps." SnRNPs are made of RNA and protein, with part of the RNA sticking out of the complex so that it can bond with RNA with complementary bases to the exposed snRNP RNA. Isolating snRNPs is difficult, but Dr. Michael Lerner, a medical doctor working toward his PhD in Dr. Steitz's lab, was able to use antibodies in the blood of patients with lupus to isolate them. Drs. Steitz and Lerner were then able to show that the snRNPs were essential for RNA splicing (CSH Oral History Collection). Figure 8.2 shows the process schematically. This work was first presented in a 1980 paper in *Nature* and is what Dr. Steitz is best known for (Lerner et al. 1980).

In addition to the research on the production of mRNA for protein coding, Dr. Steitz has studied a number of additional RNA features. She examined how multiple protein products can be formed from a single gene and has studied the function of new forms of RNA her lab discovered, including microRNAs and long noncoding RNAs, both of which play a role in the regulation of gene expression (Wang and Farhana 2023; The Jackson Lab 2017). She has also studied RNA stability and degradation. Her work has led to an understanding of the role of RNA in diseases, particularly diseases that are the result of mutations or deficiencies in splicing of mRNA.

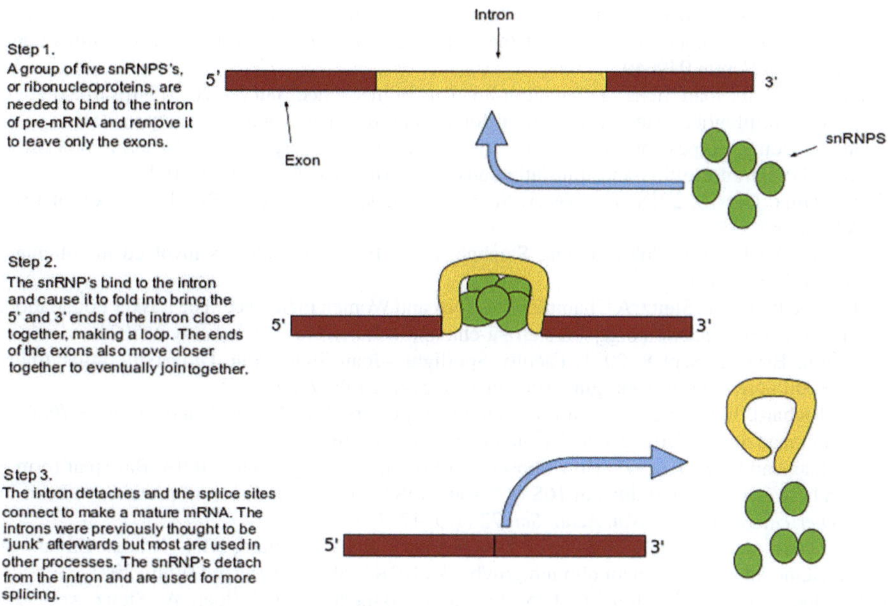

Fig. 8.2 The process of RNA splicing. (Wikipedia Commons, CC BY-SA 4.0)

8.3 Conclusion

Dr. Joan A. Steitz is a molecular biologist credited with several important discoveries related to RNA. She is best known for showing that the process of production of proteins begins with a ribosome bonding to mRNA at locations dictated by complementary base pairing, as in DNA. She also showed that mRNA is spliced by snRNPs so that only the parts of the mRNA that code for protein production are preserved. Dr. Steitz was awarded the National Medal of Science in 1986.

References

Connecticut Women's Hall of Fame, Joan Steitz. https://www.cwhf.org/inductees/joan-steitz,. Accessed July 9, 2025.

CSH Oral History Collection., Joan Steitz. https://library.cshl.edu/oralhistory/interview/scientific-experience/becoming-scientist/steitz-becoming-a-scientist/. Accessed July 9, 2025.

Greider, Carol, Sheltzer, Jason, et al. 2019. Increasing gender diversity in the STEM research workforce. *Science* 366 (6466):692–695.

Heath, Joan, Cory, Suzanne, and Steitz, Joan. 2022. Joan Heath interviews Suzanne Cory and Joan Steitz: a female perspective of the swinging '60s, *Dis Model Mech.* 15(5). https://doi.org/10.1242/dmm.049649.

The Jackson Lab, Joan Steitz: a champion for women in science. Jan 25, 2017. https://resources.jax.org/jax-blog/joan-steitz-a-champion-for-women-in-science. Accessed July 9, 2025.

Khan Academy, Types of RNA, https://www.khanacademy.org/test-prep/mcat/biomolecules/x04f6bc56:transcription-and-translation/a/types-of-rna. Accessed July 9, 2025.

Kolata, Gina. Oct 10, 2018. Thomas A. Steitz, 78, Dies; Illuminated a Building Block of Life. *NY Times.*

Lerner, Michael, Boyle, John, mount, Stephen, et al. 1980. Are snRNPs involved in splicing?, *Nature* 283: 220–4.

Leslie, Mitchell. Joan Steitz: A Champion for RNA and Women in Science, The Lasker Foundation. https://laskerfoundation.org/joan-steitz-a-champion-for-rna/. Accessed July 9, 2025.

Naughton, Brigitte. Sept 8, 2021. Faculty Spotlight—Joan Steitz, Yale University, https://mbb.yale.edu/news/faculty-spotlight-joan-steitz. Accessed July 9, 2025.

Panek, Richard. Jul/Aug 2019. Don't listen to the naysayers. *Yale Alumni Magazine.* https://yalealumnimagazine.org/articles/4923-joan-steitz, Accessed July 9, 2025.

Steitz, Joan and Jakes, K. 1975. How ribosomes select initiator regions in mRNA: Base pair formation between the 3' terminus of 16S rRNA and mRNA during initiation of protein synthesis in *Escherichia coli, Proc. Nat. Acad. Sci.* 72 (12): 4734–8.

Wang, David and Farhana, Aisha. 2023. Biochemistry, RNA Structure, NIH National Library of Medicine. https://www.ncbi.nlm.nih.gov/books/NBK558999/. Accessed 9 July 2025.

Wikipedia, Joan A. Steitz (Jun 27, 2025). https://en.wikipedia.org/wiki/Joan_A._Steitz. Accessed 9 July 2025.

Chapter 9
Rita Levi-Montalcini

For a major breakthrough in neurobiology by her discovery of the nerve growth factor and its effect on the growth of the sympathetic nervous system which set the stage for worldwide studies of the molecules involved in normal and malignant growth

Keywords Nerve growth factor · Neuroimmunology · Growth factors · Neuroembryology · Nerve ganglia · Silver staining · Stanley Cohen · Viktor Hamburger · Giuseppe Levi

Dr. Rita Levi-Montalcini, shown in Fig. 9.1, discovered nerve growth factor, a naturally occurring protein that promotes the growth of nerve cells. This research opened the door to understanding how cells communicate and led to the field of neuroimmunology. It was the first cell growth factor identified, although more than one hundred cell growth factors in animals have now been identified. For her work discovering nerve growth factor, she and collaborator Dr. Stanley Cohen were awarded a Nobel Prize in Physiology or Medicine in 1986. She received a National Medal of Science in 1987.

J. S. Tietjen et al., *Women Laureates 1970 – 2007*, Women in Engineering and Science, https://doi.org/10.1007/978-3-032-23116-1_9

Fig. 9.1 Dr. Rita Levi-Montalcini. (From Quirinale.it)

9.1 Life (1941–2012)

Rita Levi-Montalcini was born in Turin, Italy, in 1909 to parents Adamo Levi and Adele Montalcini. She and her twin sister, Paola, were the youngest of the four children in the family. The family was well off and happy in their surroundings. Rita's father worked as an electrical engineer and made all the significant decisions for the family. Rita's mother was a painter (Rita Levi-Montalcini, NobelPrize.org 2025).

Rita's father did not believe that women should be highly educated. He reasoned that it might interfere with their wifely and motherly duties. As a result, Rita did not initially complete high school. However, when she hit the age of twenty, Rita was able to convince her father that the life he envisioned for her was neither what she desired, nor what was likely to happen. Reluctantly, he agreed to let Rita pursue a medical degree, a pursuit Rita was inspired to do when the family governess was stricken with cancer. Having not completed her primary schooling, Rita then quickly caught up on required courses over a period of eight months, graduated from high school, and then entered medical school at the University of Turin (NobelPrize.org (b) 2025). She graduated with highest distinction in medicine and surgery in 1936.

She was briefly engaged while in medical school, but never had any long-term romances subsequently (Abbott 2009).

During medical school, Dr. Levi-Montalcini was mentored by Dr. Giuseppe Levi, a professor and researcher in histology. Significantly, Dr. Levi taught Dr. Levi-Montalcini how to stain nerve cells with silver, which then made them clearly visible under a light microscope. Dr. Levi-Montalcini thought that a career in research might appeal to her more than practice as a physician. She entered a three-year specialization program in neurology and psychiatry in 1936 with this in mind. However, in 1938, Mussolini's Race Law prohibiting non-Aryans from having an academic career or holding a profession was enacted, and Dr. Levi-Montalcini, a Jew, was kicked out of the program (NobelPrize.org (b) 2025).

Determined to make progress toward her goal as a research scientist, Dr. Levi-Montalcini left Italy for Brussels to study at a neurological institute. She was forced to flee again in 1940, when it seemed that Germany was about to invade Belgium. Dr. Levi-Montalcini returned to Turin and her family. She turned a bedroom in the estate into a laboratory using items she had obtained over the years, which she fashioned from materials she could still purchase, or that she obtained on the black market. For instance, she created scalpels from sewing needles and used the tiny scissors employed by ophthalmologists and the forceps used by watchmakers (NobelPrize.org (a) 2025).

By 1941, Turin was subjected to heavy bombing by the Allied forces. Dr. Levi-Montalcini and her family moved to their vacation home in the Piedmont region of the country, where the war was mostly a distant concern. In 1943, the Germans invaded Italy, and the family traveled under assumed names to Florence, where they lived underground with the help of friends until the Germans left the city in 1944 (NobelPrize.og (b) 2025). During this time, Dr. Levi-Montalcini was in contact with and supported members of the Action Party, a liberal, socialist political party.

Throughout the moves in Italy, Dr. Levi-Montalcini conducted research in a makeshift lab. Inspired by a paper by Dr. Viktor Hamburger in the USA, she began to study how nerve cells grew in chick embryos. For part of the time, Dr. Giuseppe Levi, her beloved mentor who had also been kicked out of the University of Turin for being Jewish, served as her assistant. When the family moved to the country to avoid the bombing of Turin, Dr. Levi-Montalcini bicycled to local farms to get fertilized hen eggs, but when the family fled to Florence, she was forced to temporarily give up her experiments (Abbott 2009; NobelPrize.org (a) 2025). Up until that moment, Dr. Levi-Montalcini had been able to publish her research results in foreign journals (as publication by Jews in Italian scientific journals was forbidden).

In 1944, as the Germans left Florence and the Allies took control of the city, Dr. Levi-Montalcini was hired as a physician to treat the influx of war refugees. This was the only time in her life that she served as a practicing medical doctor, and when the war finally ended in 1945, she was happy to return to Turin and to her academic pursuits (NobelPrize.org (b) 2025). Dr. Giuseppe Levi returned to his lab as well, and brought Dr. Levi-Montalcini in as his assistant, thus reversing the roles they had recently had.

In 1946, Dr. Hamburger contacted Dr. Levi-Montalcini and invited her to spend a term with him at Washington University in St. Louis, Missouri. Dr. Hamburger was aware that his research led him to conclusions that Dr. Levi-Montalcini disagreed with, and he hoped that working together they could find the scientific truth. At this point, Dr. Levi-Montalcini was also feeling awkward about the reversal of roles between her and Dr. Levi, so she gladly accepted the invitation. Commenting later, she said, "I felt at home the day I landed" (NobelPrize.org (a) 2025). Her plan to stay for as long as one year later changed, and she ultimately stayed at Washington University for thirty years. She was made an Associate Professor in 1956 and promoted to Full Professor in 1958.

In 1962, Dr. Levi-Montalcini set up a research unit in Rome and began splitting her time between Rome and St. Louis. From 1969 to 1978, she served as the Director of the Institute of Cell Biology of the Italian National Council on Research and built it into one of the largest biological research labs in Italy. In 1977, she retired from Washington University, moving full-time to Rome. Starting in 1979, Dr. Levi-Montalcini served as a Guest Professor at the Institute of Cell Biology in Switzerland (NobelPrize.org (b) 2025).

Dr. Levi-Montalcini received many awards for her work on nerve cell growth. In 1963, she became the first woman to receive the Max Weinstein Award of the United Cerebral Palsy Association for outstanding contributions in neurological research. In 1969, she received an award from the National Academy of Sciences in Rome. In 1975, Dr. Levi-Montalcini became the first woman to be installed in the Pontifical Science Academy. In the 1980s, she received awards from Brandeis University and Columbia University, as well as the Albert Lasker Basic Medical Research Award. In 1986, Dr. Levi-Montalcini and her research collaborator, Dr. Stanley Cohen, were recipients of the Nobel Prize in Physiology or Medicine, and in 1987, Dr. Levi-Montalcini received the National Medal of Science. In 2009, she received the Leonardo da Vinci Award of the European Academy of Sciences. Dr. Montalcini was a member of the American Academy of Arts and Sciences, the National Academy of Sciences, and the National Academy of Sciences of Italy (Dubinsky). She received numerous honorary doctorates from universities around the world. She is shown receiving the Nobel Prize in Fig. 9.2.

Toward the end of her career, Dr. Levi-Montalcini turned her attention to ensuring that others would not have to endure the obstacles to their research work that she had endured. In 1999, she was one of four people named to be an ambassador for the United Nations' Food and Agriculture Organization in their campaign to fight world hunger. In 2002, she founded the European Brain Research Institute (EBRI), and she later established the Rita Levi-Montalcini Foundation (RLMF) to provide African women "with the tools for a full development of their capabilities" (NobelPrize.org (b) 2025). RLMF supported the education of over 6000 women in the sciences as of 2009.

In 2001, she became a Senator for Life in Italy, a very high honor, and she used this position to protect science funding. In 2006, at the age of 97, Dr. Levi-Montalcini was the deciding vote on the budget in the Italian Parliament. She insisted she would

Fig. 9.2 Dr. Levi-Montalcini receiving the Nobel Prize in Physiology or Medicine in 1986. (Courtesy of NobelPrize.org)

only support the budget if the science funding, which had been cut in the proposed budget, was restored. In typical fashion, this was an argument she won. The Leader of the Right Party and part of the opposition coalition had stated publicly in the week before the vote that he would send her crutches, suggesting she was too old to vote and was being used as a crutch by the failing government (Abbott 2009).

Dr. Levi-Montalcini was a prolific writer, producing twenty-one books in her lifetime. While most of these were technical books, typically with her as the sole author, she also produced a couple of books about her life: *In Praise of Imperfection: My Life and Work* (Levi-Montalcini 1988) and *The Hourglass of Life: A Nobel Laureate Reflects on Her Life* (Levi-Montalcini and Tripodi 2010). She also wrote a book with biographies of marginalized women in science and social movements, entitled *Le Tue Antenate*, which roughly translates to "Your Ancestors" (Abikhzer et al. 2024). Dr. Levi-Montalcini figured prominently in a 1995 science documentary called *Death by Design: The Life and Times of Life and Times* (Friedman and Brunet 1995).

Dr. Levi-Montalcini was not only the first Nobel Laureate to reach the age of 100, but she was still working when she attained that age. She would spend mornings at EBRI and afternoons at the education foundation for African women (Abbott 2009). She was known to be a tireless worker, calling people early in the morning and late at night to discuss research findings.

Looking back on her life and career, Dr. Levi-Montalcini said, "If I had not been discriminated against or had not suffered persecution, I would never have received the Nobel Prize (Rita Levi-Montalcini, NobelPrize.org 2025)." On the occasion of her 100th birthday, she also said

> If I were to die tomorrow or in a year's time, it would be the same: what counts is the message you leave behind you. It's the only way to prevent that our passing on Earth doesn't exhaust in a big nothing (Ottria 2013).

Dr. Levi-Montalcini passed away in 2012, at the age of 103. In 2016, an orchid was named in her memory.

9.2 Nerve Growth Factor

Dr. Levi-Montalcini's first research project was to try to replicate the results obtained by Dr. Viktor Hamburger in a paper she read in the late 1930s. Dr. Hamburger had removed the budding limbs of a chick embryo as a means of determining how nerve cells emerged from the spinal cord to infiltrate the limbs. Dr. Hamburger found that removal of the budding limbs caused the nerve ganglia (group of nerve cell bodies) to decrease in size, from which he concluded that the limb tissue must be emitting some sort of inductive factor causing precursor cells to proliferate and then differentiate into nerve cells. Once the limbs were removed, the inductive factor was missing, and thus the nerve ganglia would start to die. But Dr. Hamburger had not been able to visualize the nerve cells clearly under a microscope. Dr. Levi-Montalcini replicated his experiments in her bedroom lab but used the silver staining method she had learned from Dr. Giuseppe Levi to make the nerve cells easy to visualize. She found that the ganglia didn't atrophy immediately, but instead proliferated, differentiated, and then started to grow before dying prior to reaching their target. From this, Dr. Levi-Montalcini concluded that it wasn't the tissue inducing precursor cells to proliferate but rather a growth factor in the limb tissue that was missing. The invitation she received from Dr. Hamburger was to work together at Washington University to determine which of their views was correct (Abbott 2009).

The deciding work was built upon the research of one of Dr. Hamburger's graduate students, Elmer Bueker, who had grafted a lump of mouse sarcoma tumor onto a chick embryo limb. Bueker found that nerve fibers grew more densely into the tumor than into the limb tissue. He theorized that the difference was explained by the surface area being greater in the tumor, but Dr. Levi-Montalcini had other thoughts. She delayed her return to Italy to test her ideas. Dr. Levi-Montalcini saw that the nerves growing into the tumor were in a seemingly random pattern and growing wildly. This convinced her further that the tumor was releasing some sort of growth factor. She repeated Bueker's experiment but with the tumor outside the sac containing the embryo so that they shared only a blood supply. Nonetheless, the nerves grew wildly in the tumor, supporting her theory that it was emitting a growth factor rather than the embryo encouraging growth. The nerves grew into tissue and even veins but not into arteries, so they were not directly connected to the embryo blood. This proved that it was a chemical in the tumor that was prompting the nerve cell growth (Abbott 2009).

Subsequent to this experiment, Dr. Levi-Montalcini sought a more reliable and stable system for experimentation than chicken embryos. She was interested in quantifying results, and this was not possible with the current set up. She put two live, tumor-ridden white mice into her handbag and flew to Rio de Janeiro, to the only lab in the world at that time capable of culturing nerve ganglia. She learned how to culture ganglia and did a key experiment. Dr. Levi-Montalcini isolated nerve ganglia and put them close to pieces of mouse sarcoma. A mere twenty-four hours later, she saw nerve cells growing from the ganglia like rays from the sun. Upon discovering this halo effect, she said, "it is like rivulets of water flowing steadily over a bed of stones." The side of the ganglia facing the tumor had the densest nerve growth (Abbott 2009; Wikipedia 2025).

Drs. Hamburger and Levi-Montalcini worked unsuccessfully for six years to isolate the growth factor they were convinced was behind the proliferation of nerve cells they had seen. Publications about nerve growth factor (NGF) during that time were met with skepticism by the scientific community, but the pair kept working. Finally, collaborator Dr. Stanley Cohen at Washington University was able to isolate the NGF. Additionally, Drs. Levi-Montalcini and Cohen were able to find an antiserum to the NGF. The antiserum, when applied to cultured ganglia and tissue, would stop nerve growth and, when injected into newborn mice, would wipe out relevant parts of the nervous system (Abbott 2009).

The work led to a series of classic papers in 1960 in Proceedings of the National Academy of Sciences (Iversen 2013). The first was a paper by Dr. Cohen on the isolation of NGF and the preparation of an NGF antiserum (Cohen 1960). The second was a paper by Dr. Levi-Montalcini and her assistant, Barbara Booker, which discussed the impact of NGF on a variety of mammalian ganglia as well as the growth-promoting effect on sympathetic ganglia when given to newborn, young, and adult mice (Levi-Montalcini and Booker 1960a). The third paper, again with co-author Barbara Booker, showed that NGF antiserum given to newborn mice, rats, rabbits, or kittens nearly completely destroyed the sympathetic nervous system without damaging anything else (Levi-Montalcini and Booker 1960b).

In 1962, Dr. Cohen reported on a growth factor for epithelial (skin) cells, and in the intervening time, hundreds of growth factors have been found, affecting almost all aspects of animal biology. Dr. Levi-Montalcini's work remained focused on NGF and its potential use to battle neurodegenerative diseases, such as dementia, Alzheimer's disease, and multiple sclerosis. She and the only student she had in her career, Ruth Hogue Angeletti, determined the structure of the NGF protein using one of the first automated sequencers in 1971 (Abbott 2009).

Interestingly, as the science community came to accept the existence of NGF, and more generally, growth factors, and the research in the area increased, Dr. Levi-Montalcini acted as though she felt her territory was being invaded. She fell out with most of the researchers in the area at one time or another, only making peace with colleagues starting in the 1980s. The controversy surrounding her increased again after she received the Nobel Prize and did not acknowledge the work of early colleagues Drs. Hamburger and Levi (Abbott 2009).

NGF is now known to be a crucial protein for nerve cells in sympathetic and sensorial nervous systems. Without NGF, brain cells die. Dr. Levi-Montalcini also showed, in a 1977 paper, that NGF affects the immune system as it is produced by a variety of lymphocytes (white blood cells in the lymphatic system) and glandular cells. The potential use of NGF in Alzheimer's treatment has been pioneered at EBRI, the institute she established (Ottria 2013).

In a 1986 experiment on young recruits doing parachute jumps, a large increase in NGF in the blood was seen shortly before the jump was made, and an even higher level immediately after the jump. This has led to the conclusion that NGF increases as anxiety rises. Further, the University of Pavia has shown that NGF levels increase in the early stages of falling in love (Ottria 2013).

9.3 Conclusion

Dr. Rita Levi-Montalcini was a neurobiologist who discovered that nerve cells are prompted to grow by a protein she named nerve growth factor. While her discovery was initially met with skepticism, it is now known that there are hundreds of growth factors important in nearly all biological processes of animals. Dr. Levi-Montalcini received the Nobel Prize in Physiology or Medicine with collaborator Dr. Stanley Cohen in 1986. She received the National Medal of Science in 1987.

References

Abbott, A (2009). Neuroscience: One hundred years of Rita, Nature 458: 564–7.

Abikhzer, Myriam, Delouya, Guila, and Taussky, Daniel. 2024. Rita Levi-Montalcini: From Persecution to the Nobel Prize and Honorary Degree from a Canadian University, *Cureus* 16(9): e69743.

Cohen, Stanley. 1960. Purification of a nerve-growth promoting protein from the mouse salivary gland and its neuro-cytotoxic antiserum. *Proc Natl Acad Sci USA* 46: 301–11.

Friedman, Peter and Brunet, Jean-Francois. 1995. Death by Design: The Life and Times of Life and Times. Available at https://www.youtube.com/watch?v=4NytzTLnyKM. Accessed 17 Jul 2025.

Iversen, Leslie. 2013. Rita Levi-Montalcini: Neuroscientist par excellence. *Proc Natl Acad Sci USA* 110(13): 4862–3.

Levi-Montalcini, Rita and Booker, Barbara. 1960a. Excessive growth of the sympathetic ganglia evoked by a protein isolated from mouse salivary glands. *Proc Natl Acad Sci USA* 46: 373–84.

Levi-Montalcini, Rita and Booker, Barbara. 1960b. Destruction of the sympathetic ganglia in mammals by an antiserum to a nerve-growth protein. *Proc Natl Acad Sci USA* 46: 384–91.

Levi-Montalcini, Rita. 1988. *In Praise of Imperfection: My Life and Work*. Basic Books.
Levi-Montalcini, Rita and Tripodi, Giuseppina. 2010. *The Hourglass of Life: A Nobel Laureate Reflects on Her Life*. Dana Press.
NobelPrize.org (a), Rita Levi-Montalcini, Nobel Prize Outreach 2025. https://www.nobelprize.org/stories/women-who-changed-science/rita-levi-montalcini/. Accessed 17 Jul 2025.
NobelPrize.org (b), Rita Levi-Montalcini Biographical, Nobel Prize Outreach 2025. https://www.nobelprize.org/prizes/medicine/1986/levi-montalcini/biographical/. Accessed 17 Jul 2025.
Ottria, L (2013). Rita Levi-Montalcini and Research, Oral Implantol 5(4): 75–6.
Wikipedia, (15 Jul 2025). Rita Levi-Montalcini, https://en.wikipedia.org/wiki/Rita_Levi-Montalcini. Accessed 17 Jul 2025.

Chapter 10
Anne Anastasi

For her work in the development of the discipline of differential psychology as a behavioral science, which illuminates the way traits are influenced by heredity and environment and the methods by which traits and human characteristics are measured

Keywords Differential psychology · Psychological testing · Psychometrics · Trait formation · Behavioral science · Test guru · Standardized tests · Test preparation

Dr. Anne Anastasi (Fig. 10.1) was a psychologist renowned for her work in differential psychology and psychological testing. She transformed differential psychology into a behavioral science that examines the formation of human traits and became known as the "test guru" for her expertise in quantifying human characteristics, such as intelligence. In 1987, Dr. Anastasi was awarded the National Medal of Science.

J. S. Tietjen et al., *Women Laureates 1970 – 2007*, Women in Engineering and Science, https://doi.org/10.1007/978-3-032-23116-1_10

Fig. 10.1 Anne Anastasi. (Barnard College, Mortarboard (1928 yearbook), Public domain, via Wikimedia Commons)

10.1 Life (1908–2001)

Anne Anastasi was born to Italian immigrant parents in New York City in 1908. Her father died when she was just one year old, leaving her mother, grandmother, and uncle to raise her. Her father had been an attendance officer for the New York City Board of Education. She referred to her grandmother as a "grande dame," her uncle as a "dilettante," and her mother as the realist of the family (Anastasi 1980). Though educated in languages and the humanities, none of the three was prepared to earn a living. Her mother taught herself how to do bookkeeping and tried her hand at business. She started a piano factory with her brother, but it closed after a few years. After a series of jobs, she landed at the Italian-language newspaper *Il Progresso Italo-Americano*, where she was an office manager until her retirement (Anastasi 1980).

Anne's education was up to her grandmother, who took the job seriously and made it enjoyable. Anne was tutored in arithmetic, history, geography, and natural science. She also read stories for children with moral lessons and was entertained by her grandmother's tales from her childhood in Italy, where she attended balls and royal receptions. Being educated at home, Anne had little exposure to other children, stating, "I had no playmates of my own age—not even an imaginary playmate—although I did have an imaginary staircase which I greatly enjoyed (Anastasi 1980)."

Anne's grandmother did not want her to attend school, fearing that she would learn bad manners from the other children, but a school teacher, Miss Dora Ireland, was hired as a tutor to supplement Anne's education. Ireland was concerned about Anne's isolated existence and persuaded the family to let Anne enroll in the school where she taught. The situation lasted only a few months because Anne was seated at the back of the class and could not see the blackboard. No one realized that she needed glasses. The result was that she was bored and frustrated, so lessons resumed with her grandmother at home. Her next foray into public education was more successful. At age nine, she was enrolled in the local elementary school and seated at the front of the class. She was an excellent student and made friends (Anastasi 1980).

When she got to high school, she faced overcrowded classrooms in a building "unfit for human use" (Anastasi 1980). Conditions were so overcrowded that students were only allowed to attend half days, resulting in morning and afternoon shifts. Anne lasted all of two months before dropping out. A family friend suggested that she find an alternative path to college. After researching, she found some colleges that would accept results from the College Entrance Examination Board (CEEB) tests as a measure of college readiness in lieu of a high school diploma. These exams covered high school courses, such as Latin, French, advanced algebra, and geometry. Anne took courses at Rhodes Preparatory School to prepare for the CEEB exams. Anne's first love was mathematics. She taught herself trigonometry and took all the available CEEB math exams. After two years at Rhodes, Anne was admitted to Barnard College at age 15 (Anastasi 1980).

Anne planned to be a math major at Barnard until she discovered psychology. What sealed the deal was realizing that she could employ principles of mathematics and statistics in psychological research. Thus, she could have the best of both worlds by majoring in psychology. After completing her degree, Anastasi enrolled in graduate school at Columbia University with the goal of earning a PhD in psychology in just two years. She was able to use her published undergraduate honors thesis as a substitute for a master's thesis and also placed out of a statistics course. This, combined with taking summer courses, enabled her to complete her PhD in 1930 at just 21 years old. While taking graduate coursework in test construction and studying gender differences in test scores, she developed an interest in differential psychology, the field of study that examines individual differences in human behavior (Anastasi 1980).

Dr. Anastasi met fellow psychologist John Porter Foley, Jr., at Columbia. They married in 1933, and in an act that was rare at the time, she did not change her name. Only a year into their marriage, Dr. Anastasi was diagnosed with cervical cancer. Radium treatment proved effective in eradicating the cancer but left her unable to have children (Wikipedia Contributors 2019).

Initially, Drs. Anastasi and Foley were unable to find jobs in the same city. It was the Great Depression, and there were not many opportunities. Dr. Anastasi began teaching at Barnard after graduation and stayed in New York City her entire career. Meanwhile, Dr. Foley taught at George Washington University in D.C. at the beginning of their marriage. Although Anastasi was offered a government job in D.C. at one point, she turned it down, fearing that she would get bogged down in bureaucratic red tape. Eventually, Dr. Foley was able to join Dr. Anastasi in New York City when he obtained a position at Psychology Corporation working in industrial psychology. Later, he opened a consulting firm (Anastasi 1980).

Dr. Anastasi taught at Barnard until 1939. She moved to Queens College from 1939 to 1947, where she served as chairman, and at the beginning, the only faculty member of the newly formed Psychology Department. The college itself was merely one year old. The department grew to six faculty. By 1946, Dr. Anastasi was frustrated with her administrative duties, some of the personalities, and what she felt were wasteful committee meetings. In 1947, she left to become a professor at Fordham University. She served as chairman of the Psychology Department at Fordham from 1968 to 1974. Although this coincided with the Vietnam War, a tumultuous time on university campuses, Dr. Anastasi had a more positive experience heading a department the second time around. She remained at Fordham until her retirement in 1979, at which time she was appointed professor emeritus (Anastasi 1980).

During her career, Dr. Anastasi published more than 150 articles, monographs, and books. *Psychological Testing,* first published in 1954, and *Differential Psychology,* first published in 1937, became classic textbooks used in university classes for decades. Both went through multiple editions and were translated into several languages. She was known for her clear and succinct writing style (Reznikoff and Procidano 2001). Dr. Anastasi was a sought-after world-renowned consultant and leader in her field. She served as President of the Eastern Psychological Association (1946–1947); President of the American Psychological Association (APA) Division 1, General Psychology (1956–1957); President of the APA Division 5, Evaluation, Measurement and Statistics (1965–1966); President of the American Psychological Foundation (1965–1967); and was the third female president of the APA (1972) (Human Intelligence n.d.).

Dr. Anastasi's contributions to psychology were recognized with five honorary doctorate degrees and many awards, including the APA Distinguished Scientific Award for the Applications of Psychology, the Educational Testing Services Award for Distinguished Service to Measurement, the American Psychological Foundation's Gold Medal for Life Achievement, and the American Education Research Association's Award for Distinguished Contributions to Research in Education (Pickren 2009). In 1987, she received the National Medal of Science from President Ronald Reagan, the first ever awarded in psychology.

Dr. Anastasi died in 2001 at the age of 92. The funds in her estate were used to establish the Anne Anastasi Foundation, which benefits psychology students.

10.2 Research

Dr. Anastasi was a pioneer in the field of differential psychology. She urged researchers to consider the interrelation between traits, environment, and heredity (Anastasi 1937). In her influential paper, "Heredity, Environment, and the Question How," Dr. Anastasi advanced the ongoing debate between nature and nurture in understanding human traits. She shifted the focus from trying to assign fixed percentages to the influence of heredity and environment, instead emphasizing the complex and dynamic interaction between the two. Rather than advocating for one side over the other, Dr. Anastasi encouraged her colleagues to explore how genetics and environmental factors collaborate in shaping traits (Anastasi 1958).

Dr. Anastasi argued that while genetic factors may predispose individuals and establish a range of possibilities for specific traits, the environment plays a crucial role in determining where an individual might fall within that range. Moreover, environmental influences can activate or suppress particular genetic tendencies, illustrating that human development is shaped not by heredity or environment alone but by the ongoing interaction between both. For instance, an outgoing child might be treated differently by peers and adults than a shy child, which in turn impacts their future development. This reciprocal relationship demonstrates that heredity and environment are intertwined in ways that defy easy separation.

In examining trait formation, Dr. Anastasi became an expert in test development and interpretation and was known as the "test guru." As a doctoral student, she tackled issues such as test bias early in her research. She and two other graduate students worked to develop an intelligence test free from cultural bias by avoiding the use of language and numbers. The challenges her team faced in this task convinced her that it was impossible to create a completely "culture-free" test. Instead, she advocated for recognizing that "Each culture demands, fosters, and rewards a different set of abilities, which constitute intelligence within the culture (Anastasi 1985)."

She also believed that intelligence tests cannot measure all types of intelligence. As such, intelligence tests must be tailored to measure "how well individuals have acquired skills and knowledge valued in [their own] culture" (Anastasi 1981).

Dr. Anastasi also warned against misusing test data, emphasizing that intelligence tests do not explain why test takers performed the way they did. The tests are a snapshot in time and can change over time based on future experiences. Therefore, intelligence tests should not be used to label students but rather to assess their current academic level and design education experiences that will increase their intelligence (Wells 2019).

In her textbooks, Dr. Anastasi introduced students to the concepts of validity and reliability. Validity relates to how well a test measures what it is intended to, whereas reliability refers to the consistency of scores by the same test-taker on multiple occasions. Variance in test scores can be reduced by maintaining consistency in test conditions such as temperature and lighting of a room, time of day the test is taken, etc. If a test lacks objective criteria for scoring, variances are introduced by the scorer. For example, with essay tests, more than one scorer is often employed to grade the test to reduce "scorer variance" (Anastasi 1985).

Given that a well-constructed test strives to minimize the variance in test scores by the same test taker, what about the claims of test coaches, particularly when it comes to aptitude tests? In answering this question, Dr. Anastasi considered the SAT, a standardized test used in college admissions. It should be noted that the SAT has undergone many changes since its introduction in 1926. Initially known as the Scholastic Aptitude Test (SAT), it was designed to measure innate qualities to predict academic success, as opposed to achievement tests, which measure knowledge. The SAT has been redesigned over the years and is no longer an aptitude test, and the name SAT is not an acronym.

Dr. Anastasi considered three types of test preparation and their impact on SAT scores during the time when the SAT was still the Scholastic Aptitude Test. If test preparation improves scores without improving the "intellectual behavior in real-life contexts," the test's predictive validity is compromised. The first test preparation method considered was coaching, or short-term drilling and cramming on items covered on the test, such as memorizing vocabulary lists. In reviewing studies on this subject, Dr. Anastasi found that,

> In general, such research indicates that intensive drill on items similar to those on the SAT is unlikely to produce score gains appreciably larger than those found when students are retested with the SAT after a year of regular high school instruction. (Anastasi 1985)

The second type of test preparation considered was "test-taking orientation," designed to familiarize the test taker with the types of questions asked and other aspects of the test to minimize the difference in scores based on previous test-taking experience. This type of preparation improves the validity of the test by reducing individual variance (Anastasi 1985).

Finally, Dr. Anastasi considered the effects of preparation on improving cognitive skills. This type of training should improve both test scores and the ability to perform intellectual tasks in the future, thus maintaining the test's validity (Anastasi 1985).

10.3 Conclusion

Dr. Anne Anastasi had a profound impact on generations of psychologists through her classic textbooks and scholarly papers. She defined the field of differential psychology through her study of group and individual differences and how heredity and environment interact in the development of traits. She was the authority on standardized testing, highlighting the concepts of test construction, reliability, validity, and biases. Her legacy is embedded in every standardized test taken and bubble sheet carefully filled in.

References

Anastasi, A. (1985). Psychological testing: Basic concepts and common misconceptions. In A. M. Rogers & C. J. Scheirer (Eds.), *The G. Stanley Hall Lecture Series* (Vol. 5, pp.87–120). Washington, DC: American Psychological Association.

Anastasi, Anne. 1937. *Differential Psychology*. New York: Macmillan.

Anastasi, Anne. 1958. "Heredity, Environment, and the Question 'How?'" *Psychological Review* 65 (4): 197–208. doi:https://doi.org/10.1037/h0044895.

Anastasi, Anne. 1980. "Anne Anastasi." In *A History of Psychology in Autobiography. Vol VII*, edited by Gardner Lindzey, 1–37. San Francisco: W. H. Freeman. https://archive.org/details/historyofpsychol0007unse/page/n31/mode/2up?q=anastasi.

Anastasi, Anne. 1981. "Diverse Effects of Training on Tests of Academic Intelligence." In Issues in Testing: Coaching, Disclosure, and Ethnic Bias, edited by Bert F. Green, 5–20. San Francisco: Jossey-Bass.

Human Intelligence. n.d. "Human Intelligence: Anne Anastasi." https://www.intelltheory.com/anastasi.shtml. Accessed April 4, 2025.

Pickren, Wade. 2009. "Anne Anastasi (1908-2001)." Apa.org. American Psychological Association. 2009. https://www.apa.org/about/governance/president/anastasi.

Reznikoff, Marvin, and Mary Procidano. 2001. "Anne Anastasi (1908-2001)." *American Psychologist* 56 (10): 816–17. https://doi.org/10.1037/0003-066x.56.10.816.

Wells, Sarah. 2019. "The Woman Who Turned Psychological Testing into a Science." Gizmodo. August 2, 2019. https://gizmodo.com/the-woman-who-turned-psychological-testing-into-a-scien-1836056411.

Wikipedia Contributors. 2019. "Anne Anastasi." Wikipedia. Wikimedia Foundation. September 2, 2019. https://en.wikipedia.org/wiki/Anne_Anastasi.

Chapter 11
Rosalyn Yalow

For her historic contributions to the discovery and development of radioimmunoassay, a technique that employs radioactive isotopes to detect and measure the levels of insulin and hormones in the blood and body tissues

Keywords Radioimmunoassay · Isotopes · Insulin · Hormones · RIA · Radioisotopes · Peptides · Diabetes · Blood banks · Drug screening · Thyroid glands · Viruses · Vitamins

Nobel Laureate Rosalyn Yalow (Fig. 11.1) was the first American-educated woman to receive a Nobel Prize in the sciences. She shared the 1977 Nobel Prize in Physiology or Medicine for her work on the development of radioimmunoassay (RIA) on peptide hormones. RIA is a technique used to screen for very tiny amounts of chemicals in biological tissues and fluids and has revolutionized biological and medical research. It led to the development of the field of neuroendocrinology. Dr. Yalow received the National Medal of Science in 1988.

J. S. Tietjen et al., *Women Laureates 1970 – 2007*, Women in Engineering and Science, https://doi.org/10.1007/978-3-032-23116-1_11

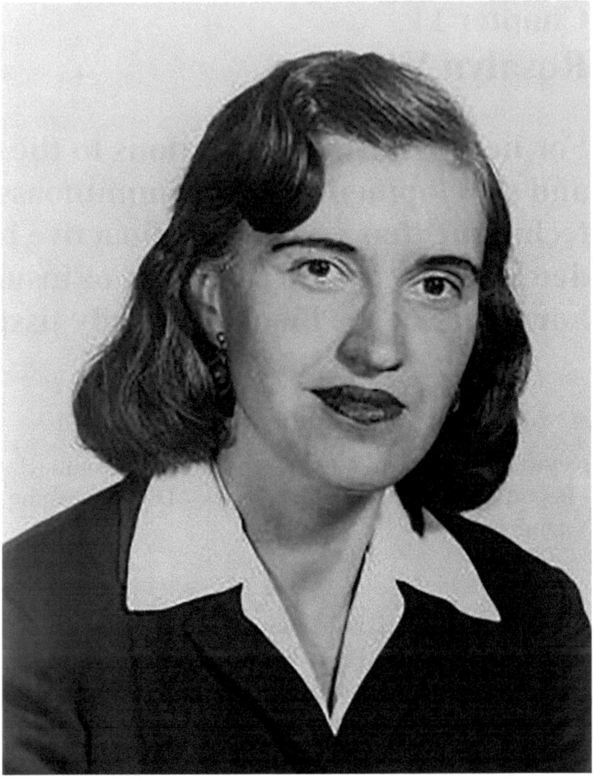

Fig. 11.1 Rosalyn Yalow. (Courtesy of the Library of Congress)

11.1 Life (1921–2011)

Rosalyn Sussman was born in 1921 in the Bronx and grew up there. In fact, she would spend most of her life there. Seemingly born a fighter, she avenged her older brother after his first-grade teacher hit him with a ruler—5 years after the fact. When that same teacher hit her with the ruler, she hit back. Her parents weren't upset—they immortalized the event—staging a picture of her with boxing gloves looking like she had just kayoed her brother in a fight. She kept that photo on her desk—"That's the attitude that made it possible for me to go into physics," she said (McGrayne 1993).

Although neither parent was educated—her mother Clara Zipper Sussman came to the USA from Germany as a small child, and her father Simon Sussman, born in New York City of Russian immigrant parents, dropped out of school after eighth grade to work—education was prized in the household. Her father read *The New York Times* and encouraged her to do whatever boys could do. "I never really got the message that girls were not as important as boys. I was very close to my father. He took me to baseball games." Her mother read all the schoolbooks that her children brought home. Rosalyn was frequenting the public library by the time she was 5 years old. By the time she was 8 years old—she already knew what her future held—she was going to be a "big-deal scientist," get married, and have children. She advised her friends about achieving these goals, "If you think now about how you're going to get married, you'll do both [have a career and a family]" (McGrayne 1993; Proffitt 1999).

At age ten, Rosalyn enrolled in an all-girls junior high school. There, her focus was narrowed to medical research. She graduated from high school in 1937 when she was 15. With her grades, she was able to enter Hunter College, which was tuition free at the time—and had selected physics as her major (McGrayne 1993).

> In the late thirties when I was in college, physics, and in particular nuclear physics, was the most exciting field in the world. A few people would sit around and talk to each other. One of them would have a great idea, go to the laboratory, and work for days or weeks, and make the discovery worthy of a Nobel Prize. It was an absolutely fantastic time. (McGrayne 1993)

Marie Curie served as a significant role model, especially after her daughter Eve Curie's biography of her mother came out. "For me, the most important part of the book was that, in spite of early rejection, she succeeded. It was in common with my background, with my being aggressive" (McGrayne 1993).

In her senior year at Hunter, she took a part-time secretarial job at the Columbia University, College of Physicians and Surgeons, planning to take some science classes there. Rosalyn wanted to go to medical school but knew she would not be admitted. Jewish men were not admitted to medical school, so being a Jewish woman and expecting to be admitted was out of the question. After she graduated from Hunter College at age 19 in January 1941, she applied to a number of universities for a physics teaching assistantship, which would pay for her graduate school. But no offer was immediately forthcoming. So, she went to secretarial school after graduating from Hunter College with high honors in physics and chemistry. She had been the first physics major at Hunter College (In Memoriam 2012; McGrayne 1993; Proffitt 1999).

A few months later, with the USA's seemingly inevitable entry into World War II, graduate schools started looking to women to fill their seats, and she became one of those women. Rosalyn received an offer to become a physics teaching assistant at the University of Illinois starting in the Fall of 1941. She continued her part-time secretarial job until June 9, quit secretarial school, and took free physics courses at New York University during the summer. In September 1941, she took a train to Illinois. She was the first woman in the University of Illinois engineering school since 1917, when World War I had been underway. She was the only woman in the College of Engineering's faculty of the four hundred professors and teaching assistants. She was now also "a stubborn-determined graduate student" (McGrayne 1993; Proffitt 1999; Shearer and Shearer 1996).

Rosalyn met her future husband, Aaron Yalow, on the first day of graduate school at the University of Illinois. They did PhD research in nuclear physics together and were married in 1943—after Aaron got a fellowship and was no longer considered faculty. They couldn't be married until one was not considered a faculty member, because otherwise nepotism rules precluded them both being employed. In 1945,

Yalow received her PhD (she had gotten her masters in 1942), the second woman to receive a PhD in physics at the University of Illinois (McGrayne 1993; Proffitt 1999; Shearer and Shearer 1996).

She returned to New York City after receiving her PhD and became the first woman engineer at the Federal Telecommunications Laboratory, the research laboratory for the International Telephone and Telegraph Corporation. A year later, when that job ended, she returned to teaching at Hunter College. But Hunter College had no research facilities, and Dr. Yalow wanted to do research. She didn't want to do high-energy physics and instead found radioisotopes, radioactive versions of atomic elements, through volunteering at a medical lab at Columbia University. In 1947, she began part-time work at the Veterans Affairs Medical Center in the Bronx. The VA wanted to set up a radioisotope lab and use the radioisotopes to treat cancer so they hired her. But she had other ideas (In Memoriam 2012; McGrayne 1993).

While teaching full-time at Hunter, Dr. Yalow developed the VA hospital's radioisotope service and started research projects. She turned a janitor's closet into a functioning radioisotope lab and published eight clinical papers with physicians at the VA. In 1950, she resigned from Hunter College to work full-time at the VA, although she taught one course for one additional semester because she had found a protégé to take under her wing—Mildred Spiewak (later Dresselhaus), who would win the National Medal of Science herself in 1990 (In Memoriam 2025; McGrayne 1993).

Also in 1950, Dr. Yalow began looking for a collaborator with whom to pursue research in medical physics. She found physician Dr. Solomon Berson, a resident in internal medicine at the VA hospital. Berson understood biology, physiology, anatomy, and clinical medicine. She understood physics, mathematics, chemistry, and engineering. They were a dynamic team and collaborated until his death in 1972 (McGrayne 1993).

Their first project together used radioisotopes to estimate blood volume. Using Dr. Yalow's expertise in nuclear counting techniques and Berson's in diabetes, their first major contribution involved an examination of how the kidneys and the thyroid gland remove iodine from the blood (In Memoriam 2012).

They then turned their attention to diabetes. Dr. Yalow had a significant personal interest in diabetes as Aaron Yalow had diabetes. She and Berson were the first to prove that a peptide such as insulin could stimulate the production of antibodies, an immunologic response. The initial response of the medical community was skepticism and rejection. Dr. Yalow would later tell schoolchildren, "Initially, new ideas are rejected. Later they become dogma, if you're right. And if you're really lucky you can publish your rejections as part of your Nobel presentation!" (In Memoriam 2012).

Using radioisotopes, they were able to measure diseases, viruses, hormones, and trace elements within the human body—a practice that came to be known as radioimmunoassay (RIA). It was for the development of RIA that Dr. Yalow would win the Nobel Prize (Berson had died before the awarding of the Nobel Prize, and Nobel Prizes are not given posthumously).

It was during the early years of their collaboration that she had her two children—Benjamin in 1952 and Elanna in 1954 (In Memoriam 2012; McGrayne 1993; Nobel Prize 2025; NSTMF 2025; Shearer and Shearer 1996). Dr. Yalow defied the VA rule that pregnant women must resign during their fifth month of pregnancy for both of her children, took very little maternity leave, had a live-in housekeeper, and benefited from her mother just happening to "drop in" most afternoons. Dr. Yalow used to joke that Benjamin was the only eight-pound two-ounce "five-month" baby in history (McGrayne 1993).

Dr. Yalow received many awards in addition to the 1977 Nobel Prize in Physiology or Medicine. She was elected a member of the National Academy of Sciences in 1975. In 1976, she was the first woman to win the Albert Lasker Basic Medical Research Award. In 1978, the Public Broadcasting Service aired a five-part series on Marie Curie, which Dr. Yalow hosted. In 1988, she received the National Medal of Science. Dr. Yalow has been inducted into the National Women's Hall of Fame (Proffitt 1999; Shearer and Shearer 1996). She was the first female and first Nobel Laureate to serve as President of The Endocrine Society (1978–1979) (In Memoriam 2012). She received numerous honorary doctorates (McGrayne 1993).

Dr. Yalow held professorships at other New York institutions, including the Albert Einstein College of Medicine at Yeshiva University, Montefiore Hospital and Medical Center, and Mount Sinai School of Medicine. She continued to work all that time at the lab, now named the Solomon A. Berson Research Laboratory, at the VA Hospital (McGrayne 1993). Dr. Yalow retired from the VA Hospital in 1991 and continued as a science activist, which included advocating for women's participation in science (Shearer and Shearer 1996). Understanding her own role as a trailblazer for women scientists, Dr. Yalow said in her Nobel Prize acceptance speech:

> We cannot expect in the immediate future that all women who seek it will achieve full equality of opportunity. But if women are to start moving towards that goal, we must believe in ourselves or no one else will believe in us; we must match our aspirations with the competence, courage and determination to succeed; and we must feel a personal responsibility to ease the path for those who come afterwards. The world cannot afford the loss of the talents of half its people if we are to solve the many problems which beset us. (Yalow 1977)

11.2 Radioimmunoassay

Radioimmunoassay (RIA) was developed as a side issue to insulin research by Drs. Berson and Yalow. When adult diabetics were injected with radioactively tagged insulin, the insulin was retained longer than expected and longer than in individuals who were serving as the controls—those without diabetes who had not been receiving regular insulin injections. In addition, one group of people had been subject to insulin treatments even though they didn't have diabetes—schizophrenics who had undergone insulin shock therapy—also retained the insulin longer than expected. Drs. Yalow's and Berson's conclusion was that people who had taken insulin had developed antibodies to it. This was at a time when the insulin that was administered to people was made from pigs and cattle (McGrayne 1993).

Unfortunately, this conclusion flew in the face of conventional wisdom. Their initial conclusions were met with skepticism. It was believed at the time that the insulin molecule was too small to stimulate the production of antibodies. The journal *Science* refused to publish their article. *The Journal of Clinical Investigation* only published it (1956) once Drs. Yalow and Berson had removed the words "insulin antibody" from the title. Interestingly, other researchers had also discovered that people who had been treated with insulin processed the hormone more slowly than those who had not been treated with insulin (McGrayne 1993).

It was what happened next that led to RIA. Drs. Yalow and Berson realized that although their technique measured the antibodies to a hormone, its inverse would measure the hormone itself. Later, other researchers would state that they had the same idea but had not done anything about it. Dr. Yalow said, "It isn't by accident that you interpret the observation correctly. That's creativity. Once you saw it one way, you saw it the other way. I still think discovery is the most exciting thing in the world" (McGrayne 1993). Drs. Yalow and Berson called it radioimmunoassay because it used radioactively tagged substances to measure antibodies produced by the immune system. RIA is very sensitive—it can detect a billionth of a gram—and it works for virtually every hormone as well as many other biologically important substances. In addition, it determines this information very quickly and very precisely (McGrayne 1993; Proffitt 1999).

The first disease for which RIA was applied was diabetes. With RIA, patients could be separated into two groups—children who lose the ability to make insulin and adults who produce insulin but lose the ability to use it (McGrayne 1993). Drs. Yalow and Berson then attacked human growth hormone—identifying which children could be helped by growth hormone treatments. This was followed by applications of RIA for parathyroid hormone, hepatitis B-antigen, and ACTH. Before their work, patients who underwent blood transfusions could end up with infected livers because of hepatitis B-antigen. With RIA, transfusion hepatitis virus is no longer a fear associated with blood banks in North America. A huge success for RIA was identifying which newborns had underactive thyroid glands. Otherwise, the symptoms are not obvious until the baby is 3 months old, by which point the brain damage is not reversible and the child will have an intellectual disorder. About this success, Dr. Yalow said, "What a bonus to these children, their families, and society!" (McGrayne 1993).

A whole new era in endocrinology was made possible by RIA, and in fact, it spawned a whole new science—neuroendocrinology. The technique enabled major advances in diagnosing and treating hormonal problems related to diabetes, fertility, thyroid function, and growth. RIA is used to measure levels of vitamins, enzymes, drugs, and viruses. Blood banks use RIA to screen blood for the hepatitis virus. It is also the technology used to screen athletes for drug use. RIA and a variation of RIA called ELISA, that uses enzymes or fluorescent markers, are used in many medical and clinical research labs. The Nobel Prize organization said that the technique "brought a revolution in biological and medical research" (In Memoriam 2012;

McGrayne 1993; Proffitt 1999; Shearer and Shearer 1996). In her Nobel Prize lecture in 1977, Dr. Yalow identified more than one hundred biological substances that were already being measured by RIA (Proffitt 1999). Drs. Berson and Yalow decided not to patent RIA but instead make it available to the medical community (Shearer and Shearer 1996).

11.3 Conclusion

Nobel Laureate Dr. Rosalyn Yalow decided as a very young girl that she would become a "big-deal scientist" and marry and have a family—and she did. The technique that she and Dr. Solomon Berson developed, radioimmunoassay (RIA), allows scientists to measure diseases, viruses, hormones, and trace elements in the human body, leading to better treatments and outcomes for millions of people.

References

In Memoriam: Dr. Rosalyn Yalow, PhD., 1921–2011. Mol Endocrinol. May 1, 2012;26(5):713–4. https://doi.org/10.1210/mend.26.5.zmg713. PMCID: PMC5417100.

McGrayne, Sharon Bertsch, *Nobel Prize Women in Science: Their Lives, Struggles, and Momentous Discoveries*, New York, New York: Carol Publishing Group, 1993.

National Science and Technology Medal Foundation (MSTMF). Rosalyn S. Yalow. https://nationalmedals.org/laureate/rosalyn-s-yalow/. Accessed July 13, 2025.

The Nobel Prize. Nobel Prize in Physiology or Medicine 1977. https://www.nobelprize.org/prizes/medicine/1977/summary/, Accessed July 14, 2025.

Proffitt, Pamela, Editor, *Notable Women Scientists*, Detroit, Michigan: The Gale Group, 1999.

Shearer, Benjamin F. and Barbara S. Shearer, Editors, *Notable Women in the Life Sciences*, Westport, Connecticut: Greenwood Press, 1996.

Yalow, Rosalyn. The Nobel Prize Banquet Speech. 1977. https://www.nobelprize.org/prizes/medicine/1977/yalow/speech/. December 10. Accessed July 14, 2025.

Chapter 12
Katherine Esau

For her distinguished service to the American community of plant biologists, and for the excellence of her pioneering research, both basic and applied, on plant structure and development, which has spanned more than six decades; for her superlative performance as an educator, in the classroom and through her books; for the encouragement and inspiration she has given to a legion of young, aspiring plant biologists; and for providing a special role model for women in science

Keywords Plant virus · Phloem · Sieve tubes · Sieve elements · Plant anatomy · Sugar beets · Curly-top virus · Plant tissue · Plant morphology · Plant physiology · Electron microscope · Plant structure · Plant biology · Xylem

The author of the bible of plant anatomy and the bible of phloem (food-conducting plant tissue), Katherine Esau (Fig. 12.1) was born in what was then Russia and today is Ukraine. Her family emigrated to Germany after World War I where she completed her college education. The family then emigrated to the USA and settled in California, where she worked in her chosen field of agriculture and then completed her PhD. The rest of her career was spent in academia, where she taught, researched, wrote six textbooks, and wrote hundreds of articles. It was said that she dominated the field of plant anatomy and morphology for several decades and set the stage for advances in plant physiology and molecular biology. Esau received the National Medal of Science in 1989 (Warren 2025).

J. S. Tietjen et al., *Women Laureates 1970 – 2007*, Women in Engineering and Science, https://doi.org/10.1007/978-3-032-23116-1_12

Fig. 12.1 Katherine Esau. (Courtesy of the Hunt Institute for Botanical Documentation, Carnegie Mellon University, Pittsburgh, Pennsylvania)

12.1 Life (1898–1997)

Katherine Esau was born in Ekaterinoslav (also Yekaterinoslav, now Dnepropetrovsk), Ukraine, in 1898 to John and Margarethe Esau. She had a younger brother Paul. Her parents were part of a community of Mennonites from Germany who sought refuge in what was then Russia starting in the late eighteenth century. Her father was a mechanical engineer who became the mayor of their city. Esau grew up speaking German and Russian, studying at a Mennonite school before entering a Russian gymnasium (a European secondary school that prepares students for college) when she was 11. Her family also secured the services of a private tutor from whom she began learning English. In 1916, she began her studies at the Golitsin Women's

Agricultural College in Moscow, but war interrupted her education (Christ 2023; Evert 1999; Shearer and Shearer 1996).

The Germans occupied her city during World War I. When they retreated, the family felt it was most prudent to flee to Germany as her father was now considered "an enemy of the people" by the Russians, as he had continued to serve as mayor during the German occupation. The family went to Berlin, where Esau enrolled in the Landwirtschaftliche Hochschule, the Berlin agricultural college, and earned a diploma in plant sciences. She continued her instruction in English there also. After 3 years in Germany, the family immigrated to the USA in 1922, having enough English to make it through Ellis Island, and settling in Reedley, California, which had a large Mennonite population (Christ 2023; Evert 1999; Shearer and Shearer 1996).

Once in the USA, Esau worked as a housekeeper before finding employment with the Sloan Seed Company. Here she learned Spanish so that she could communicate with the workers. She worked for the Spreckels Sugar Company from 1924 to 1927, which had hired her to breed sugar beets that were resistant to the curly-top virus. This was her first encounter with a plant virus (Christ 2023; Evert 1999; Ogilvie and Harvey 2000; Shearer and Shearer 1996).

A visit to Spreckels from Professor W.W. Robbins, the chairman of the Botany and Truck Crops Division at the University of California at Davis, in 1927 led to Esau seeking further education. When she asked Dr. Robbins how she could do further research in agricultural botany, he arranged for her to have a graduate assistantship. She moved to the UC-Davis campus but was officially enrolled at the University of California—Berkley since UC-Davis didn't have a graduate school at the time. Her first order of business was preparing for the general examination required for a PhD in the US educational system. This meant three years of scientific training before she could take the exam. By 1930, she had completed the course requirements and passed the exam (Christ 2023; Cleveland 1997; Shearer and Shearer 1996).

In 1931, she received her PhD and was elected to Phi Beta Kappa (Shearer and Shearer 1996). Her dissertation research was a comparison of the anatomical structure of normal sugar beets and those infected with the curly-top virus. That she completed her PhD research in 1 year prompted Dr. George Ledyard Stebbins, botanist and geneticist who wrote her memoir, to say,

> I stand in awe of this extraordinary feat. I have never heard of another candidate who has completed his or her thesis in less than three years after passing the qualifying examination... The success of her lifetime career as a plant anatomist could therefore be predicted from this performance. (Christ 2023)

She joined the faculty at UC-Davis upon the completion of her PhD (Christ 2023; Shearer and Shearer 1996).

One of her first publications reported the results of her work that the curly-top virus spread through a sugar beet plant via the phloem, or food-conducting tissue. Her further research demonstrated that this was also the case for the tobacco plant. The tobacco plant was also susceptible to the curly-top virus (Cleveland 1997).

Thus, phloem and viruses became one of her primary research focuses over the course of her career. Phloem is the plant tissue that transports sugars and other metabolic products of photosynthesis downward from the leaves. She examined how plant viruses affect the structure and development of the phloem. She was able to distinguish between primary and secondary viral symptoms such that studies of viral damage to specific plant tissues could be performed. She elucidated the development phases of plant tissues, especially the sieve tubes—which move the metabolic products throughout a plant. Dr. Esau provided the impetus for much phloem study in the USA. Said of her, she "absolutely dominated the field of plant anatomy and morphology for several decades. She set the stage for all kinds of modern advances in plant physiology and molecular biology." The Guggenheim Foundation recognized her contributions with a one-year fellowship, granted in 1940 (Christ 2023; Cleveland 1997; Proffitt 1999; Shearer and Shearer 1996).

Dr. Esau used microscopes to conduct her research. She outfitted a darkroom in her home with equipment that she purchased and produced the microphotographs herself in the 1940s and 1950s. The illustrations of plant structures for her seminal "The Bible of Plant Anatomy" 1953 book, *Plant Anatomy*, were produced in this manner (Christ 2023; Evert 1999; Proffitt 1999; Shearer and Shearer 1996).

One of her graduate students said of this book,

> Dr. Esau's Plant Anatomy took a dynamic, developmental approach designed to enhance one's understanding of plant structure. The book Plant Anatomy brought to life what previously had seemed to me to be a rather dull subject. I was not the only one so affected. Plant Anatomy had an enormous impact worldwide, literally bringing about a revivification of the discipline. (Cleveland 1997)

Her work on plant structure underpins most of the current research on plant function (Evert 1999).

Others of her books were also exceedingly influential.

> "Katherine Esau's work [The Anatomy of Seed Plants] is the Webster's of plant biology—it's encyclopedic. Her prose is elegant and precise; each word is carefully chosen. When you read her publications, you're at the microscope with her—you see what she's seeing. If my students and I have a disagreement about cell definition, I turn to Esau's work to settle it." This professor of plant biology also said that her book The Phloem is still the bible on that tissue (Cleveland 1997).

Dr. Esau progressed through the ranks at UC-Davis from instructor and junior botanist to full professor, a rank she achieved in 1949. She taught plant anatomy, systematic botany, and morphology of crop plants, among other subjects. She studied celery, tobacco, carrot, and pear. In 1963, Dr. Esau moved to the University of California at Santa Barbara, when her colleague and co-researcher on phloem moved there as Chancellor. Although she officially retired in 1965, she continued to do research and came into her office until 1994. The National Science Foundation gave her a grant for her own electron microscope after she retired—in 1969! Her last lecture (on plasmodesmata) was given at UC-Davis in 1982. Her last paper was published in 1991 when she was 93. She published six textbooks and more than 160 articles over the course of her career. She even learned to use a personal computer with which to update *Plant Anatomy* in the 1980s (Christ 2023; Cleveland 1997; Evert 1999; Ogilvie and Harvey 2000; Shearer and Shearer 1996).

Dr. Esau was active in professional organizations and served as President of the Botanical Society of America in 1951 (Evert 1999). She received a 1956 Merit Award from that organization as part of its celebration for its 40th anniversary. The citation for the award read: "Katherine Esau, plant anatomist and histologist, for her numerous contributions on tissue development of vascular plants and in particular for her outstanding studies on the structure, development, and evolution of phloem" (Evert 1999; Shearer and Shearer 1996).

She received many awards over the course of her career, with the first being the 1946 Faculty Research Lecture, at UC-Davis (Shearer and Shearer 1996). She was elected to the National Academy of Sciences (the sixth woman) in 1957 (Cleveland 1997). She was also elected to the American Academy of Arts and Sciences (Evert) and the American Philosophical Society, 1964 (Christ 2023). Dr. Esau received honorary degrees from Mills College (1962) and the University of California (1966) (Evert). Her other awards include the 11th International Botanical Congress Medal (1969) and election to the Swedish Royal Academy of Science (1971) (Shearer and Shearer 1996). Dr. Esau received the National Medal of Science in 1989 (Christ 2023). An international symposium in her honor was held at UC-Davis in 1992 (Ogilvie).

Upon her election to the National Academy of Sciences, Esau said,

> I never worried about being a woman. It never occurred to me that that was an important thing. I always thought that women could do just as well as men. . . My surprise at being elected to the National Academy of Sciences was not because I was a woman, but because I didn't think that I had done enough to be elected. (Evert 1999)

In 1988, Dr. Esau donated $648,000 to UC-Davis to establish an endowment to fund plant research fellowships in perpetuity. In 2022, The Sciences Laboratory Building on the UC-Davis campus was named Katherine Esau Science Hall, a posthumous honor (Sicard 2023).

Dr. Esau never married and was beloved by her many students and colleagues. She had a wicked sense of humor as manifested by the title of one of her lectures, "The Saga of Vladimir-the-Virus and the Sad Fate of Norman-the-Nucleus." Since many of her lectures began with "Once upon a time. . ." they became known as "Esau's fables" (Cleveland 1997).

12.2 Phloem

Xylem and phloem (Fig. 12.2) are the two tissues that comprise a plant's transport system. Xylem tissue provides mechanical strength to plant parts and also serves to conduct water and minerals. Xylem functions in only one direction, from the roots to the stem and leaves (Carillo-Lopez and Yahia 2019).

Phloem is a bi-directional transportation system within plants composed of vascular tissue that transports sugars, amino acids, and signaling molecules. It brings

Fig. 12.2 Plant phloem and xylem. (Courtesy of Wikipedia)

the sugars from sites of photosynthesis, such as the leaves, to where the sugars need to be consumed—the so-called sink tissues: roots, flowers, and fruits. Thus, not only does the phloem provide the mechanism to fuel the growth of sink tissues, it also provides the plant with the ability to respond to both internal and external stimuli (Truernit 2022).

The phloem is comprised of various types of cells. The sieve elements are long cells that are arranged in series to form sieve tubes. The sieve tubes are the component of the plant that allows water and other materials to move. Phloem also appears to transport peptides, proteins, and RNA (Thompson 2017).

Sieve elements were first described by Theodor Hartig in 1837. The use of electron microscopy to study subcellular features of the phloem was reinvigorated by Katherine Esau's introduction of transmission electron microscopy into the field to study phloem ultrastructure. "Her work paved the way for a better understanding of phloem structure and function, culminating in her 1969 publication, *The Phloem*, which has come to be considered a bible of phloem research" (Heo et al. 2014).

12.3 Conclusion

The author of two books, each regarded as a "bible" of plant physiology, Dr. Katherine Esau dominated the field of plant anatomy and morphology for several decades and set the stage for advances in plant physiology and molecular biology. The recipient of the National Medal of Science in 1989, she was a beloved teacher and mentor.

References

Carillo-Lopez, Armando and Elhadi M. Yahia. 2019. Chapter 6 – Morphology and Anatomy. Postharvest Physiology and Biochemistry of Fruits and Vegetables. Pages 113-130. Sawston, United Kingdom: Woodhead Publishing. https://www.sciencedirect.com/science/article/abs/pii/B9780128132784000063. Accessed August 31, 2025.

Christ, Marian L. 2023. Katherine Esau, 1898-1997. https://www.amphilsoc.org/blog/katherine-esau-1898-1997. September 11. Accessed August 23, 2025.

Cleveland, Debra. 1997. Remembering Katherine Esau. Katherine Esau Postdoctoral Fellowships – University of California – Davis. October. http://esau.plb.ucdavis.edu/about.htm. Accessed August 23 2025.

Evert, Ray F. Katherine Esau 1898-1997 – A Biographical Memoir. 1999. https://www.nasonline.org/wp-content/uploads/2024/06/esau-katherine.pdf. National Academy of Sciences. Accessed August 23, 2025.

Heo, Jung-ok, Pawel Roszak, Kaori M. Furuta, and Ykä Helariutta. 2014. Phloem development: Current knowledge and future perspectives. American Journal of Botany. September 1. https://doi.org/10.3732/ajb.1400197. Accessed August 31, 2025.

Ogilvie, Marilyn and Joy Harvey, Editors. The Biographical Dictionary of Women in Science: Pioneering Lives from Ancient Times to the Mid-20th Century. New York, New York: Routledge. 2000.

Proffitt, Pamela, Editor. Notable Women Scientists. Detroit, Michigan: Gale Group. 1999.

Shearer, Benjamin F. and Barbar S. Shearer, Editors. 1996. Notable Women in the Life Sciences: A Biographical Dictionary. Westport, Connecticut: Greenwood Press.

Sicard, Clémentine. 2023. A Gift Grows On: The Legacy of Katherine Esau. UC-Davis College of Biological Sciences. https://biology.ucdavis.edu/news/gift-grows-legacy-katherine-esau. Accessed August 23, 2025.

Thompson, G.A. and H.L. 2017. Wang. Encyclopedia of Applied Plant Sciences (Second Edition). Volume 1. Pages 100-118. https://www.sciencedirect.com/science/article/abs/pii/B978012394807600071X. Accessed August 31, 2025.

Truernit, Elisabeth. 2022. Sieve elements and their cell neighbours in the Arabidopsis root – Roles and relationships. Journal of Plant Physiology. Volume 268. January. https://www.sciencedirect.com/science/article/pii/S017616172100208X. Accessed August 31, 2025.

Warren, Rachel. Katherine Esau. National Science & Technology Medals Foundation. https://nationalmedals.org/laureate/katherine-esau/. Accessed August 23, 2025.

Chapter 13
Helen T. Edwards

For the contributions to the design, construction, and initial operation of the TEVATRON particle accelerator. The scientific instrument was designed to explore the fundamental properties of matter. The innovative design and successful operation of the TEVATRON have been crucial to the design of the Superconducting Super Collider, the planned next-generation particle accelerator

Keywords Fermilab · Tevatron · Elementary particles · Superconducting magnets · Top quark · Tau neutrino · Particle accelerator · Energy doubler · Bc meson

It all started with a Big Bang! In the mid-twentieth century, scientists sought to recreate the conditions of the Big Bang to discover the tiniest particles that make up matter and further their understanding of the nature of the universe. Dr. Helen T. Edwards (Fig. 13.1) played a pivotal role in the development of the Tevatron, the most powerful particle accelerator of its time, which ultimately led to the discovery of the top quark and the tau neutrino, two fundamental particles of matter. For this work, Dr. Edwards was awarded the National Medal of Technology in 1989.

J. S. Tietjen et al., *Women Laureates 1970 – 2007*, Women in Engineering and Science, https://doi.org/10.1007/978-3-032-23116-1_13

Fig. 13.1 Helen T. Edwards. (Reidar Hahn, Fermilab)

13.1 Life (1936–2016)

Helen T. Edwards, née Thom, was born in 1936 in Detroit, Michigan, and grew up in nearby Pontiac and on a ranch outside Metamora, Michigan. Helen was the youngest of five children. Her father was a stockbroker, and her mother was a homemaker (Stephens 2020). Her time on the ranch instilled a love of animals and nature. Helen enjoyed mechanical tasks and excelled in math and science in school. She struggled with reading and spelling, but it was not until she was an adult that she was diagnosed with dyslexia, which was not a well-known condition when she was a child (Shearer and Shearer 1997).

In 1950, as a high school sophomore, Helen was sent to join two of her older sisters at the all-girls Madeira School in McLean, Virginia. There, she participated in sports and student government and demonstrated an "unusual ability" (Stephens 2020) in science and math. For college, Helen applied to the most competitive schools for science and engineering, including MIT, Cornell, and Duke. Having attended high school without boys, Helen had not experienced the gender bias and discrimination that her parents feared she would encounter in college as a woman pursuing a science degree. Helen wanted to attend MIT, but the engineering and physics programs consisted almost solely of men, and her parents thought the isolation would be too much. They were more comfortable with her attending Cornell, which had a few more women (Shearer and Shearer 1997).

Helen enrolled at Cornell and majored in physics, graduating with a BA in 1957. There, she found an environment where she could succeed, even though she did sometimes encounter the attitude that "educating women in the sciences was a waste of time because they would choose marriage over a research career" (Shearer and Shearer 1997). She remained at Cornell for graduate studies. While working in the Laboratory of Nuclear Studies there, she met Donald Edwards, a doctoral student. In 1963, Helen received her master's degree and married Donald. Helen T. Edwards continued doing research at Cornell's Laboratory of Nuclear Studies and completed her PhD in Experimental Physics in 1966. She stayed on at the lab for another 4 years as a research associate. At the lab, she and her husband worked with particle accelerators under the lab's director, Dr. Robert Wilson. Particle accelerators are used to study the fundamental building blocks of matter by speeding up charged particles, such as protons and electrons, and then smashing them into each other or another target to determine whether any new subatomic particles emerge. Dr. Donald Edwards was responsible for the design and construction of the 10 GeV Synchrotron, a particle accelerator that accelerated electrons to very high speeds by moving them around a circular path using magnetic and electric fields (GeV, or gigaelectron Volt, is a unit of energy). Dr. Helen T. Edwards was responsible for bringing the 10 GeV Synchrotron into operation (Stephens 2020).

Dr. Wilson left Cornell in 1967 for the National Accelerator Laboratory (later renamed the Fermi National Accelerator Laboratory (Fermilab)) in Batavia, Illinois, and 2 years later, he hired Drs. Helen and Donald Edwards. Cornell's policy prohibiting married couples from being on the faculty of the same department may have contributed to the couple's decision to leave. At Fermilab in 1970, Dr. Helen T. Edwards became associate head of the Booster Section (and later head of the Accelerator Division) and was involved in every aspect of creating a new, much higher-energy synchrotron called the Tevatron, overseeing its design, construction, commissioning, and ultimately its operation (Hesla and Salles 2016). This was the defining project of her career. The Tevatron, which became operational in 1983, reigned as the world's most powerful particle accelerator for 25 years, until the Large Hadron Collider became operational in Europe. The Tevatron was a new class of accelerator, which used superconducting magnets made of materials that enabled them to be kept extremely cold. In this environment, the electricity running through the accelerator encounters very little resistance, resulting in a stronger magnetic field. Particles in the accelerator could achieve speeds near the speed of light, and their collisions led to the monumental discovery of two of the tiniest particles of matter: the top quark in 1995 and the tau neutrino in 2000 (Chandran 2022).

In 1987, Dr. Edwards became head of Fermilab's Accelerator Division (MacArthur Foundation 2016). She also became involved in the project to build the Superconducting Super Collider (SSC), an even larger and higher-energy accelerator. The SSC was to be built underground and have a circumference of 54 miles with a budget of $4.4 billion. In 1989, when the government decided to build the SSC in Texas, Dr. Edwards and her husband relocated near Dallas, Texas, where Dr. Helen T. Edwards served as head of the Superconducting Super Collider Laboratory from 1989 to 1992, designing key features of the SSC (Hesla and Salles 2016). However, the project was a significant government expense with no contributions from other countries. When it ran over budget, Congress pulled the funding, and the project was scrapped in 1993. This "ended more than four decades of American leadership in high-energy physics" (Chandran 2022).

Dr. Edwards stayed at Fermilab as a guest scientist from 1992 to 2010, although her home base was in Montana, where she and her husband had relocated to be closer to nature. During this period, she also spent time in Germany, collaborating with scientists at the Deutsches Elektronen Synchrotron (DESY), where she designed a superconducting linear accelerator. In this accelerator, particles move in a straight line, as opposed to a circular path.

Although she left Fermilab in 2010, Dr. Edwards was invited back in 2011 when the Tevatron was retired. It was only fitting that she be the one to turn the machine off.

In her career, Dr. Edwards developed a reputation as a force to be reckoned with. She worked long hours, was determined to get things done, and had no patience for bureaucratic red tape. She could be intimidating, but was also a great mentor who loved teaching and created opportunities for her students to succeed (Chandran 2022).

Of her scientific prowess, University of Maryland Physics Professor Timothy Koeth, who studied under Dr. Edwards, said,

> I was scientifically mesmerized by her. She had this intuitive and innate grasp of the material, and she was always absolutely right—she was never wrong in the 20 years I knew her. (Hesla and Salles 2016)

In addition to the National Medal of Technology in 1989, Dr. Edwards received the US Department of Energy's E.O. Lawrence award in 1986, the MacArthur Foundation Fellowship in 1988, and the American Physical Society's Robert R. Wilson Prize in 2003. She was a Fellow of the American Physical Society and a member of the National Academy of Engineering (Higgins 2022).

Dr. Helen T. Edwards died in 2016 from cancer at the age of 80. The Integrated Engineering Research Center (IERC) at Fermilab was officially renamed the Helen T. Edwards Engineering Research Center by an act of Congress signed into law by President Joe Biden on December 23, 2024, to recognize her significant contributions to particle physics (Marc 2025).

13.2 Particle Accelerators, the Tevatron, and the Big Bang

The Big Bang theory is the leading scientific explanation for the origin of the universe. It states that the universe was created approximately 13.8 billion years ago from an extremely hot and dense state. In the beginning, the universe consisted of tiny fundamental particles, light, and energy. As the universe expanded and cooled, these fundamental particles combined to form protons, neutrons, and electrons, which over time, formed atoms. Atoms combined to create matter, leading to the eventual formation of stars, galaxies, and the building blocks of life.

During the very first moments of the Big Bang, not only were the fundamental particles of matter created, but so were their corresponding antimatter particles. Each particle has an antiparticle with the same mass but opposite electrical charge. When a particle and its antiparticle meet, they annihilate each other, releasing energy. For reasons that scientists have yet to understand, after the Big Bang, instead of all the matter and antimatter destroying each other entirely, a small amount of matter remained. Today, tiny amounts of antimatter are still naturally produced when cosmic rays from space enter the Earth's atmosphere or during radioactive decay of some elements (CERN 2019).

In the 1970s, physicists proposed the Standard Model, which defined the fundamental building blocks of matter—the basic particles that cannot be broken down into smaller components. Scientists have used particle accelerators to replicate the high-energy conditions of the Big Bang to confirm the existence of many of the subatomic particles predicted by the Standard Model. By producing high-energy collisions, accelerators can create antimatter and collide it with matter, breaking the matter apart to gain a better understanding of its inner structure.

Particle accelerators can be thought of as race tracks for subatomic particles. How fast can a car go? The driver steps on the accelerator to find out. Particle accelerators are measured not by the speeds they achieve, but by the energy created. In a particle accelerator such as the Main Ring built at Fermilab before the Tevatron, particles are accelerated around a ring, reaching speeds near the speed of light. As the particles approach this limit, further acceleration increases their energy rather than their speed.

The Main Ring was an underground loop nearly four miles in circumference, designed by Dr. Edwards's mentor, Dr. Robert Wilson. It used magnets to steer and accelerate beams of protons to energies of 200 GeV, its initial goal, in 1972, and later far surpassed its goal with energies of 500 GeV by 1976 (Fermilab n.d.).

Dr. Wilson conceived the idea of creating another accelerator in the same underground tunnel as the Main Ring, utilizing superconducting magnets. Dr. Helen T. Edwards oversaw the work on this project, initially known as the Energy Double and later as the Tevatron, named for its aim of doubling the energy of the Main Ring, which would be 1000 GeV, or one tera electron-volt. The design would use the Main Ring to feed particles to the Tevatron. It took 11 years from the beginning of the design phase in 1972 to the Tevatron becoming operational (Fermilab n.d.). One of the significant challenges of the Tevatron was the design and construction of the superconducting magnets, which had never been used in particle accelerators before. 774 superconducting magnets would be used to guide the particles through the 3.9-mile loop. After years of research and development, it was decided to use a niobium-titanium alloy embedded in copper for the magnetic coils. Each coil was encased in a stainless steel tube with liquid helium flowing through to cool the magnets and maintain a temperature of approximately −450 °F. At this temperature, the magnets become superconducting, resulting in virtually no electrical resistance. This creates the powerful magnetic fields needed to steer the subatomic particles around the Tevatron at nearly the speed of light (The American Society of Mechanical Engineers 1993).

To build the large number of magnets required, Fermilab had to corner the market on niobium-titanium. The Tevatron was the first large-scale application of mass-produced superconducting magnets. Manufacturers of niobium-titanium wire scaled up their production to meet demand, resulting in lower costs, improved performance and reliability, as well as broader applications. As a direct result, commercial applications using niobium-titanium became more viable, most notably magnetic resonance imaging (MRI) machines (Engineering and Technology History Wiki n.d.).

Installing and maintaining the magnets was also a challenge since small misalignments could cause the magnet to cease superconducting, a phenomenon known as "quenching." To avoid this problem, Dr. Edwards and her team developed a method of using collars, which became standard for particle accelerators. The stainless-steel collars are clamped around the magnetic coils, which serve to pre-stress the coils and provide a rigid support to keep the magnets in place (Prestemon and Gourlay 2018).

The last of the 774 magnets was installed in 1983, and the Tevatron was turned on. A series of steps must take place before the particles enter the Tevatron. Figure 13.2 shows an aerial view, and Fig. 13.3 depicts a schematic of the accelerator chain.

The steps in the accelerator chain are as follows:

1. The process begins with a bottle of hydrogen and the Cockcroft-Walton generator. The generator produces H^- ions, containing one proton and two electrons (Fermilab 2004).
2. The ions travel through the linear accelerator (Linac), approximately 500 feet long, which increases their speed to about 70% of the speed of light. The ions then pass through a carbon foil, which strips the electrons away, leaving positively charged protons (Fermilab 2014).
3. The protons travel 20,000 times around the circular booster, which uses an electric field to raise the protons' energy to about 8 GeV (Fermilab 2014).

Fig. 13.2 Aerial view of Fermilab with components of accelerator chain in red. (Fermilab)

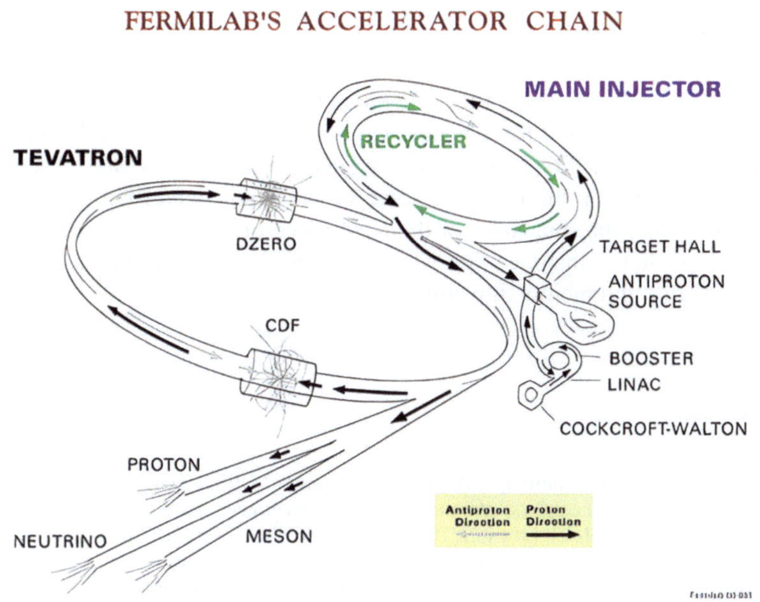

Fig. 13.3 Fermilab accelerator chain. (Fermilab)

4. The protons then enter the main ring, referred to on the diagram as the main injector, where further acceleration occurred to 150 GeV (Fermilab 2014).
5. Some of the protons are then injected into the Tevatron, while others are used to create antiprotons by colliding proton beams with a fixed target of nickel. The antiprotons were stored and cooled before joining the protons in the Tevatron (Fermilab 2014).
6. In the Tevatron, the superconducting magnets allow the protons and antiprotons to reach speeds close to the speed of light and energies near 1000 GeV. The protons and antiprotons are traveling in opposite directions and crashing into each other. Like two cars having a head-on collision on a race track, the protons and antiprotons break apart, but unlike the racecars, the "debris" from the crash is so small that special detectors, the Collider Detector at Fermilab (CDF) and DZero, had to be built to analyze the debris and resulting fundamental particles (Fermilab).

On March 2, 1995, the Tevatron system discovered the subatomic particle called the top quark (Fermilab 2001). "The decades-long search for one of the last missing pieces in the Standard Model of particle physics had come to an end" (Liss and Tipton 1997).

Three years later, in 1998, the Bc meson was discovered, not a fundamental particle itself, but a composite of two quarks held together by a strong nuclear force (Jackson 1998). And, in 2000, the tau neutrino, another elementary particle of the Standard Model, was observed (Fermilab 2000).

13.3 Conclusion

Dr. Helen T. Edwards furthered the understanding of how the universe began. She was the brains and driving force behind the design, construction, and commissioning of the Tevatron, the first superconducting synchrotron. She advanced accelerator science, leading to the discovery of some of the fundamental building blocks of matter, and setting the standards for the next generation of particle accelerators. In addition to her fierce intellect, she was renowned for her exceptional work ethic, leadership skills, and mentorship abilities.

References

CERN. 2019. "The Matter-Antimatter Asymmetry Problem." CERN Accelerating Science. 2019. https://home.cern/science/physics/matter-antimatter-asymmetry-problem.

Chandran, Anita. 2022. "Helen Edwards: Pioneer of Fermilab's Tevatron." Physics World. July 26, 2022. https://physicsworld.com/a/helen-edwards-pioneer-of-the-tevatron/.

Engineering and Technology History Wiki. n.d. "Milestones: Superconducting Magnet System for the Fermilab Tevatron Accelerator/Collider, 1973-1985." ETHW.org. https://ethw.org/Milestones:Superconducting_Magnet_System_for_the_Fermilab_Tevatron_Accelerator/Collider,_1973-1985. Accessed June 30, 2025.

Fermilab. 2000. "Physicists Find First Direct Evidence for Tau Neutrino at Fermilab." U.S. Department of Energy Fermi National Accelerator Laboratory. July 20, 2000. https://news.fnal.gov/2000/07/physicists-find-first-direct-evidence-tau-neutrino-fermilab/.

Fermilab. 2001. " Discoveries at Fermilab - the Top Quark." Fnal.gov. U.S. Department of Energy Fermi National Accelerator Laboratory. February 26, 2001. https://www.fnal.gov/pub/inquiring/physics/discoveries/top_quark.html.

Fermilab. 2004. "Accelerator Report No. 1: The Cockcroft-Walton." Www.fnal.gov. U.S. Department of Energy Fermi National Accelerator Laboratory. September 9, 2004. https://www.fnal.gov/pub/news04/update_archive/update_9-10.html.

Fermilab. 2014. "Tevatron Accelerator." Www.fnal.gov. U.S. Department of Energy Fermi National Accelerator Laboratory. August 5, 2014. https://www.fnal.gov/pub/tevatron/tevatron-accelerator.html.

Fermilab. n.d. "A Brief History of Fermilab." History.fnal.gov. U.S. Department of Energy Fermi National Accelerator Laboratory. https://history.fnal.gov/history-fermi.html. Accessed June 30, 2025.

Hesla, Leah, and Andre Salles. 2016. "Helen Edwards, Visionary behind Fermilab's Tevatron, Dies." Fnal.gov. U.S. Department of Energy Fermi National Accelerator Laboratory. June 27, 2016. https://history.fnal.gov/historical/people/edwards_obituary.html.

Higgins, Valerie. 2022. "Helen T. Edwards: A Trailblazer." U.S. Department of Energy Fermi National Accelerator Laboratory. https://www.fermilabcommunity.org/wp-content/uploads/2022/09/Edwards_Talk_20220922_final.pdf.

Jackson, Judy. 1998. "CDF Corrals the Last of the Mesons." *FermiNews* 21 (6): 1–2. https://www.fnal.gov/pub/ferminews/FermiNews98-03-20.pdf.

Liss, Tony M., and Paul L. Tipton. 1997. "The Discovery of the Top Quark." *Scientific American* 277 (3): 54–59. https://doi.org/10.1038/scientificamerican0997-54.

MacArthur Foundation. 2016. "Helen T. Edwards." Macfound.org. 2016. https://www.macfound.org/fellows/class-of-1988/helen-t-edwards.

Marc, Tracy. 2025. “Fermilab Building Named for Renowned Scientist, Dr. Helen Edwards.” U.S. Department of Energy Fermi National Accelerator Laboratory. January 14, 2025. https://news.fnal.gov/2025/01/fermilab-building-named-for-renowned-scientist-dr-helen-edwards/.
Prestemon, Soren, and Steve Gourlay. 2018. “Superconducting Accelerator Magnets, Unit 14 Construction Methods and Support Structures II.” Lawrence Berkeley National Laboratory. https://uspas.fnal.gov/materials/18MSU/U14-2018.pdf.
Shearer, Benjamin F. and Barbara S. Shearer. 1997. *Notable Women in the Physical Sciences: A Biographical Dictionary*. Westport, Conn.; London: Greenwood Press.
Stephens, Nastassja. 2020. *The Life and Legacy of Helen Thom Edwards*. University of Maryland.
The American Society of Mechanical Engineers. 1993. “The Fermilab Tevatron Cryogenic Cooling System.” https://www.asme.org/wwwasmeorg/media/resourcefiles/aboutasme/who%20we%20are/engineering%20history/landmarks/169-cryogenic-cooling-system-fermilab-tevatron.pdf.

Chapter 14
Mildred S. Dresselhaus

For her studies of the electronic properties of metals and semimetals, and for her service to the nation in establishing a prominent place for women in physics and engineering

Keywords Mildred Dresselhaus · Nanotubes · Carbon · Queen of Carbon · Gene Dresselhaus · Buckyballs · Graphene · Graphite · Andre Geim · Konstantin Novoselov · Enrico Fermi · Robert Curl · Richard Smalley · Sir Harry Kroto · Frederick Reif · Rosalyn Yalow · Raman spectroscopy · Armchair nanotube · Chiral nanotube · Zigzag nanotube · Kavli Prize · MIT · Lithium-ion batteries

Dr. Mildred Dresselhaus (Fig. 14.1) was the "Queen of Carbon." For 50 years, she studied carbon, unlocking the mysteries of graphite, fullerenes, and nanotubes. Her fundamental research laid the groundwork for modern nanomaterials, including smaller, stretchable, and biodegradable electronics, as well as lithium-ion batteries. As a child, her family relied on welfare to get by, and Dr. Dresselhaus pursued scholarships to fund her education. She juggled family and career, becoming a powerful advocate for gender equity. She felt strongly about giving back by mentoring students, volunteering her time on committees, and providing opportunities for young scientists, particularly women. She received the National Medal of Science in 1990.

J. S. Tietjen et al., *Women Laureates 1970 – 2007*, Women in Engineering and Science, https://doi.org/10.1007/978-3-032-23116-1_14

Fig. 14.1 Mildred S. Dresselhaus. (Department of Energy)

14.1 Life (1930–2017)

Mildred Dresselhaus, nee Spiewak, was born in Brooklyn, New York, in 1930. Her father, Meyer, was a Polish Jewish immigrant who arrived in the USA in the early 1920s with limited education. Her mother, Ethel, was also a Polish Jew, but her family had moved to Holland during World War I. She had more education than her future husband, but did not attend high school. Ethel immigrated to the USA in 1926, and the two married. Their son, Irving, was born in 1928, followed by Mildred "Millie" 2 years later (Sherkow 1976).

During the Great Depression, jobs were scarce, and Meyer's lack of skills and education meant that he had to settle for day jobs when he could find them. The family survived with the help of public assistance, which only afforded them a subsistence diet. At one time, Millie had to make do with just one set of clothes and a pair of shoes.

Millie's brother was a musical prodigy who played the violin. Millie's father came from a musical family, and he valued musical ability, so when his five-year-old son was offered a scholarship for music lessons, the family moved from Brooklyn to the Bronx to be nearer to the violin teacher. Unfortunately, the teacher died a few months later, but Irving received a scholarship to the Greenwich House's Music School in Greenwich Village in Manhattan. Millie tagged along to her brother's music lessons, and the teachers took notice of her musical ability, too. Although not as gifted as her brother, she also won a scholarship to Greenwich House and began taking violin lessons there, learning to read music at the age of four or five (Weiss 2009). Millie made the one-hour subway ride to Greenwich House two or three afternoons a week and, on Saturdays, all the way through her high school years.

For her academic education, Millie attended public school in the Bronx through junior high school. She described where she lived as a "rough" neighborhood with tenements (Sherkow 1976). She recalled getting attacked by gangs while walking to elementary school, and the teachers spending most of their time trying to manage the students rather than teaching (Weinstock 2022). On top of school and music, Millie also had responsibilities at home. Her father was ill during much of her childhood and was in and out of the hospital, and her mother worked nights at an orphanage, so Millie had to be independent and took on some of the cooking responsibilities. The family was barely getting by, so Millie began working as soon as possible. Her first job was as a tutor when she was just 8 for a child with severe developmental and learning disabilities. It was hard work that paid little, and it was both challenging and frustrating; however, she learned a great deal (Sherkow 1976). In junior high, she had a job in a zipper factory and also helped her mother with sewing work that she brought home to earn extra cash (Weinstock 2022).

The time Millie spent at the music school was like being in another world. Most of the other students came from higher socioeconomic backgrounds, and their families were more educated. She quickly made the connection between education and the possibility of higher-paying jobs and an escape from struggling to stay fed and clothed. It was at music school that she and her brother learned about the excellent public specialized high schools in New York City, where admission was determined through entrance exams rather than by neighborhood. Her brother went to the renowned Bronx High School of Science. There was only one of these schools open to girls, Hunter College High School. Millie's junior high in the Bronx offered no encouragement or assistance, so Millie had to prepare for the entrance exam on her own. She took the initiative to write away for copies of old exams and used books from the public library to try to fill in the many gaps in her knowledge. She found teaching herself math relatively easy, but English proved to be much more challenging for her. Still, she passed the exam and was admitted. Later, Millie would say that getting into Hunter College High School was a turning point in her life. She was around kids whose parents were college-educated and had high aspirations for the children. The environment was conducive to learning, and the teachers were encouraging, which set her on the path to college (Sherkow 1976).

Millie thrived at Hunter College High School. Math and science came easily to her, and she was able to improve in English to earn A's in all her classes. She was active in clubs, played in the orchestra, and made many friends. She still needed to have some income, so she began tutoring. She taught her students study skills in addition to the academic subjects. Word got around about her tutoring prowess, and she built up a nice business. She tutored throughout her college years, earning enough money to cover her expenses (Sherkow 1976).

In high school, the girls who did not have the means to go to the more selective colleges were funneled into Hunter College. Millie received a scholarship to Cornell University, but she turned it down, fearing she would not be able to afford the uncovered expenses and feeling that it would limit her ability to contribute to the family finances. Millie entered Hunter College, which specialized in educating future teachers, in 1947 (Hunter College Campus Schools 2017).

Although Millie enjoyed math and science, she did not think of them as career options, and with all her tutoring experience, she felt that teaching was a good fit (Sherkow 1976). She enjoyed school but did not have career goals. She just wanted to be sure that she could support herself in a job that did not require enduring the conditions and drudgery of factory work. She wanted to have a family, and most college professors were not supportive of women having careers (Sherkow 1976). However, during her sophomore year of college, she had a female adjunct professor, future Nobel Prize and National Medal of Science winner Rosalyn Yalow, for a physics class, who encouraged her to go to graduate school and pursue a career in physics.

Millie had not thought about physics as a career, but she did not consider physics to be the domain of men. Hunter College had been an all-women's school, but it began admitting men in 1946. By the time Millie attended the school, there were a number of men in her classes. She noted,

> The boys in the science classes were toward the bottom of the class... They always used to come to me for help.... That might be somewhat significant in my story, because I never got the idea in college that science was a man's profession. (Sherkow 1976)

Mildred Dresselhaus graduated from Hunter College with an undergraduate degree in liberal arts in 1951. Looking back, Dresselhaus did not feel that the science education she got at Hunter College was strong, but she learned other valuable lessons about being part of a community, saying, "It's not enough only to take, but you have to give (Sherkow 1976)."

After graduation, Dresselhaus spent a year in England on a Fulbright Fellowship, which paid all of her expenses. She attended the University of Cambridge and pursued additional undergraduate-level physics studies to prepare for graduate school. She received a fellowship for graduate study and chose to go to Radcliffe College in 1952 to be with her boyfriend, who was attending Harvard (Sherkow 1976). This was presumably her first husband, physicist Frederick Reif. Dresselhaus did not speak publicly about their marriage and its duration. The Radcliffe College Catalog of Names 1952–1953 lists her as Mildred Reif (Radcliffe College 1952). Dresselhaus received her Master of Science in Physics from Radcliffe in 1953. She and Reif then went to the University of Chicago, where she pursued a PhD in physics, and Reif joined the faculty. The two divorced sometime before 1958.

The physics program at the University of Chicago was shaped by the renowned physicist Enrico Fermi, was quite rigorous, and demanded independent research without much faculty input. About three-quarters of the incoming PhD students would drop out (Tietjen and Ozturk 2026). Dresselhaus's advisor did not believe that women should pursue graduate education and was upset when she won fellowships, feeling that she was taking opportunities away from men. As a result, she avoided her advisor, and he was not even aware of what her research topic was. He was not the only person Dresselhaus encountered who thought women did not belong in the lab.

> I got a lot of negative input from people who believed that a woman's place is in the home. I heard so much of this that I believed it. Another thing I remember as a student was all the nude pinup girls on the walls. These were things you had to put up with in a lab to get your experiments working. It was not an environment in which women felt comfortable. But if you couldn't handle that, you knew you couldn't have a career in science. (Weinstock 2022)

Not all the faculty and students felt this way. Dresselhaus considered Enrico Fermi to be a role model. Although they were only together at the University of Chicago for 1 year due to Fermi's death in 1954, she got to know him and his family well, and credits Fermi with teaching her to think like a physicist (Kavli 2012). In 1956, she met her future husband, physicist Gene Dresselhaus. Having completed his PhD at the University of California, Berkeley, Dr. Gene Dresselhaus was a postdoctoral student and instructor at the University of Chicago for 1 year. He then went to Cornell University as an Assistant Professor (Wikipedia Contributors 2025).

Mildred Dresselhaus remained in Chicago for another 2 years, finishing her thesis on superconductivity, the property where some materials achieve near-zero resistance and create a strong magnetic field at very low temperatures. Dresselhaus finished her PhD thesis in 1958 and was awarded a postdoctoral fellowship, funding that she could take with her to any institution. She chose to go to Cornell to be with Gene, and the two married in 1958 (Sherkow 1976). Gene was encouraging of Millie's desire to have a career and a family. The two worked together at Cornell on interpreting experiments she had done for her thesis, and also started a family with the birth of their first child, Marianne, in 1959. Dr. Mildred Dresselhaus immediately returned to work with her weeks-old baby in tow. She was fortunate to be able to take Marianne to the lab and to have the departmental secretaries be willing to look after her when necessary (Weinstock 2022).

In 1960, Dr. Mildred Dresselhaus's fellowship was finished, but the Dresselhauses wanted to keep working together. Due to nepotism rules at Cornell, which were not uncommon in academia at the time, employment on the faculty there was not an option. As a result, Drs. Dresselhaus accepted positions at the Massachusetts Institute of Technology's (MIT) Lincoln Laboratory. During her 7 years at Lincoln Lab, the Dresselhauses worked together, independently, and with other collaborators, at times. They also had three more children: Carl in 1961, Paul in 1963, and Eliot in 1964 (Weinstock 2022).

At Lincoln Lab, Dr. Mildred Dresselhaus was given a great deal of freedom in her research, but she was asked to pick an area other than superconductivity. The Lab's management did not feel that there were significant discoveries still to be made in this area. After some work with semiconductors and semimetals, she settled on carbon. Gene had worked on carbon in graduate school and suggested it as an area that would be of interest. At the time, carbon was not thought to be an exciting field of study, so there was little competition. Dr. Dresselhaus considered this an opportunity to achieve a better work-life balance (Weinstock 2022).

Dr. Dresselhaus worked with carbon for the rest of her career, but by 1967, she felt the atmosphere at Lincoln Lab had become too bureaucratic. With four kids at home, she wanted more flexibility than the rigid 8:00 a.m. start to the workday afforded by her job (Weinstock 2022). She was happy to be appointed as the Abby Rockefeller Mauzé Visiting Professor at MIT. The Rockefeller family established the fund to bring "distinguished women scholars to campus" (Wiesner 1967). In 1968, MIT offered Dr. Dresselhaus a full professorship in the Department of Electrical Engineering and Computer Science, making her the first female full professor at the university (Lahiri 2019). She remained at MIT for nearly 50 more years, becoming the first female MIT Institute Professor in 1985 (an MIT Institute Professor is a lifetime appointment held by no more than twelve active MIT professors) (Tietjen and Ozturk 2026).

The "Queen of Carbon's" accomplishments in the lab led to a revolution in electronics. Her research on graphene vibration enabled the discovery of ways to capture heat from vibrating particles and turn it into usable electricity (Dwyer 2017).

Her research was key to the development of lithium-ion batteries and smaller and faster computer chips.

> She devised carbon fibers that are stronger than steel at a fraction of steel's weight. Her research helped usher in the age of nanotechnology, the wildly popular effort to downsize electronic circuits, medical devices and a host of other products to molecular dimensions (Dwyer 2017)

Dr. Dresselhaus practiced what she preached when it came to giving back. She was a strong advocate for women in science. At Lincoln Laboratory, there was only one other female scientist on the staff. She felt that she was judged harshly by the men every time she had to arrive late or leave early due to her family responsibilities. When she began teaching at MIT, just 4% of the undergraduates were women (Lahiri 2019). For many years, she led an engineering seminar at MIT for first-year students, designed to boost the confidence of female students, which also attracted a large number of men. In 1972, Dr. Dresselhaus and Dr. Emily Wick planned the first seminar to discuss the status of women at MIT. The event attracted about 100 women affiliated with MIT, including students, faculty, staff, postdocs, alumni, and even faculty wives. This evolved into the MIT Women's Forum, which significantly raised the profile of the environment for women at MIT and led to the appointment of a special assistant to the president and chancellor for women and work (MIT Women in Mathematics 2022). A year later, in 1973, she won a grant from the Carnegie Foundation to encourage women to enter science and engineering. She was a mentor to men and women, advising more than 75 PhD students (Yeh 2017). In 1999, Dr. Dresselhaus received the Dwight Nicholson Medal for Outreach from the American Physical Society:

> For being a compassionate mentor and lifelong friend to young scientists; for setting high standards as researchers, teachers and citizens; and for promoting international ties in science. (MIT Department of Physics 2017b)

Her generosity and dedication to young scientists were exemplified when she won the prestigious $1 million Kavli Prize in 2012 for her work in nanoscience. She used the money to create the Mildred S. Dresselhaus fund to support women or junior members of the faculty, saying that she received the award because of, "The investment MIT made in me many years ago through the generosity of others" (School of Engineering 2013).

Dr. Dresselhaus's leadership roles extended beyond MIT. She served as the Director of the Office of Science at the US Department of Energy from 2000 to 2001, the President of the American Physical Society (APS) in 1984, the President of the American Association for the Advancement of Science (AAAS) in 1998, the Treasurer of the US National Academy of Sciences, and the Chair of the Governing Board of the American Institute of Physics (AIP) from 2003–2007 (Lahiri 2019).

In addition to the National Medal of Science and the Kavli Prize, Dr. Dresselhaus received numerous awards. Some of her highest honors include: the L'Oreal-UNESCO for Women in Science Award in 2007; the US Department of Energy Enrico Fermi Award in 2012, the second female recipient; the Presidential Medal of Freedom in 2014; and the IEEE Medal of Honor in 2015, the first female recipient. She was elected to membership in many prestigious societies, including the National Academy of Sciences, the National Academy of Engineering, and the American Academy of Arts and Sciences (Wikipedia Contributors 2022). She was also recognized with 40 honorary degrees (Weinstock 2022). Dr. Dresselhaus published more than 1700 papers and co-wrote eight books (Dwyer 2017). She had dozens of patents and was inducted into the National Inventors Hall of Fame in 2014 (Busch-Vishniac et al. 2024).

14.2 Research

In the 1960s, at Lincoln Laboratory, Dr. Dresselhaus began experimenting with graphite, which is composed of a layered, lattice structure of carbon atoms (Fig. 14.2). Her goal was to understand better how electrons move within and between layers and to determine the electronic band structure of graphite. The electronic band structure of a solid, such as graphite, refers to the energy levels that electrons may occupy, and just as important, the energy ranges they may not occupy (Wikipedia Contributors 2019b). Dr. Dresselhaus performed magneto-optical experiments by exposing graphite to a strong magnetic field, which forced the electrons into discrete energy states called Landau levels. Then, by shining laser light on the sample, she studied how the graphite absorbed specific wavelengths, revealing how electrons moved between those states. These experiments provided the first precise experimental map of graphite's electronic band structure and overturned some long-held beliefs about the arrangement and behavior of its electrons (MIT Department of Physics 2017a).

In the 1970s, Dr. Dresselhaus, working with her husband and other collaborators, investigated graphite intercalated compounds (GICs). GICs are formed by inserting different elements between the graphite's carbon layers (Fig. 14.3). They demonstrated that this process could fundamentally change the graphite's electrical

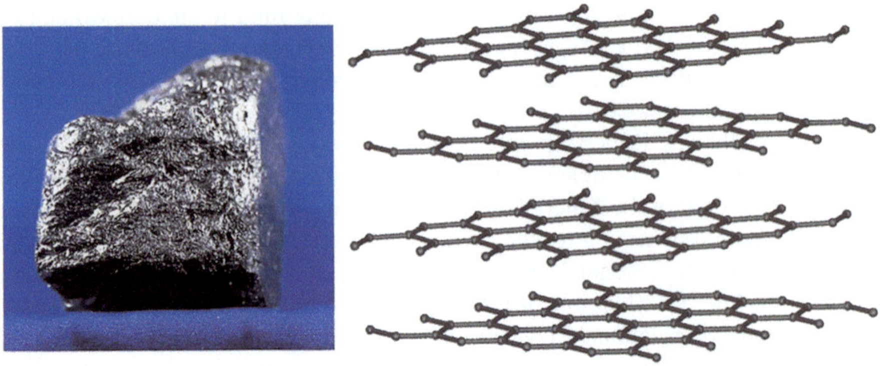

Fig. 14.2 Graphite and its lattice structure. (Itub, CC BY-SA 3.0, via Wikimedia Commons)

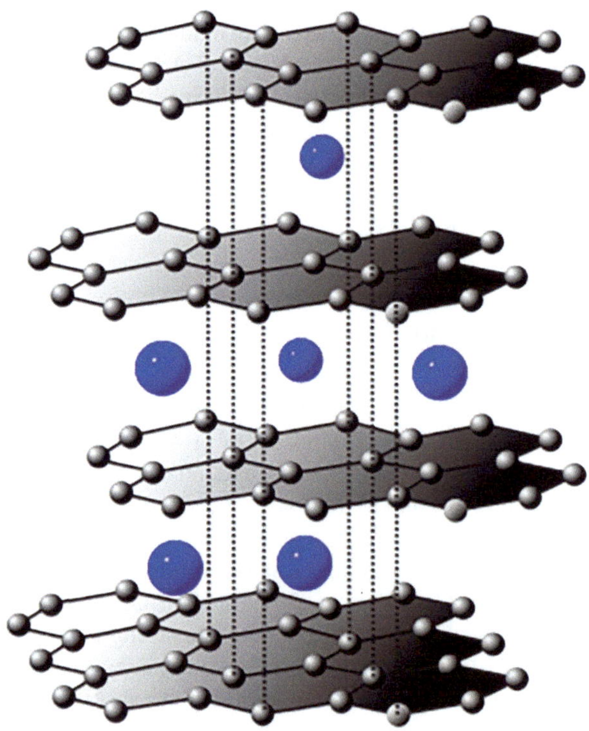

Fig. 14.3 Graphite intercalated compounds (GICs). (Anton, CC BY-SA 2.5, via Wikimedia Commons)

conductivity and optical properties, depending on the inserted element. In some cases, such as with potassium, intercalation even produced superconductivity. This work demonstrated that graphite could serve as a platform for designing materials with tailored electronic properties, launching a revolution in carbon research (Lahiri 2019).

Dr. Dresselhaus continued her research in the 1980s, bombarding graphite with powerful lasers and observing that large chunks of carbon atoms were expelled from the surface, a surprising result initially doubted by other scientists (Hargittai 2015). She and her team at MIT analyzed these carbon clusters and found that they reorganized themselves into unusually stable clusters, rather than the layered sheets that comprise graphite. Dr. Dresselhaus encouraged researchers at Exxon to use this laser ablation technique to study how carbon atoms organize themselves (Fleming 2022).

> These discussions directly contributed to work which led to the publication two years later of the first observation of the buckminsterfullerene, the famous football-shaped "bucky-ball" molecule made up of 60 carbon atoms (C60) with a range of interesting and useful properties such as superconductivity and light absorption. Three of the paper's authors, Robert Curl, Richard Smalley, and Sir Harry Kroto, were jointly awarded the Nobel Prize in Chemistry for the discovery in 1996. (Fleming 2022)

Figure 14.4 shows the structure of a buckyball. In the USA, the buckyball is said to resemble a soccer ball instead of a football.

In 1990, Dr. Dresselhaus and Richard Smalley proposed that fullerenes, the general term for a hollow molecule composed of carbon atoms like the buckyball, could assume a narrow, tubular shape. In 1991, Sumio Iijima of Japan was the first to observe these hollow cylindrical tubes of carbon atoms, with diameters of just a few nanometers, which came to be called carbon nanotubes (CNTs). By 1993, single-walled nanotubes (SWNTs), which can be thought of as a single sheet of carbon atoms (known as graphene) rolled into a cylindrical structure (Fig. 14.5), had been independently created by groups in Japan and at IBM.

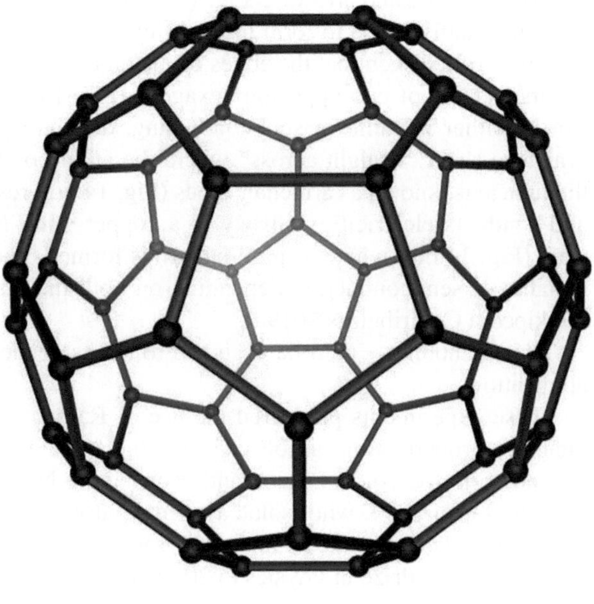

Fig. 14.4 Buckyball. (Designism, CC BY-SA 4.0 via Wikimedia Commons)

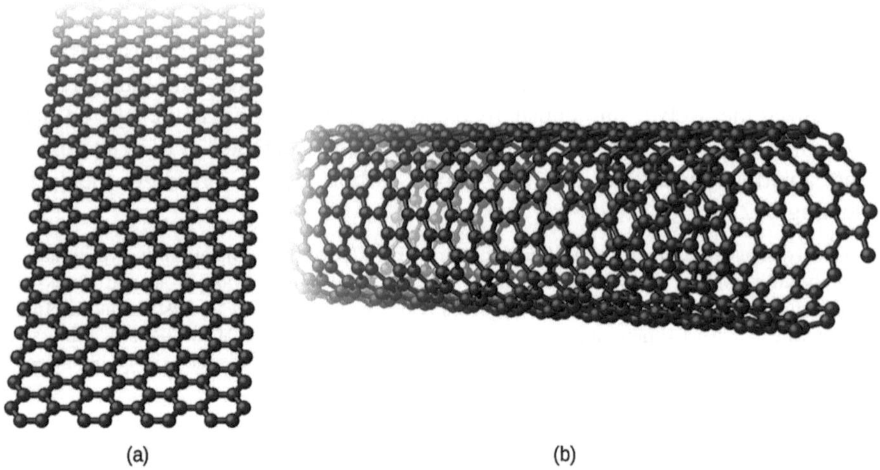

Fig. 14.5 Graphene sheet and nanotube. (OpenStax, CC BY 4.0, via Wikimedia Commons)

Dr. Dresselhaus quickly recognized the importance of these materials and became one of the earliest and most influential researchers to study their properties. She set out to prove that the angle at which the sheet is rolled would determine whether a SWNT behaves as a semiconductor or a metal, explaining why nanotubes of similar size exhibit different properties. She and her team applied Raman spectroscopy, a technique that reveals how light interacts with atomic vibrations. The blue dashed lines and arrows in Fig. 14.6a illustrate how a flat sheet of graphene can be conceptually rolled to form these different cylindrical structures. She found that graphene rolled such that the edges of the tube run perpendicular to a line connecting the centers of two opposing hexagons (Fig. 14.6b) forms a zigzag pattern and can be either metallic or semiconducting, depending on the diameter. Rolling a graphene sheet "straight across" where the atom bonds line up perpendicularly to the tube axis, known as armchair tubes (Fig. 14.6c), results in a tube that is metallic and conducts electricity similarly to a copper wire. Finally, graphene rolled at a slant (Fig. 14.6d), where a spiral pattern is formed, known as chiral tubes, is either metallic or semiconducting depending on both the diameter and the rolling angle (Wikipedia Contributors 2019a).

Thus, nanotubes could be designed to act as needed for a myriad of electronic applications.

These experiments pioneered the use of Raman spectroscopy in exploring the quantum properties of carbon nanostructures (Fleming 2022). They also laid the groundwork for Andre Geim and Konstantin Novoselov at the University of Manchester in 2004, who isolated a single-atom-thick sheet of carbon atoms in a honeycomb lattice, called graphene, from bulk graphite. This achievement won them the Nobel Prize in Physics in 2010 (The University of Manchester n.d.). After

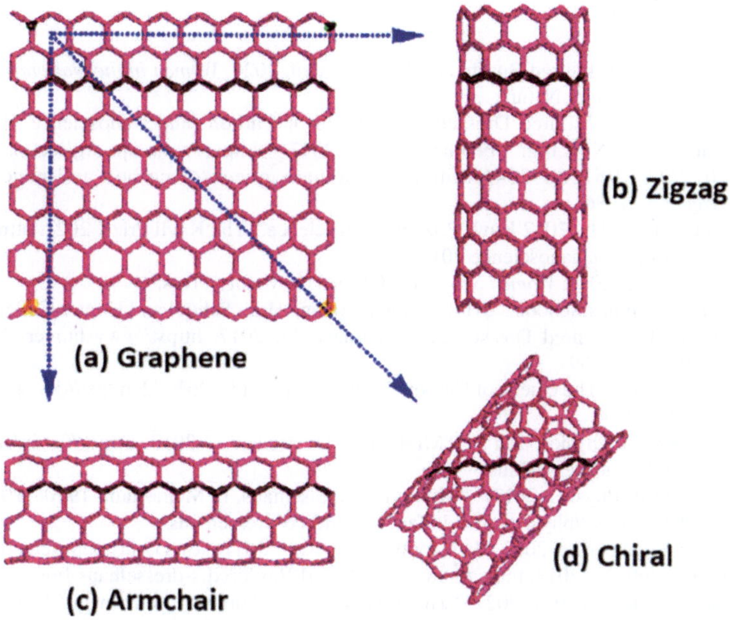

Fig. 14.6 Types of nanotubes. (Lesnar1234, CC0, via Wikimedia Commons (annotated))

this 2004 discovery, Dr. Dresselhaus researched the vibrational and electronic behavior of the two-dimensional graphene, exploring its fundamental physics. She continued to go to her lab at MIT until 2 weeks before she died in 2017 (The Ogden Trust 2020).

14.3 Conclusion

Dr. Mildred Dresselhaus was science royalty for her pioneering research on carbon. During her long career, the "Queen of Carbon" studied the thermal and electrical properties of graphite, graphene, fullerenes, and carbon nanotubes, laying the groundwork for modern nanotechnology. Her research into the behavior of carbon at the atomic scale underpins many of the technologies we rely on today, from smaller and faster electronics and stronger materials to more efficient batteries. She encouraged women in science and engineering and paved the way for women at MIT to have more opportunities and a supportive work environment. Her influence on women in STEM went far beyond MIT, with national leadership roles that enabled her to advocate for inclusive policies and promote gender equity in science and engineering.

References

Busch-Vishniac, Ilene, Lauren Busch, and Jill S. Tietjen. 2024. *Women in the National Inventors Hall of Fame*. Springer Nature.

Dwyer, Colin. 2017. "Mildred Dresselhaus, 'Queen of Carbon' and Nanoscience Trailblazer, Dies at 86." NPR.org. February 24, 2017. https://www.npr.org/sections/thetwo-way/2017/02/24/517004254/mildred-dresselhaus-queen-of-carbon-and-nanoscience-trailblazer-dies-at-86.

Fleming, Nic. 2022. "The 2012 Kavli Prize in Nanoscience." The Kavli Prize. 2022. https://www.kavliprize.org/prizes/nanoscience/2012.

Hargittai, Magdolna. 2015. *Women Scientists*. Oxford University Press.

Hunter College Campus Schools. 2017. "Hunter College High School and Hunter College Mourn the Loss of Dr. Mildred Dresselhaus." February 24, 2017. https://www.hunterschools.org/news-detail?pk=914097.

Lahiri, Jayeeta. 2019. "The Queen of Carbon!" *Resonance* 24 (3): 263–72. https://doi.org/10.1007/s12045-019-0778-4.

MIT Department of Physics. 2017a. "Mildred S. Dresselhaus." 2017. https://fi.edu/en/awards/laureates/mildred-s-dresselhaus.

MIT Department of Physics. 2017b. "Mildred S. Dresselhaus, in Memoriam: 1930 - 2017." MIT Physics. 2017. https://physics.mit.edu/faculty/mildred-dresselhaus/.

MIT School of Engineering. 2013. "In It for the Long Run." MIT News | Massachusetts Institute of Technology. July 31, 2013. https://news.mit.edu/2013/mildred-s-dresselhaus-fund.

MIT Women in Mathematics. 2022. "The MIT Women's Forum – Women in Math." Mit.edu. 2022. https://math.mit.edu/wim/2022/01/14/the-mit-womens-forum/.

Radcliffe College. 1952. "Student Directory Radcliffe College 1952-1953." 1952. https://iiif.lib.harvard.edu/manifests/view/drs:428023154.

Sherkow, Shirlee. 1976. "Mildred Dresselhaus Edited Transcript 1976 June–August" Interview by Shirlee Sherkow. Massachusetts Institute of Technology, Oral History Program, Women Scientists and Engineers Oral History Collection (MC-0086). Massachusetts Institute of Technology. Libraries. Department of Distinctive Collections.

The Ogden Trust. 2020. "Women in Physics Research Cards." The Ogden Trust. 2020. https://www.ogdentrust.com/wp-content/uploads/2021/10/Women-in-Physics-Research-Cards-1.pdf.

The University of Manchester. n.d. "Discovery of Graphene - Graphene - the University of Manchester." Www.graphene.manchester.ac.uk. The University of Manchester. https://www.graphene.manchester.ac.uk/learn/discovery-of-graphene/. Accessed September 18, 2025.

Tietjen, Jill S., and Mehmet C. Ozturk. 2026. *Pioneering Women of Solid-State Physics – Breaking Boundaries*. Springer Nature.

Weinstock, Maia. 2022. *Carbon Queen*. MIT Press.

Weiss, Paul S. 2009. "A Conversation with Prof. Mildred Dresselhaus: A Career in Carbon Nanomaterials." *ACS Nano* 3 (9): 2434–40. https://doi.org/10.1021/nn901147q.

Wiesner, Jerome B. 1967. "News Release." *MIT*, September 20, 1967. https://cdn.libraries.mit.edu/dissemination/diponline/AC0069_NewReleases/NewsRelease_1960/AC0069_1967/AC0069_196709_009.pdf.

Wikipedia Contributors. 2019a. "Carbon Nanotube." Wikipedia. Wikimedia Foundation. April 13, 2019. https://en.wikipedia.org/wiki/Carbon_nanotube.

Wikipedia Contributors. 2019b. "Electronic Band Structure." Wikipedia. Wikimedia Foundation. June 23, 2019. https://en.wikipedia.org/wiki/Electronic_band_structure.

Wikipedia Contributors. 2022. "Mildred Dresselhaus." Wikipedia. November 14, 2022. https://en.wikipedia.org/wiki/Mildred_Dresselhaus.

Wikipedia Contributors. 2025. "Gene Dresselhaus." Wikipedia. Wikimedia Foundation. March 14, 2025. https://en.wikipedia.org/wiki/Gene_Dresselhaus.

Yeh, Nai-Chang. 2017. "Mildred S. Dresselhaus (1930–2017): A Pinnacle of Scholarship." *ACS Nano* 11 (6): 5215–16. https://doi.org/10.1021/acsnano.7b03868.

Chapter 15
Gertrude B. Elion

For her basic research which enhanced the sciences of chemistry and medicine by elucidating fundamental principles of drug-receptor interactions for nucleic acid antagonists

Keywords Purines · Acyclovir · Rational drug design · Allopurinol · Kidney transplants · Leukemia · Imuran · Azathioprine · AZT · Mercaptopurine · Burroughs Welcome · GlaxoSmithKline · Gout · Herpes · Epstein-Barr · Chickenpox · Shingles · Thioguanine · Anticancer drugs · Pharmacology · Nobel Prize in Physiology or Medicine · Nucleic acid antagonists · Antagonists

Gertrude B. Elion, shown in Fig. 15.1, was a biochemist who changed how new drugs are designed by focusing on the molecular level differences between pathogen and human cell replication. Her work was responsible for the first effective drug to combat leukemia, the first drug to suppress immune systems so that organs could be transplanted, and the first anti-viral drug. Her work laid the foundation for the first effective drug to combat AIDS and contributed to work leading to medicines for malaria, meningitis, sepsis, gout, and bacterial infections. Elion was awarded a National Medal of Science in 1991 and received the Nobel Prize in Physiology or Medicine in 1988 along with colleagues Dr. George Hitchings and Dr. James Black.

J. S. Tietjen et al., *Women Laureates 1970 – 2007*, Women in Engineering and Science, https://doi.org/10.1007/978-3-032-23116-1_15

Fig. 15.1 Gertrude B. Elion. (Courtesy of GlaxoSmithKline plc)

15.1 Life (1918–1999)

Gertrude Belle Elion was born in New York City on a January night in 1918 so cold that the pipes froze and burst in the apartment of her parents, Berta Cohen Elion and Robert Elion. Berta Elion was originally from a part of Russia that later became Poland and arrived in the US at 14 years of age. Robert Elion had emigrated to the US from Lithuania when he was 12 years old and was a 1914 graduate of the NYU School of Dentistry. Their living quarters adjoined the dental office (Elion nd).

When Gertrude was 3 years old, her grandfather emigrated from Russia and moved into an apartment near her family's apartment. They spent a great deal of time together. Six years after Gertrude was born, her brother Herbert arrived, and the family moved to the Bronx (Elion nd; The Nobel Prize 1988).

Gertrude has described her childhood as happy. She attended the local public school and took delight in learning. As a gifted student, she was moved up in school so that she was two grades ahead of her age peers by the time she was twelve. She graduated from Walton High School at the age of fifteen, the same year that her beloved grandfather died of stomach cancer (Elion nd; Wikipedia 2025; Elion 1993). Looking back on that time, she has said: "I was fifteen years old when my grandfather died of cancer, and right then I decided that my goal would be to find a cure for this terrible disease" (National Inventors Hall of Fame 2025). She decided to pursue a degree in chemistry and noted:

> What was remarkable about my childhood was that I had such supportive parents who never discouraged me from having ambition. They never told me that girls weren't supposed to go into chemistry (National Inventors Hall of Fame 2025).

Gertrude's father invested in the stock market as his finances improved, but the stock market crash of 1929 wiped out the family's fortune, and Robert Elion declared bankruptcy. This meant that when Gertrude graduated from high school a few years later, she needed to limit her sights to colleges that would be free. She chose to go to Hunter College and to major in chemistry. She graduated summa cum laude in 1937 (National Inventors Hall of Fame 2025).

Upon graduating, Elion aimed to go to graduate school and applied to fifteen places for graduate assistantships, but she was turned down by all of them. What support for graduate work existed at the time was reserved for deserving men, and Elion was told that her presence in the lab would be distracting. Commenting on this period, Elion has said, "I hadn't been aware that any doors were closed to me until I started knocking on them" (American Chemical Society 1918–1999).

Over the next several years, Elion held jobs for which she was overqualified and made slow progress toward an advanced degree. She took a six-week secretarial school course and worked for a time as a secretary. She also got a one-term job teaching nursing students at New York University (NYU) the biochemistry they needed to know. Then she volunteered as an assistant in the chemistry lab of Dr. Alexander Galat. After her first six months, she was paid and managed to save enough after 18 months to pursue a master's degree (Elion nd; Wikipedia 2025; Elion 1993).

In 1939, Elion entered graduate school at NYU. She was the only woman in her class in the chemistry department. Elion finished her required coursework in eighteen months and took a job as a teacher-in-training and then a substitute teacher in the New York City schools. She conducted her research at night and on weekends, graduating with her MS degree in 1941 (Elion nd).

During this time, she also met and spent time with Leonard Canter, a statistics student at City College of New York. Before they could marry, Leonard was stricken with bacterial endocarditis, an infection of the heart valves. He died in 1941 (Wikipedia 2025).

By the time of Elion's MS completion, World War II was well underway and there was a shortage of chemists in industry, as most of the young men were off at war. This meant that laboratory positions were opened to women for the first time. Elion said:

> World War II did what I had been unable to do on my own: It opened the doors for women to work in chemistry laboratories. While the men were away, employers had to take the "risk" (Elion 1993).

She landed a job in the quality control area of grocery firm A&P testing the acidity of pickles and the color of the egg yolks being used in mayonnaise. While she grew bored with the repetitiveness of the job, she was grateful for the chance to learn how to use advanced chemical instrumentation (Elion nd, Wikipedia 2025).

Elion left after eighteen months and joined Johnson and Johnson (J&J) in New Jersey. At J&J, she was part of a growing pharmaceutical division, and the work was rewarding. Unfortunately, just 6 months after starting at J&J, a new vice president was appointed who opted to close the pharmaceuticals area. Elion was offered the opportunity to remain, but her new work was to test the strength of sutures—not at all aligned with her personal goals (Elion 1993).

In 1944, Elion left J&J to join Burroughs Welcome (now GlaxoSmithKline) in Tuckahoe, New York and work as an assistant to Dr. George Hitchings. Sulfa drugs had recently been discovered, and Dr. Hitchings was inspired by them. Rather than directly killing bacteria, the sulfa drugs worked by preventing bacteria from making folic acid, a nutrient they required for cell growth and replication. Dr. Hitchings wondered if other compounds might be used to interfere with the metabolism of various microbes and be developed into drugs. Dr. Hitchings focused on the idea of developing compounds that would interrupt the replication cycle of cancer cells. Elion's training in organic chemistry positioned her well to contribute to such a molecular level investigation. And as she learned more about microbiology and the instrumentation, Dr. Hitchings let Elion assume greater responsibility for the research (Wikipedia 2025).

While working for Burroughs Welcome, Elion decided to pursue the PhD she had been forced to forego for financial reasons. She attended Brooklyn Polytechnic Institute at night after work and on weekends for 2 years. As she finished her coursework, she was told she needed to cease being a part-time student and attend full time, a change that would require her to leave her job. Elion chose instead to leave her PhD program so she could continue to work as a research chemist at Burroughs Welcome, a position she would enjoy for the rest of her professional career (Elion nd; Elion 1993).

Dr. Hitchings and Elion spent many years as colleagues working on the development of new drugs. Alone and with Dr. Hitchings, Elion produced cancer drugs, immunosuppression medicine for transplant patients, and antiviral drugs. In each of these areas, the work reflected a complete transformation of drug design based on knowledge of biomolecular behavior rather than trial and error. Today, the approach Hitchings and Elion pioneered is referred to as rational drug design (Andersun and Eboni 2025; Wikipedia 2025).

When Burroughs Welcome moved to the Research Triangle Park in North Carolina, Elion moved to Chapel Hill, North Carolina and continued to work for them. The move enabled the lab to expand and to get more modern equipment. When Dr. Hitchings was promoted to Vice President of Research in 1967, Elion was promoted to the position he had held as Head of the Department of Experimental Therapy (Science History Institute n.d.). She held that position until her retirement in 1983.

From 1971–1983 Elion was also an adjunct professor of pharmacology and experimental medicine at Duke University. Upon retiring from Burroughs Welcome, her title at Duke University changed to research professor, and she regularly took on mentoring a third-year medical student who wanted to dip a toe into research. At the same time, she was listed as a scientist emeritus and consultant at Burroughs Welcome (National Inventors Hall of Fame 2025).

Dr. Elion chose her retirement date carefully:

> In 1983 I had to make a decision whether to retire or go on working. I chose to do both. With acyclovir now launched, I thought it might be a good time to retire as head of the Department before a new drug turned up that I couldn't bear to leave. (Elion 1993)

In addition to her retirement activities at Duke University and Burroughs Welcome, Elion became quite active in national societies. She was President of the American Association for Cancer Research in 1983–1984 and served on its Board of Directors. She served on advisory committees and the National Cancer Advisory Board of the National Cancer Institute. Elion also served on advisory committees for the American Cancer Society and the Leukemia Society of America. She chaired the Steering Commission on the Chemotherapy of Malaria for the Tropical Disease Research Division of the World Health Organization (Elion nd).

Elion was recognized with many awards throughout her career including the Garvan-Olin Medal in 1968, the Sloan-Kettering Institute Judd Award in 1983, the American Chemical Society Distinguished Chemist Award, and the Cain Award of the American Association for Cancer Research in 1985. In 1988, Elion, Dr. Hitchings, and Dr. James Black were awarded the Nobel Prize in Physiology or Medicine for their new approach to drug development. When she received the phone call informing her of the Nobel Prize, she said "quit your kidding" as she didn't quite believe she was being recognized (Onco Daily 2025). She and Dr. Hitchings are pictured in Fig. 15.2.

In 1990, Elion was awarded the American Cancer Society Medal of Honor. In 1991, she received the National Medal of Science and became the first woman inducted into the National Inventors Hall of Fame (Busch-Vishniac et al. 2025). She was also inducted into the National Women's Hall of Fame in 1991. In 1995, she was made a foreign member of the Royal Society, and in 1997 received the Lemelson-MIT Lifetime Achievement Award. Elion received 23 honorary doctorates and was inducted into the National Academy of Sciences (1990), the Institute of Medicine (1991), and the American Academy of Arts and Sciences (1991). She was one of seven people featured in a 1999 documentary entitled *Me and Isaac Newton* (Apted 1999), which described advances in the sciences.

The American Association for Cancer Research has created the Gertrude B. Elion Cancer Research Award, which "encourages and supports tenure-eligible junior faculty to conduct research in cancer and etiology, diagnosis, treatment, or prevention" (American Association for Cancer Research 2025). There is another award named in her honor as well—the Gertrude B. Elion Mentored Medical Student Research Award of the Triangle Community Foundation, which provides $10,000 to support female medical students interested in health-related research (Triangle Community Foundation 2025).

Elion was named on more than 45 US patents and published over 225 papers. Her first publication was produced in 1939, and her last in 1998. She died in 1999 after suffering a stroke (Altman and Lawrence 1999; Avery 2000).

Fig. 15.2 Elion and Hitchings in 1988. (Will & Deni McIntyre, CC BY 4.0 <https://creativecommons.org/licenses/by/4.0>, via Wikimedia Commons)

15.2 Rational Drug Design

Prior to the work of Elion and Dr. Hitchings, drug development was largely a trial-and-error procedure. New compounds would be created and tested on a variety of diseases with little attempt to engineer compounds for specific pathogens. Elion and Dr. Hitching changed that by developing rational drug design, the development of drugs based on disrupting the life cycle of the pathogen without causing ill effects to host cells. Dr. Hitching had understood that sulfa drugs did this to bacteria and wanted to determine if other compounds could be found to similarly interrupt the metabolism of other pathogens (American Chemical Society 1918–1999).

When Elion joined Dr. Hitchings's lab, the focus was on purines. Purines are compounds made using adenine and guanine, two of the bases found in DNA. They are produced in the body during cell replication and are found in high concentrations in meats, seafood, and some vegetables. The reasoning behind a focus on purines was that they were essential for replication of cells, so fast-growing cells such as bacteria and tumor cells require plentiful supplies of purines. If drugs could be found that replaced natural purines with synthetic purines that were not quite the same, maybe they could trick fast-growing cancer cells or bacteria cells into taking up the synthetic purines, which could prevent replication and stop the disease from spreading. Elion has described this period: "Each series of studies was like a mystery story in that we were trying to deduce what the microbiological results meant, with little biochemical information to help us" (Elion nd).

In 1950, Elion and Hitchings produced two compounds which entice metabolic enzymes to attach to them rather than natural purines: diaminopurine and thioguanine. These drugs were shown to be effective in treating cancer, but too toxic for patients. They were able to show remission in two adults with leukemia, with one of the remissions lasting 2 years. Two other patients didn't respond to the treatment, and most of the patients had serious side effects. Elion replaced one oxygen atom in the compound with a sulfur atom, and the net result was 6-mercaptopurine, known as 6-MP (Elion 1993; American Chemical Society 1918–1999).

In August 1953, a paper by Dr. Clarke, Elion and others showed that 6-MP could inhibit tumor growth in mice (Clarke et al. 1953). In November of the same year, collaborator Dr. Burchenal published a paper showing that 6-MP produced long-term remission of leukemia in children (Burchenal et al. 1953). This result was transformative. At that time, the life expectancy for children diagnosed with leukemia and untreated was 3 months; 6 months if treated. Using 6-MP, Dr. Burchenal extended the median life expectancy to 12 months (Elion 1993), although complete remission was seen in some children. The result was so important that the FDA approved 6-MP for use to treat childhood leukemia that same year.

The fact that leukemia remission was not routinely permanent weighed heavily on Elion and pressed her to continue her work.

> Within a year of the time 6-MP became an accepted treatment for acute childhood leukemia, it became apparent that this drug was not going to be the cure we had all hoped for. Relapses occurred and patients often no longer responded to 6-MP or methotrexate. These were heart-breaking times. (Elion 1993)

Eventually, Elion and her colleagues learned that combining 6-MP with other drugs significantly improved the situation, with significant increases in childhood leukemia cures.

Cancer drugs were merely the beginning of an astonishingly prolific drug creation career. In 1961, Elion developed azathioprine (marketed as Imuran), a drug to suppress immune response, making it safe to transplant kidneys for the first time. In 1963, she developed allopurinol, a drug to reduce the production of uric acid and thus to treat gout. Again, these drugs were designed by focusing on the biomolecular characteristics of cells and introducing antagonist compounds to interrupt a process (The Nobel Prize 1988; Wikipedia 2025).

In 1977, Elion and colleagues had yet another transformative discovery. They applied rational drug design and life cycle disruption to treat a virus for the first time. Prior to their discovery, it had been believed that drugs could not disrupt virus replication without significantly damaging host cells. The drug they produced is acyclovir. It was found to be an effective treatment for herpes, Epstein-Barr, chickenpox, and shingles, and was approved for use in 1982. This work also formed the basis for the development of AZT, the first drug to fight AIDS. AZT was produced by her lab the year after Elion retired (Elion et al. 1977; Schaeffer et al. 1978; Larsen 2009).

In addition to these drugs, Elion played a role in the development of pyrimethamine for treatment of malaria; trimethoprim for the treatment of meningitis, sepsis, and other bacterial infections; and nelarabine for the treatment of cancer. She kept a file of letters from grateful patients, and when asked about her reaction to receiving the Nobel Prize, said, "What we were aiming at was getting people well, and the satisfaction of that is much greater than any prize you can get" (The Nobel Prize 1988).

15.3 Conclusion

Gertrude B. Elion was a biochemist who had a remarkably prolific and impactful career in drug development. She was responsible for the first successful drug to fight childhood leukemia, the first immunosuppressive drug to permit safe organ transplants, and the first antiviral drug. She received the National Medal of Sciences in 1991 and the Nobel Prize in Physiology or Medicine in 1988.

References

American Association for Cancer Research, Gertrude B. Elion Cancer Research Award,. https://www.aacr.org/professionals/research-funding/funded-research/career-development-awards/aacr-gertrude-b-elion-cancer-research-award/. Accessed Jul 30, 2025.

Altman, Lawrence. Feb 23, 1999. Gertrude Elion, Drug Developer, Dies at 81. *NY Times*. https://www.nytimes.com/1999/02/23/us/gertrude-elion-drug-developer-dies-at-81.html. Accessed Jul 30, 2025.

American Chemical Society, Gertrude Elion (1918–1999). https://www.acs.org/education/whatischemistry/women-scientists/gertrude-elion.html. Accessed Jul 30, 2025.

Andersun, Eboni. May 23, 2025. Gertrude Belle Elion (1918 – 1999), Embryo Project Encyclopedia, Arizona State University. https://embryo.asu.edu/pages/gertrude-belle-elion-1918-1999. Accessed Jul 30, 2025.

Apted, Michael (director). 1999. Me and Isaac Newton. *Clear Blue Sky Productions*.

Avery, Mary. 2000. Gertrude B. Elion. In *Biographical Memoirs: Vol 78*. National Academies Press. https://nap.nationalacademies.org/read/9977/chapter/3. Accessed Jul 30, 2025.

Burchenal, J, Murphy, M, et al. 1953. Clinical Evaluation of a New Antimetabolite, 6-Mercaptopurine, in the Treatment of Leukemia and Allied Diseases. *Blood* 8:965-99.

Busch-Vishniac, Ilene, Tietjen, Jill and Busch, Lauren. 2025. *Women in the National Inventors Hall of Fame*. Springer Nature, 2025.

Clarke, D, Philips, S et al. 1953. 6-Mercaptopurine Effects in Mouse Sarcoma 180 and in Normal Animals. *Cancer Res* 13: 593-604.

Elion, Gertrude. 1993. The Quest for a Cure. *Annual Rev. of Pharmacology and Toxicology*. 33:1-25. https://www.annualreviews.org/content/journals/10.1146/annurev.pa.33.040193.000245. Accessed Jul 30, 2025.

Elion, G, Furman, P, et al. 1977. Selectivity of Action of an Antiherpetic Agent, 9-(2-Hydroxyethoxymethyl)guanine. *Proc. Nat'l Acad Sci* 74:5716-20.

Larsen, Kristine. Feb 27, 2009. Gertrude Elion, Shalvi/Hyman Encyclopedia of Jewish Women, Jewish Women's Archive. https://jwa.org/encyclopedia/article/elion-gertrude-belle. Accessed on Jul 30, 2025.

National Inventors Hall of Fame, Gertrude Belle Elion: Anti-Leukemia Drugs, https://www.invent.org/inductees/gertrude-belle-elion. Accessed Jul 30, 2025.

The Nobel Prize, Gertrude B. Elion: The Nobel Prize in Physiology or Medicine 1988. https://www.nobelprize.org/stories/women-who-changed-science/gertrude-elion/. Accessed Jul 30, 2025.

Onco Daily. June 2025. Gertrude B. Elion: The Woman Who Rewrote the Rules of Cancer Treatment. https://oncodaily.com/opinion/gertrude-b-elion-the-woman-who-rewrote-315534. Accessed Jul 30, 2025.

Schaeffer, H, Beauchamp, L et al. 1978. 9-(2-H)guanine Activity Against Viruses of the Herpes Group, Nature 272: 583-5.

Science History Institute, (n.d.) George Hitchings and Gertrude Elion. https://www.sciencehistory.org/education/scientific-biographies/george-hitchings-and-gertrude-elion/. Accessed Jul 30, 3035.

Triangle Community Foundation, Gertrude B. Elion Mentored Medical Student Research Award. https://trianglecf.org/award/gertrude-b-elion-mentored-medical-student-research-award/. Accessed Jul 30, 2025.

Wikipedia. Jul 29, 2025. Gertrude B. Elion. https://en.wikipedia.org/wiki/Gertrude_B._Elion. Accessed Jul 30, 2025.

Chapter 16
Mary Ellen Avery

For her discovery of the major cause of respiratory distress syndrome of premature infants and the strategies for treatment and prevention

Keywords Mary Ellen Avery · Mel Avery · Respiratory Distress Syndrome · RDS · Neonatal Respiratory Distress Syndrome · NRDS · Surfactant · Hyaline Membrane Disease · HMD

The sound of a newborn gasping for its first breath is a relief to the new parents. Still, for many premature infants, breathing becomes increasingly difficult due to a condition called respiratory distress syndrome (RDS). Dr. Mary Ellen Avery (Fig. 16.1) was a pioneering pediatrician who dedicated her life to the health of infants, transforming neonatal care in the process. She discovered the cause of RDS and collaborated with other researchers on developing treatments and prevention strategies, substantially reducing the mortality rate of premature infants. For this work, she was awarded the National Medal of Science in 1991.

J. S. Tietjen et al., *Women Laureates 1970 – 2007*, Women in Engineering and Science, https://doi.org/10.1007/978-3-032-23116-1_16

Fig. 16.1 Mary Ellen Avery. (History of Modern Biomedicine Research Group, CC BY-SA 4.0, via Wikimedia Commons)

16.1 Life (1927–2011)

Mary Ellen "Mel" Avery was born in 1927 in Camden, New Jersey, and grew up in Moorestown, New Jersey. Her mother had been a high school vice-principal who quit working after getting married. Her father, who left school in eighth grade, was a cotton goods salesman when Mel, the younger of his two daughters, was born. The family struggled financially during the Great Depression, and times became more difficult when Mel's father lost most of his eyesight from an unknown cause. Concerned about his employability, he took matters into his own hands and borrowed money to start manufacturing cotton goods, particularly canvas products, such as tents and awnings. The business became a success, with a workforce of up to 100 employees. Although Mel did not consider the family wealthy, they were able to move out of their apartment and into a small house, send the girls to private school, and eventually buy a summer home in Maine (Gartner 1998).

Both parents valued education, with her mother having been a teacher and her father wishing he had more formal education. Mel enjoyed reading to her father, whose near blindness left him unable to read. She read the business pages of the newspaper to him so that he could stay informed about the stock market and cotton futures. As a result, Mel developed an interest in the stock market and became a successful investor in her adulthood. Overall, she described her childhood as being filled with love and support (Gartner 1998).

Mel's attraction to a career in medicine began as a teenager. Next-door neighbor and family friend, Dr. Emily Bacon, was a professor of pediatrics at the Woman's Medical College of Pennsylvania (now part of the Drexel University College of Medicine) and the first pediatric specialist in Philadelphia (Shore 2018). Dr. Bacon became a role model for Mel and took her on rounds to see premature babies at Woman's Medical College (Gartner 1998).

Mel felt that her mother was unfulfilled as a homemaker, but of Dr. Bacon, she said,

> I saw her life as more exciting and meaningful than most of the women I knew who were my mother's friends, for example, who were busy doing good works, raising children, and I admired them greatly; but I still thought Emily Bacon had something going for her in terms of reaching out to all children. But mainly, she reached out to me, and I'm eternally grateful. (U.S. National Library of Medicine 2015)

For college, Mel considered women's colleges, stating, "That's what you did in those days" (Gartner 1998). She was nervous about leaving home and going to the big city, so she chose a small rural school where she felt she would be a big fish in a small pond. This was Wheaton College, in Norton, Massachusetts. She majored in chemistry and gained a lot of confidence in her abilities, so that when it was time to apply to medical school, she wanted to go to the best medical school possible. She graduated from Wheaton in 1948 summa cum laude and entered Johns Hopkins Medical School. Hopkins was Dr. Bacon's alma mater and the first school that accepted Mel (Gartner 1998).

There were only four women in her class of 90 at Hopkins, but all the female students lived together in the "hen house" and formed a supportive community. She enjoyed medical school and graduated in 1952. She decided to stay at Hopkins for an internship in pediatrics because she continued to feel wanted there. One reason she chose pediatrics was that she liked the idea of "working with a younger age group where what you do might have a 70-year payoff" (Joint Committee on the Status of Women 2025).

Unfortunately, 1 month into her internship, a routine tuberculosis screening discovered that Dr. Avery had the disease. She was shipped off to a sanatorium in New York and ordered to stay in bed flat on her back for 6 months. Unable to get a satisfactory explanation for the treatment and being asymptomatic, Dr. Avery declined to check into the sanatorium and went home to her parents' house. There, she took the medicine she had been prescribed and stayed in bed for a few months before deciding to head to Europe with a friend. During her 3 months in Europe, she took her medication and made sure not to overdo it, sightseeing for 12 h a day and sleeping the other 12. After the trip, she was able to return to Hopkins and finish her residency there.

Dr. Avery's personal experience with tuberculosis and bed rest fueled her interest in the lungs.

> Every day in bed I thought about LUNGS … and I realized at the time that a lot of medicine was preached with great conviction and with very little scientific basis, so I decided that maybe there was room for somebody to put some underpinnings in there. (Hostetter 2005)

Dr. Avery's desire to work with newborns, particularly premature infants, stemmed from her experiences during her residency, where she saw many premature infants with hyaline membrane disease (HMD), now known as respiratory distress syndrome (RDS), for which the cause, treatment, and cure were unknown. Almost half of the infants with this lung condition died within a few days of birth (Stark 2014).

But before working on respiratory diseases in newborns, Dr. Avery felt that she needed to learn much more about physiology and the mechanics of breathing in infants. She secured a fellowship at the Harvard School of Public Health in Boston, under the guidance of Dr. Jeremiah (Jere) Mead, a leader in the field of respiratory mechanics, and Dr. Clement Smith, who was working with newborns at the Boston Lying-In Hospital (now Brigham and Women's Hospital). Dr. Smith's fellows were doing a study on the Apgar score, a method for evaluating the health of a newborn that must be measured at 1 min and 5 min after birth. Dr. Avery volunteered to perform the Apgar tests on the night shift so that she could be present in the delivery room and witness babies take their first breaths. She observed that some premature infants struggled to reinflate their lungs after exhaling (Marquard 2012). She went to the autopsies of the infants that had died from HMD and noticed that these infants did not have air in their lungs, unlike babies who died from other causes (Gartner 1998). Her breakthrough came when further research led her to the discovery that these infants' lungs lacked pulmonary surfactant, the foamy coating that lines the lungs' air sacs and prevents them from collapsing during exhalation, while facilitating the expansion of the air sacs during inhalation (Martin 2012).

Drs. Avery and Mead published the findings in 1959, but the paper was met with a lackluster response because it went against the prevailing view that HMD was due to an obstruction caused by a glassy membrane seen in the air sacs of babies that died of the disease. Avery and Mead had instead proposed that surfactant deficiency was the primary cause of the babies' respiratory distress, and the formation of the membrane was a secondary effect. Their hypothesis would prove correct and lead to lifesaving treatment (Martin 2012). It also led to the name change from HMD to the more accurate RDS.

When her Harvard fellowship was completed in 1959, Dr. Avery returned to Hopkins. The following year, she became an assistant professor of pediatrics in charge of the newborn nurseries (U.S. National Library of Medicine 2015). In 1969, she went to McGill University in Montreal to become a full professor, the first woman chair of the pediatrics department (and as such, the first woman to chair any department at McGill's medical school), and Physician-in-Chief at McGill's Montreal Children's Hospital. She returned to Boston in 1974, where she served as a professor of pediatrics at Harvard Medical School and as physician-in-chief at Boston Children's Hospital. She beat out 30 men for the physician-in-chief position (Marquard 2012), making her the first woman to head a clinical department at Harvard Medical School (Stark 2014).

In the late 1970s, Dr. Tetsuro Fujiwara of Japan built upon Dr. Avery's research and developed a surfactant replacement to treat RDS. Dr. Avery, in turn, collaborated with Dr. Fujiwara and conducted trials using surfactant replacement in Boston. The success of these trials and other studies resulted in the Food and Drug Administration's (FDA) approval of the treatment in 1990. Dr. Avery also helped advance research and perform clinical trials begun by Drs. Graham Liggins and Ross Howie in the early 1970s in New Zealand. These trials showed that giving steroids to women at risk of preterm birth sped up the lung maturation and surfactant production of the fetus and was an effective preventive measure against premature infants developing RDS.

Dr. Avery stepped down from her position as physician-in-chief at Children's Hospital in Boston in 1985 and devoted more time to global health issues with UNICEF. She traveled the world promoting polio vaccination and oral rehydration therapy to treat dehydration, particularly in cases caused by diarrhea (Stark 2014).

In the mid-1950s, the death toll from RDS was about 10,000 per year in the USA alone, but by 2011, due to Dr. Avery's work, this number had dropped to 700, even though there was a significant increase in premature births (NIH National Heart, Lung, and Blood Institute 2011). At the time of her death in 2012, the *New York Times* estimated that she had saved over 800,000 lives, a number that continues to grow (Martin 2012). She was recognized with many honors, including the E. Mead Johnson Award for pediatric research in 1968 (Wikipedia Contributors 2025), the Trudeau Medal from the American Lung Association in 1984, the National Medal of Science in 1991, and the American Pediatric Society's highest award, the John Howland Medal, in 2005 (Pincock 2012). She was a member of the National Academy of Sciences, the Institute of Medicine, and the American Academy of Arts and Sciences. She was the first female president of the Society for Pediatric Research (1972–1973). She served as president of the American Pediatric Society (1990–1991) and was the first pediatrician to serve as president of the American Association for the Advancement of Science (2004) (Hostetter 2005).

Dr. Avery's work was known around the world. She published over 140 academic papers (Torday 2014) and co-authored seven textbooks (Gartner 1998). Her books, *The Lung and Its Disorders in the Newborn Infant* and *Diseases of the Newborn,* became classics in the field.

In 2011, Dr. Mary Ellen Avery died in West Orange, New Jersey, from complications of Alzheimer's disease. She was 84 (Langer 2012). She left behind adoring nieces and a nephew whom she often took on her world travels. Additionally, she left her mark on colleagues and the more than 75 doctors she mentored, who were inspired by her scientific achievements in pediatric healthcare and her ability to break the glass ceiling in academia (Stark 2014).

16.2 Respiratory Distress Syndrome and Dr. Avery's Discovery

In the uterus, the fetus gets oxygen through the umbilical cord. When the cord is cut at birth, the baby must breathe on its own. Still, in premature births, the lungs may not have developed enough to perform this task, and at the time that Dr. Avery was an intern at Johns Hopkins, nearly half of the newborns with this condition, now known as respiratory distress syndrome (RDS), died within a few days of birth.

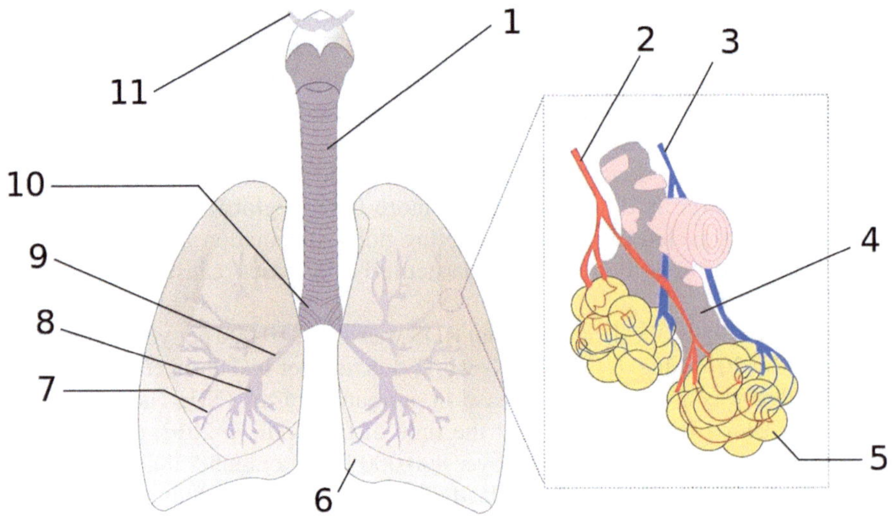

Fig. 16.2 Diagram of lungs and alveoli. (Rastrojo with the biological support of Retama, CC BY-SA 4.0, via Wikimedia Commons). Legend: (1) Trachea, (2) Pulmonary artery, (3) Pulmonary vein, (4) Alveolar duct, (5) Alveoli, (6) Cardiac notch, (7) Bronchioles, (8) Tertiary bronchi, (9) Secondary bronchi, (10) Primary bronchi, (11) Larynx

Figure 16.2 illustrates a healthy lung with inflated alveoli, the tiny air sacs where gas exchange takes place. During breathing, inhaled air fills the alveoli, and oxygen from that air enters the bloodstream. At the same time, the carbon dioxide waste gas moves from the blood into the alveoli and is then exhaled. If the alveoli are not functioning correctly, they will inflate less with each breath and eventually collapse, making it impossible to breathe.

In the 1950s, Dr. John Clements, an army doctor, was researching the effects of nerve gas on lungs and the mechanism by which lungs inflate and deflate during breathing. He studied the alveoli, which like balloons, expand as they fill with air and shrink when air is let out. He observed what he named "pulmonary surfactant," a foamy substance that lines the alveoli. Clements built a device to measure how the surface tension of this surfactant changes during breathing. He found that the surface tension was high when the alveoli were stretched during inhalation and dropped to near zero when the alveoli were small during exhalation. This lower surface tension results in less force resisting inflation, so when the next breath comes in, the alveoli can expand more easily (Torday 2014).

Dr. Avery learned of Clements's work and put two and two together. She hypothesized that infants with RDS had no air in their lungs at death because a lack of pulmonary surfactant led to high surface tension in the alveoli and made breathing too difficult. Dr. Avery met with Clements, who showed her how to build a device to measure surface tension. Dr. Avery and Dr. Mead measured the surface tension in the lungs of babies who died of RDS and confirmed that the tension was abnormally high compared to babies who died from other causes. They published the findings, but they had only proven a correlation between RDS and surface tension, not a causation. Many scientists in the field at the time still believed that the lack of surfactant was a result of RDS, rather than its cause (Stark 2014).

In the late 1970s, Dr. Tetsuro Fujiwara developed a pulmonary surfactant replacement using surfactant taken from cows' lungs. In clinical trials in Japan, premature infants with RDS who received the replacement showed immediate improvement in breathing, thus proving that surfactant deficiency was indeed the cause of RDS, not just a symptom. Dr. Avery met with Fujiwara during a trip to Japan, and, like Clements, the two became friends rather than competitors. Dr. Avery arranged for clinical trials of Fujiwara's surfactant replacement in the U.S. By 1990, surfactant replacement was commercially available, and the mortality rate from RDS, the leading cause of death for premature infants, declined significantly (Stark 2014).

16.3 Conclusion

As a teenager, Dr. Mary Ellen Avery sought to ensure that she had a more fulfilling life than what was typically available to women of her generation. Inspired by her neighbor, pediatrician Dr. Emily Bacon, she also became a pediatrician. She transformed neonatal care when she discovered the cause of respiratory distress syndrome, saving hundreds of thousands of lives. She collaborated with scientists worldwide to promote infant healthcare around the globe. Through her textbooks and mentorship, she influenced generations of pediatricians. Her leadership in the field of pediatrics, dominated by male physicians, inspired other women in medicine. Not only did she more than meet her goal of leading a fulfilling life, but she also allowed parents of many premature infants to breathe easier and enabled their tiny children to survive.

References

Gartner, Lawrence. 1998. "Mary Ellen Avery, MD" Interview by Lawrence M. Gartner. Pediatric History Center, American Academy of Pediatrics, Oral History Project.

Hostetter, Margaret K. 2005. "Introduction of the American Pediatric Society's 2005 John Howland Award Recipient, Mary Ellen Avery, M.D." *Pediatric Research* 58 (6): 1311–16. https://doi.org/10.1203/01.pdr.0000189312.67186.d8.

Joint Committee on the Status of Women, "Oral history interview with Mary Ellen Avery (video and transcript)," *OnView*, https://collections.countway.harvard.edu/onview/items/show/6003. Accessed July 13, 2025.

Langer, Emily. 2012. "Mary Ellen Avery, Pediatrician." The Washington Post. January 20, 2012. https://www.washingtonpost.com/national/health/mary-ellen-avery-pediatrician/2012/01/17/gIQAHLQ9BQ_story.html.

Marquard, Bryan. 2012. "Mary Ellen Avery, 84: Found Key Factor to Save Preemies." *The Boston Globe*, January 11, 2012.

Martin, Douglas. 2012. "Mary Ellen Avery, Premature Babies' Savior, Dies at 84." *The New York Times*, January 12, 2012. http://www.nytimes.com/2012/01/12/health/research/mary-ellen-avery-premature-babies-savior-dies-at-84.html.

NIH National Heart, Lung, and Blood Institute. 2011. "Respiratory Distress Syndrome: Brighter Prospects for Premature." Nih.gov. August 2011. https://www.nhlbi.nih.gov/news/2011/respiratory-distress-syndrome-brighter-prospects-premature-babies.

Pincock, Stephen. 2012. "Mary Ellen Avery." *The Lancet* 379 (9816): 610. https://doi.org/10.1016/s0140-6736(12)60259-1.

Shore, Eleanor G. 2018. "Mary Ellen Avery: Brief Life of a Groundbreaking Neonatologist ." Harvard Magazine. February 7, 2018. https://www.harvardmagazine.com/2018/02/dr-mary-allen-avery.

Stark, Ann R. 2014. "My Tribute to Mary Ellen Avery." *Frontiers in Pediatrics* 2 (June). https://doi.org/10.3389/fped.2014.00050.

Torday, John Steven. 2014. "Mary Ellen Avery's Research Career - Remembrance of Things Past." *Frontiers in Pediatrics* 2 (April). https://doi.org/10.3389/fped.2014.00034.

U.S. National Library of Medicine. 2015. "Changing the Face of Medicine: Dr. Mary Ellen Avery." Nih.gov. 2015. https://cfmedicine.nlm.nih.gov/physicians/biography_17.html.

Wikipedia Contributors. 2025. "Mary Ellen Avery." Wikipedia. Wikimedia Foundation. June 21, 2025.

Chapter 17
Grace Hopper

For her pioneering accomplishments in the development of computer programming languages that simplified computer technology and opened the door to a significantly larger universe of users

Keywords COBOL · FLOW-MATIC · Harvard Mark 1 · UNIVAC · Howard Aiken · J. Presper Eckert · John Mauchly · Charles A. Phillips · Computer compiler · Computer software · Computer bug · Debugging · Standardization · Portability · FORTRAN · Computer science · Computer programming languages · Machine language

"Amazing Grace," as Dr. Grace Hopper (Fig. 17.1) was called with admiration, had a remarkable career in both the military and industry. After more than a decade teaching college mathematics, she joined the Navy during World War II and eventually rose to the rank of Rear Admiral. Her work on early computer systems for the Navy shaped the rest of her life. Dr. Hopper became a visionary in computer science, foreseeing the commonplace use of computers and setting out to help make that vision a reality. She created the first computer compiler and was instrumental in the development of COBOL, which was an English-based programming language that became a standard for business and government applications throughout the world. She was awarded the National Medal of Technology in 1991.

J. S. Tietjen et al., *Women Laureates 1970 – 2007*, Women in Engineering and Science, https://doi.org/10.1007/978-3-032-23116-1_17

Fig. 17.1 Grace Hopper. (James S. Davis, Public domain, via Wikimedia Commons)

17.1 Life (1906–1992)

Grace Hopper was born Grace Brewster Murray in 1906 in New York City. Her father, Walter, was an insurance broker, and her mother, Mary, was a homemaker. The Murrays had three children: Grace was the oldest, followed by her sister, Mary, and brother, Roger (Billings 1989).

Grace's childhood was filled with books and music. She played basketball, field hockey, and water polo. She attended private schools for girls, and summers were spent at the family's cottage in New Hampshire (Billings 1989).

John Van Horne, Grace's maternal grandfather, was a civil engineer, and Grace's mother enjoyed tagging along with him when he performed land surveys. She loved mathematics, but because she was a girl in the 1800s, she was unable to go beyond the study of geometry. Grace also had the opportunity to go on surveying trips with Van Horne, and she, too, loved math (Billings 1989).

While the Murray children were young, their father had both legs amputated due to hardening of the arteries. The Murrays wanted all three children to go to college, and they encouraged their daughters to disregard gender norms. Perhaps because of his precarious health, Walter wanted his daughters to work for at least a year after finishing college to ensure that they could support themselves should something happen to their future husbands (Billings 1989).

When it came time for college, Grace chose Vassar College in Poughkeepsie, New York, although she was not accepted on her first try. She flunked the Latin portion of the entrance exam, but she was admitted the following year, 1924. She graduated in 1928 with a Bachelor of Arts degree in mathematics and physics and went on to earn a master's degree in mathematics from Yale in 1930, thanks in part to a fellowship granted by Vassar (Billings 1989).

In June of 1930, Grace married Vincent Hopper, whom she met one summer in New Hampshire. Vincent had a master's degree from Princeton and was teaching English at New York University. The couple settled in New York City, but the following year, Grace Hopper returned to Vassar to teach math. Their marriage became a commuting marriage, with her husband traveling to Poughkeepsie on the weekends. She remained on the faculty of Vassar from 1931 to 1943. During that period, she also studied for a doctorate in mathematics at Yale, which she received in 1934 (Billings 1989).

By the early 1940s, the Hoppers were separated, and their divorce was finalized in 1945. When the USA entered World War II in December 1941, Dr. Grace Hopper wanted to serve her country by joining the Navy Reserves. However, the Navy rejected her for multiple reasons. First, she was considered too old. Second, as a mathematician, she was deemed more valuable to the war effort as a civilian, and finally, she was underweight. She was able to get the Navy to make an exception for all three of these obstacles and officially joined the Navy Reserve and was placed on active duty in December 1943 (Billings 1989).

In June 1944, Dr. Hopper was assigned by the Navy to the Bureau of Ordnance Computation Project at Harvard University. This project was using the Harvard Mark 1, also known as the Automatic Sequence Controlled Calculator (ASCC), a "calculating machine" which was the brainchild of Howard Aiken of Harvard. This early computer, built by IBM in collaboration with Harvard, could perform long calculations automatically. The massive machine weighed over 5 tons, was 2 feet deep and 8 feet high, and had a 51-foot-long rotary shaft (Wikipedia Contributors 2018). This electromechanical device could perform basic arithmetic as well as compute sines and logarithms. It could complete three addition operations per second, and multiplication took 6 s. Input was done through punching codes onto a paper tape, and output was printed using IBM electric typewriters. Electromechanical counters, called registers, stored numbers during calculations. Dr. Hopper worked under Aiken as one of three programmers, or coders, on the Mark 1, making calculations to create mathematical tables that the Navy used in the design of torpedoes, underwater detection systems, and other equipment (The Collection of Historical Scientific Instruments n.d.).

Although it was Thomas Edison who, in the 1870s, coined the term "bug" for a technical problem with his inventions, Dr. Hopper popularized the term "debugging" when it came to fixing a problem with a computer. In the summer of 1945, Dr. Hopper was working on the Mark II, the successor to the Mark I. One day, the machine malfunctioned, and the cause was found to be a moth in the hardware. Dr. Hopper and the team removed the moth and taped it into the log book with the description, "First actual case of bug being found." From then on, whenever Aiken checked in on the group and found that they were not in the midst of making calculations, he was told they were "debugging the computer."

As the team of coders wrote more programs, they began saving lines of code that performed functions that they often needed. They also borrowed groups of code from each other that they knew worked, so as not to reinvent the wheel.

By 1946, World War II was over, and Dr. Hopper requested a transfer to the regular Navy, but was denied due to her age. She remained in the Navy Reserve, no longer on active duty, while pursuing a civilian career (Foundation for Women Warriors 2020). She was offered a faculty position to return to Vassar, but she felt working with computers would be more fun. She stayed at Harvard as a research fellow with a three-year contract (Pantages 1980), programming the Mark II and then the Mark III (Schneider and Schneider 1998). Aiken began thinking about the uses of computers beyond military applications, particularly in the business sector. He had Dr. Hopper create and run a data processing program for Prudential Insurance Company.

When her contract with Harvard was up, Dr. Hopper decided that she wanted to focus on the commercial sector, so she took a job with Eckert-Mauchly Computer Corporation in Philadelphia. J. Presper Eckert and John Mauchly had built the ENIAC (Electronic Numerical Integrator and Calculator), the first general-purpose computer. The ENIAC was built at the University of Pennsylvania for the US Army during World War II. Eckert and Mauchly also began work on the EDVAC (Electronic Discrete Variable Automatic Computer), which used a binary numerical system instead of decimal, before they left Penn in 1946 to establish their own firm. When Dr. Hopper joined Eckert-Mauchly, the company was preparing to deliver its computer, the Binary Automatic Computer (BINAC), which operated using a binary system, to the Northrop Aircraft Corporation. Dr. Hopper was on the team that went to Northrop to train its employees to use the computer. The BINAC was the first general-purpose electronic computer for commercial use. Unfortunately, it was not a financial success, and Northrop claimed it never really worked. Northrop blamed Eckert-Mauchly, who pointed the finger back at Northrop (McCartney 1999).

Eckert-Mauchly had already begun development of a more advanced computer, the Universal Automatic Computer (UNIVAC) I, but the financial debacle caused by the BINAC forced the sale of the company to Remington Rand in 1950. Eckert and Mauchly stayed on to continue development of the UNIVAC I, which became the first mass-produced commercial electronic computer. Dr. Hopper was part of the team developing code for the UNIVAC I. Following the failure of the BINAC, a great deal was riding on the UNIVAC I.

> We used to say if the UNIVAC I didn't work, we were going to throw it out one side of the factory, which was a junkyard, and we were going to jump out the other side, which was a cemetery! (Billings 1989)

Fortunately, the first UNIVAC I became operational in 1951 and proved to be a success.

During the early days of computers, not everyone predicted their widespread use. However, Dr. Hopper was a visionary who wanted to make programming easier and more accessible to a broader audience. She thought that computers could be programmed to understand commands that used symbols and English words. Dr. Hopper is credited with inventing the compiler. This intermediate program translates commands written in a higher-level language into the low-level machine language, such as binary code, that computers require. Over the years, Dr. Hopper had been collecting bits of code written by her and her peers that were commonly used in their programming. These routines were recorded on tapes and assigned identifiers so that the computer could find them. Dr. Hopper's first compiler, the A-0, was completed in 1952. The A-0 was able to translate code into computer instructions that could access the stored routines, eliminating the need for programmers to write repetitive low-level code by hand.

The industry did not immediately accept the concept of the compiler. Computers were mainly thought of as giant calculators rather than as machines capable of interpreting symbols and human language.

> I had a running compiler and nobody would touch it, because, they carefully told me, computers could only do arithmetic, they could not write programs. It was a selling job to get people to try it. (Centre for Computing History n.d.)

Dr. Hopper and her group at Remington Rand made improvements to the compiler, and version A-2, completed in 1953, was distributed to UNIVAC users. In a historical context, the A-0 is considered to be a "link/loader" because it linked symbolic names to stored subroutines. The A-2 was the first operational compiler. It had significantly more advanced capabilities in translating symbolic instructions into low-level machine code and allowed for user-defined subroutines.

In the mid-1950s, IBM also recognized that expanding the use of computers required advancements in both hardware and software. They developed FORTRAN, which became the dominant scientific and engineering programming language and remained so for years. Because FORTRAN was intended for those with a technical background, it lacked the ease of use and readability that Dr. Hopper felt necessary for business users. Dr. Hopper focused on business applications, such as data processing for UNIVAC's users, and led the development of FLOW-MATIC in the mid-1950s, the first programming language to use English words as commands. She introduced commands such as DIVIDE, MOVE, and REPLACE, whose functions were intuitive and straightforward.

By the late 1950s, an issue was arising in the computer industry: software was being developed that could only run on the specific hardware for which it was designed. For example, FLOW-MATIC worked exclusively on UNIVAC systems. Business leaders and government agencies, particularly the Department of Defense (DoD), sought to avoid being locked into a particular computer manufacturer and advocated for standardization and portability in programming languages. To address the problem, a consortium initially known as the Short-Range Committee on Data Systems Languages, later becoming CODASYL (Conference on Data Systems Languages), was formed in 1959. The goal was to create a universal programming language for business and data processing that could run on different computers.

Dr. Hopper was part of the committee that helped shape the standards for this new language, and she strongly pushed for the use of English-like syntax. Although features were taken from IBM's COMTRAN and RCA's 501 business language, the resulting language, COBOL (Common Business-Oriented Language), was most significantly influenced by Dr. Hopper's FLOW-MATIC (Beyer 2012).

Several companies began developing COBOL compilers to translate COBOL code into machine language. At the end of 1960, Remington Rand (which had since become Sperry Rand and would later become Unisys Corporation) and RCA demonstrated that test COBOL programs running on a UNIVAC and an RCA 501 computer yielded the same results. By 1962, IBM had also adopted COBOL and made it a primary development language. Dr. Hopper and Charles A. Phillips, the first Chairman of CODASYL and at the time employed by the DoD, were two of COBOL's most influential advocates. Although no single person created COBOL, Dr. Hopper made foundational contributions to COBOL's design and helped coordinate its development. She and Phillips both pushed for the widespread adoption of the language by government agencies. In the early 1960s, DoD began using COBOL for its data applications and encouraged private industry to do the same. By the beginning of the 1970s, COBOL had become the most widely used programming language globally (Beyer 2012). Although there were many contributors to COBOL's success, Dr. Hopper became known as the "mother of COBOL," and later, the "grandmother of COBOL."

Dr. Hopper was particularly instrumental in having the Navy adopt COBOL. As a naval reservist, she served as a technical consultant on computing and programming. She was also known as a great communicator and teacher who could explain complex topics in layperson's terms. She taught classes and seminars to naval personnel on computers and data processing. She also taught courses at the University of Pennsylvania and George Washington University.

In 1966, at the age of 60, Dr. Hopper, now a Commander, was informed that she had to retire from the Navy Reserve. She said her retirement day was the saddest of her life. Wisely, the Navy decided they could not do without her and called her back to active duty just 7 months later. Her orders were for a six-month assignment, but when that time period was up, the Navy asked her to stay on indefinitely. She continued working on the proliferation of COBOL in the Navy. She served as the Director of the Navy Programming Languages Group, where she and her group developed COBOL training materials, a certifier to test COBOL compilers, and programs to translate code from one version of COBOL to a newer version.

Dr. Hopper retired from Sperry Rand in 1971, but stayed in the Navy Reserve and continued to advance through the ranks into her 70 s. She became a captain in 1973. In 1983, an act of Congress made her a Commodore (Dickason 2025).

The bill's sponsor, Rep. Philip Crane, said,

> It is time the Navy recognized the outstanding contributions made by this officer recalled from retirement over a decade and a half ago and promote her to the rank of Commodore. (Dickason 2025)

In 1985, Dr. Hopper was promoted to the rank of Rear Admiral (Lower Half). The following year, at age 79, she was the oldest active-duty commissioned officer, and the Navy forced her to retire again ("Remembering Grace Murray Hopper: A Legend in Her Own Time" 2025). This Navy retirement stuck, but Dr. Hopper wanted to continue working, so she became a consultant in public relations for Digital Equipment Corporation, giving lectures around the country. One point that she liked to stress was the importance of change, saying, "The most damaging phrase in the language is 'We've always done it this way'" (Dickason 2025).

She believed in fighting resistance to change. Her most famous and widely used quote drives this point home: "It's always easier to ask forgiveness than it is to get permission (Dickason 2025)".

Dr. Hopper died in 1992 at age 85. Throughout the years, she collected many honors. Her Navy service medals included the Legion of Merit in 1967, the Meritorious Service Medal in 1980, and the Defense Distinguished Service Medal in 1986 (Wikipedia Contributors 2019). Hopper Hall, the Naval Academy's Center for Cyber Security Studies, is the first academic building named after a woman at one of the primary US service academies (Skidmore et al. 2025), and the USS Hopper Navy destroyer was launched in 1996.

Her many civilian honors include: the Society of Women Engineers Achievement Award, the society's highest honor, in 1964; the first-ever Data Processing Management Association Man of the Year award in 1969, election to the National Academy of Engineering, in 1973; the first woman and first American to be made a Distinguished Fellow of the British Computer Society, in 1973; the National Medal of Technology, in 1991; named a Fellow of the American Academy of Arts and Sciences, in 1991; induction into the National Women's Hall of Fame, in 1994; and the Presidential Medal of Freedom, awarded posthumously in 2016 (Wikipedia Contributors 2019).

Dr. Hopper was awarded 40 honorary degrees from universities worldwide, and numerous scholarships and awards have been named in her honor (Yale University Office of the President 2017).

17.2 Machine Code and COBOL

When Dr. Hopper first started programming on the Harvard Mark 1, she had to write code in a binary format and then transfer the code to paper tape by punching holes to represent ones and zeros. Each line of code had three sections of eight digits each. A simple mathematical operation is illustrated in Fig. 17.2, with the three lines of code highlighted.

The left-most section indicates where the result of the command was to be stored, the middle section indicates where the input was stored, and the right-most section tells the computer what operation to perform.

Fig. 17.2 Addition example for the Mark 1

	Result	Source	Operation
Line 1	00000011	01000001	00000001
Line 2	00000011	01000010	00000010
Line 3	00000011	00000000	00000100

The operations are defined as follows:

00000001—Move
00000010—Add
00000100—Print

Line 1 instructs the computer to take the number stored in computer register 65 (01000001 in binary = 65) and move it to storage counter 3 (00000011 in binary = 3).

Line 2 instructs the computer to take the number stored in computer register 66 (01000010 in binary = 66) and add it to the number in storage counter 3.

Finally, line 3 instructs the computer to print the number in storage counter 3. (The middle section is not used for this instruction.)

As an example, to add 42 and 15, the Mark 1 operator would manually set the computer register 65 to the number 42 and computer register 66 to the number 15 by turning the mechanical dials.

After the first line of code is executed, storage counter three contains the number 42.

After the second line of code is executed, storage counter three contains the number 57.

Line 3 prints the number 57 on the attached IBM typewriter.

This type of programming was subject to a lot of errors, not at all intuitive, and therefore, only accessible to highly skilled mathematicians, engineers, and scientists. There were many years and developmental steps in between, such as the creation of compilers and languages like FLOW-MATIC, before the first version of COBOL was introduced in 1960. COBOL was highly structured and easier to read and understand. The portions of a COBOL program that define the variables and the logic required to add 42 and 15 in COBOL follow.

The statements shown in Fig. 17.3 define the names and type of three variables, NUM1, which is set to 42, NUM2, which is set to 15, and RESULT-NUM, which is not assigned a value. PIC and VALUE are words defined by COBOL. PIC defines the variable type, with 9 indicating a numeric variable, and the number in

Fig. 17.3 Variable declaration example in COBOL

```
NUM1        PIC 9(02)  VALUE 42.
NUM2        PIC 9(02)  VALUE 15.
01  RESULT-NUM      PIC 9(03).
```

Fig. 17.4 Example in COBOL to add two numbers

```
ADD NUM1 TO NUM2 GIVING RESULT-NUM.
DISPLAY RESULT-NUM.
```

Fig. 17.5 COBOL example of adding two numbers with algebraic symbol

```
COMPUTE RESULT-NUM = NUM1 + NUM2.
```

parentheses is the number of digits allowed. In the example, PIC 9(02) defines NUM1 and NUM2 as two-digit numbers, while RESULT-NUM is defined as a three-digit number to allow for results that exceed two digits. VALUE initializes the variable.

The code to add the numbers and print the result is shown in Fig. 17.4, where ADD, TO, GIVING, and DISPLAY are words defined by COBOL. The code reads more like a sentence and instructs the computer to add NUM1 to NUM2, give the result to RESULT-NUM, and then display the result.

COBOL was known for being verbose, but programmers also had the flexibility to write the command to add the numbers with algebraic symbols, as shown in Fig. 17.5.

COMPUTE is a reserved word in COBOL, telling the computer that the line of code is an algebraic function (Conference on Data Systems Languages 1960).

By the 1970s, COBOL was the dominant programming language for business computing. A 1997 report by the Gartner Group estimated that 80% of businesses worldwide used COBOL, with more than 200 billion lines of code in existence and an estimated 5 billion lines of new code being written annually (Robinson 2009).

COBOL has continued to be updated and modernized over the years, but other languages, such as Java, C#, and Python, have become the favored languages for new systems, and COBOL is rarely taught in today's computer programming classes. Although some legacy systems have been converted to other languages, this is an expensive and potentially risky process. In 2017, Reuters found that 43% of banking systems are built on COBOL and 95% percent of ATM swipes rely on COBOL code. In total, they estimated that $3 trillion in daily commerce used COBOL systems (Irrera 2017). In addition, there are many government systems still written in COBOL. In the USA, the Social Security Administration still maintains 60 million lines of COBOL code. Other US government agencies that still rely heavily on COBOL include the Department of the Treasury, the Department of Education, and the Department of Health and Human Services. 45 of the 50 states in the USA are also still running COBOL systems (Vaughan-Nichols 2025).

17.3 Conclusion

Dr. Grace Hopper was a pioneer in computer programming whose work broadened the view of what computers could do and who could use them. Her work on FLOW-MATIC and COBOL helped democratize computing by making it more accessible and extending its use to the business world. She championed software standardization and education, influencing the development of computer languages and generations of computer programmers. Hundreds of billions of lines of computer code were written in COBOL as its use proliferated around the world. Rear Admiral Hopper was perhaps most proud of her military service, which spanned 43 years and resulted in a namesake destroyer. She was beloved in the Navy, in industry, and as a teacher, and left an indelible mark on both computer science and the military.

References

Beyer, Kurt. 2012. *Grace Hopper and the Invention of the Information Age*. Cambridge, Mass.; London: MIT Press.

Billings, Charlene W. 1989. *Grace Hopper: Navy Admiral and Computer Pioneer*. Hillside, N.J., U.S.A.: Enslow Publishers.

Centre for Computing History. n.d. "Grace Hopper Completes the A-0 Compiler - Event - Computing History." Www.computinghistory.org.uk. https://www.computinghistory.org.uk/det/5487/Grace-Hopper-completes-the-A-0-Compiler/. Accessed September 4, 2025.

Conference on Data Systems Languages. 1960. "COBOL Report: Report to the Conference on Data System Languages." Washington, D.C.: U.S. Department of Defense.

Dickason, Elizabeth. 2025. "Remembering Grace Murray Hopper: A Legend in Her Own Time." Navy.mil. 2025. https://www.history.navy.mil/content/history/nhhc/research/library/bibliographies/hopper-grace-admiral-select-bibliography/remembering-grace-m-hopper-legend-her-own-time.html/.

Foundation for Women Warriors. 2020. "Grace Murray Hopper, Rear Admiral, U.S. Navy - Foundation for Women Warriors." March 8, 2020. https://foundationforwomenwarriors.org/grace-hopper/.

Irrera, Anna. 2017. "Banks Scramble to Fix Old Systems as IT 'Cowboys' Ride into Sunset." *Reuters*, April 10, 2017. https://www.reuters.com/article/technology/banks-scramble-to-fix-old-systems-as-it-cowboys-ride-into-sunset-idUSKBN17C0CN/.

McCartney, Scott. 1999. *ENIAC, the Triumphs and Tragedies of the World's First Computer*. New York: Walker.
Pantages, Angeline. 1980. "Oral History of Captain Grace Hopper." https://archive.computerhistory.org/resources/text/Oral_History/Hopper_Grace/102702026.05.01.pdf.
Robinson, Brian. 2009. "Cobol Remains Old Standby at Agencies despite Showing Its Age -- FCW." *FCW the Business of Federal Technology*, July 9, 2009. https://web.archive.org/web/20140427010019/http:/fcw.com/Articles/2009/07/13/TECH-COBOL-turns-50.aspx.
Schneider, Carl J, and Dorothy Schneider. 1998. *Grace Murray Hopper*. Las Cruces, NM: Software Press.
Skidmore,Owings&Merrill.2025."U.S.NavalAcademy–HopperHall."SOM.August7,2025.https://www.som.com/projects/u-s-naval-academy-grace-hopper-center-for-cyber-security-studies/.
The Collection of Historical Scientific Instruments. n.d. "Harvard IBM Mark I - Use." Chsi.harvard.edu. https://chsi.harvard.edu/harvard-ibm-mark-1-use. Accessed September 4, 2025.
Vaughan-Nichols, Steven. 2025. "If COBOL Is so Problematic, Why Does the US Government Still Use It?" ZDNET. 2025. https://www.zdnet.com/article/if-cobol-is-so-problematic-why-does-the-us-government-still-use-it/.
Wikipedia Contributors. 2018. "Harvard Mark I." Wikipedia. Wikimedia Foundation. 2018. https://en.wikipedia.org/wiki/Harvard_Mark_I.
Wikipedia Contributors. 2019. "Grace Hopper." Wikipedia. Wikimedia Foundation. 2019. https://en.wikipedia.org/wiki/Grace_Hopper.
Yale University Office of the President. 2017. "Biography of Grace Murray Hopper." February 10, 2017. https://president.yale.edu/biography-grace-murray-hopper#_edn3.

Chapter 18
Maxine F. Singer

For her outstanding scientific accomplishments and her deep concern for the societal responsibility of the scientist

Keywords Nucleotide triplets · LINEs · Carnegie Institution of Washington · National Cancer Institute · First Light · Carnegie Academy for Science Education · Protein coding · Gordon Conference · Asilomar Conference · Paul Berg · Women in science · Magellan telescopes · Capital Science Evening Lecture · Chromatin · Gene splicing · National Medal of Science · Chromatin

Dr. Maxine F. Singer (Fig. 18.1) was one of a rare group of scientists who excel not only in their chosen technical field but also as a scientific enterprise manager and as an advocate for social responsibility and women in science. Dr. Singer is known for her studies identifying the importance of triplet strings of nucleotides in RNA, enabling each to be identified with the protein it codes for. She also discovered long sequences of nucleotides in DNA that can jump around and showed that these sequences can cause mutations as they migrate. Dr. Singer served as the President of the Carnegie Institution of Washington and in this role created programs to promote science and technical careers and to improve education. She was also a leader in the movement to have scientists consider the social and ethical ramifications of DNA editing, ultimately leading to a voluntary pause on recombinant DNA research in 1975. Dr. Singer received the National Medal of Science in 1992.

J. S. Tietjen et al., *Women Laureates 1970 – 2007*, Women in Engineering and Science, https://doi.org/10.1007/978-3-032-23116-1_18

Fig. 18.1 Dr. Maxine F. Singer. (Courtesy of the Library of Congress)

18.1 Life (1931–2024)

Maxine Frank was born in New York City on February 15, 1931, to first-generation American parents Henrietta (Perlowitz) and Hyman Frank. Her father worked as a lawyer, and her mother was a homemaker. Maxine attended public schools in Brooklyn, graduating from Midwood High School in 1948. Along the way, she was inspired by a chemistry teacher, "I had one marvelous chemistry teacher in high school. She was an exciting teacher, interested in me because I was interested in what she taught, and very demanding" (Moyers 1988).

Maxine entered Swarthmore College to pursue a degree in chemistry. She had become interested in the interface between chemistry and biology, but at the time, Swarthmore only permitted chemistry majors to minor in physics or mathematics. Maxine appealed and ultimately was given special permission to minor in biology.

At Swarthmore, Maxine became friends with five other women who were excellent in science, noting that none of the men in her classes "were any better than these six women" (Gellene 2024). Maxine has credited these women with keeping her going in pursuit of a scientific career. Upon graduation, Maxine earned a National Science Foundation (NSF) Fellowship to Yale University. Four of the remaining group of five friends similarly earned NSF Fellowships (accounting for over 15% of the fellowships awarded to women that year), and the fifth went to medical school (Speaker 2024).

Maxine married Daniel Singer immediately after they both graduated from Swarthmore. The Singers traveled to Yale together, with Maxine entering the biochemistry department to earn a PhD while Daniel pursued a law degree. At Yale, Maxine studied protein chemistry and became interested in nucleic acids. It had been shown only a decade earlier that DNA was the basis for our genetic makeup, and the helical structure of DNA was revealed while she was a graduate student.

In 1956, Singer went to the National Institutes of Health (NIH) to work with Dr. Leon Heppel, one of the few scientists working on nucleic acids. She hadn't quite completed her PhD degree but was nonetheless hired as a postdoctoral fellow at the National Institute for Arthritis and Metabolic Diseases (now the National Institute for Arthritis and Musculoskeletal and Skin Diseases). Singer completed her PhD in 1957 and was offered and accepted a staff position at NIH (Cave 2024). Between 1959 and 1964, she was "essentially pregnant all the time" but recalled encountering no objection to her continuing scientific endeavors (Ferry 2024). The Singers welcomed four children during this time: Ellen, Amy, David, and Stephanie.

In the early 1960s, Dr. Singer continued to work with Dr. Heppel on the role of enzymes that regulate the synthesis of nucleic acids. Their work led to the identification of what nucleic acid sequences code for the twenty amino acids (protein building blocks) found in humans. Dr. Singer enjoyed her work and found it rewarding, but she was having trouble recruiting postdoctoral fellows to work with her. Apparently, applicants feared that working with a woman would ruin their future career prospects. This was Dr. Singer's first encounter with serious gender bias, and it propelled her to advocate for women in science (Speaker 2024). Her advocacy much later included a guest editorial in *Science,* which was based on a report issued by the National Academy of Sciences' Committee on Science, Engineering and Public Policy, which she had chaired. The report outlined persistent cultural, professional, and institutional obstacles to the advancement of women in science, engineering, and math. The editorial argued that universities should encourage women to pursue science, mathematics, and engineering careers instead of wasting their talents (Singer 1991).

Dr. Singer's research flourished and expanded in the 1970s, but as she considered the exponentially growing work on recombinant DNA and RNA, i.e., organisms that use DNA or RNA from multiple sources, she grew concerned that it might have potentially hazardous or unethical health and environmental consequences. In June 1973, Dr. Singer co-chaired the Gordon Conference on Nucleic Acids held in New Hampshire, aimed at discussing the risks of gene editing. After the meeting, Dr. Singer, representing the attendees, wrote to the chair of the National Academy of Sciences suggesting that a committee be formed to investigate further. She also recommended that the committee appoint Dr. Paul Berg as its chair. This led directly to the Asilomar Conference on recombinant DNA in 1975, organized largely by Drs. Singer and Berg (Cave 2024).

The Asilomar Conference attracted over 150 scientists from thirteen countries, a few lawyers, and sixteen members of the press. It produced a statement seeking the establishment of guidelines for recombinant DNA work. The conference categorized and assessed each type of recombinant DNA experiment they could envision, assigning it a risk level. They agreed that research with low levels of risk should continue, but some others should require strict containment or should be banned outright. The conference produced an agreement to voluntarily pause research on recombinant DNA using antibiotic-resistant strains or cancer-causing viruses until guidelines were established. This was the first voluntary moratorium in the history of science (Speaker 2024; Jewish Women's Archive 2009; Gellene 2024). The report on the meeting appeared in *Nature* (Asilomar Conference on DNA Recombinant molecules 1975). Bans on research have eased as the technology and its consequences have become better understood.

Following these two meetings, Dr. Singer served as an advisor to NIH Director Donald Fredrickson on the development of guidelines for recombinant DNA research. By 1976, rules were issued by NIH. These resulted in pushback from environmental activists who felt that the rules were not strict enough, as they relied on scientists to self-govern. Dr. Singer was a staunch defender of scientists being self-regulated and participated in many public hearings and debates on the topic, including testimony to Congress (Speaker 2024; Cave 2024).

From the mid-1970s onward, Dr. Singer held leadership positions in various scientific organizations. In 1975, she was made chief of the Nucleic Acid Enzymology Section of the National Cancer Institute's Laboratory of Biochemistry. From 1980 to 1987, she was the head of that Laboratory of Biochemistry. There she worked to understand sequences of nucleotides that seemed able to move and the impact that has on diseases, in addition to managing a large scientific operation.

In 1988, Dr. Singer became a scientist emeritus at the National Cancer Institute and continued working there 2 days a week while serving as President of the Carnegie Institution of Washington. She held these two positions until her retirement in 2002. The Carnegie Institution was established in 1902 by Andrew Carnegie to advance scientific knowledge. It has historically supported work in astronomy, biology, and earth and planetary sciences. In her role, Dr. Singer oversaw and financed the construction of twin 6.5 m Magellan telescopes in Chile. She created the first new department at the Carnegie Institution in over 80 years—a department of global ecology (Speaker 2024; Isaacs and Mulchaey 2024).

Dr. Singer also created a series of outreach and education programs during her term as President. Her aim was to explain that "Science is not an inhuman or superhuman activity. It's something that humans invented, and it speaks to one of our great needs—to understand the world around us" (Isaacs and Mulchaey 2024). Among these programs was First Light, aimed to bring third- to eighth-grade students and teachers in the area to weekend science immersion programs. Shown in Fig. 18.2 is Dr. Singer interacting with some of the First Light students. The success of the First Light program led to its expansion as the Carnegie Academy for Science

Fig. 18.2 Dr. Singer interacting with students at Carnegie Institution's First Light program. (Courtesy of Carnegie Science)

Education (CASE). These programs served thousands of teachers and students in the Washington, DC, area. Dr. Singer also created the Capital Science Evening Lecture, inviting members of the community in to learn about science research at the Institution. That program was in its 35th season at the time of Dr. Singer's death (Cave 2024; Isaacs and Mulchaey 2024; Speaker 2024).

Dr. Singer's educational outreach was not limited to programs established while President of the Carnegie Institution of Washington. She published three books with Dr. Paul Berg, one with Dr. Robert Hazen, and one on her own: *Genes and Genomes* (Singer and Berg 1991), *Dealing with Genes* (Berg and Singer 1993), *Why Aren't Black Holes Black?* (Hazen and Singer 1997), *George Beadle: An Uncommon Farmer* (Berg and Singer 2003), and *Blossoms and the Genes That Make Them* (Singer 2018). These vary from textbooks to overviews for lay audiences.

Dr. Singer also served on a variety of boards. She was a Fellow (trustee) of the Yale Corporation from 1975 to 1990, a member of the Board of Governors and

Science Advisory Council of the Weizmann Institute of Science (Israel) from 1978 to 2024, Chair of the Editorial Board of Proceedings of the National Academy of Sciences from 1985 to 1988, Member of the Board of Directors of Johnson and Johnson from 1975 to 1990, and a director of the Whitehead Institute from 1985 to 1994 (Speaker 2024; Jewish Women's Archive 2009).

Dr. Singer received numerous medals and awards for her service, including election to the American Academy of Arts and Sciences in 1978, election to the National Academy of Sciences in 1979, appointment to the Pontifical Academy of Science in 1986, and election to the American Philosophical Society in 1990. She was awarded the Distinguished Presidential Rank Award, the highest award for government service, by President Reagan in 1988; the National Medal of Science in 1992; the American Association for the Advancement of Science William D. Carey Lectureship Award for Leadership in Science Policy in 1996; the National Academy of Sciences Public Welfare Medal in 2007; and the 2008 Public Service Award of the American Society for Cell Biology. In 1999, she was the first woman to receive the National Science Board's Vannevar Bush Award for public service to the nation (Isaacs and Mulchaey 2024; Cave 2024; Spradling and Halpern 2024). She received numerous honorary degrees from universities, including Harvard and Brandeis.

Dr. Singer died in 2024 of COPD and emphysema. She had published over 100 scientific papers. In her final years, looking back, she said, "I've been part of an extraordinary time in biology. There hasn't been a day when I've wanted to do anything else" (Moyers 1988). A charcoal drawing of Dr. Singer hangs in the National Portrait Gallery in Washington, DC.

18.2 Triplets, LINEs, and Chromatin

Dr. Singer's first significant research result stemmed from work she conducted with Dr. Heppel while she was a postdoctoral fellow. The two of them worked with the enzyme polynucleotide phosphorylase, which allowed them to join random nucleotides into RNA (single-stranded) sequences. Dr. Singer created synthetic RNA of known nucleotide chains in this manner and shared them with Drs. Nirenberg and Matthaei, also of NIH, who created a database of the sequences and what resulted from them. In this manner, they were able to match the twenty amino acids to the RNA sequences that code for them. For this work, Dr. Nirenberg was awarded a Nobel Prize in Physiology or Medicine in 1968. In addition to the matching of coding to specific sequences, was the observation by Dr. Singer that three nucleotides in a row, a triplet, seemed to be the important unit (Wikipedia 2025).

Subsequent to the work on identifying what codes for proteins, Dr. Singer worked on isolating and manipulating fragments of DNA. At one point, she even combined fragments from two different species to see what would happen. It was while considering this work that she grew concerned that there might be consequences that were not being considered in such work, and she abandoned this set of experiments (Cave 2024).

Dr. Singer turned to consideration of long sequences in DNA that seemed to occur very often but that moved about in location in the RNA and DNA. These are called LINEs for long, interspersed, nucleotide elements. Dr. Singer specifically focused on a common LINE, which was labeled LINE-1, and found that insertion could induce mutations in nearby genes. Such mutations could explain some genetic diseases. For instance, hemophilia can be caused by insertion of LINE-1 into the gene coding for blood clotting (Speaker 2024).

Dr. Singer later conducted important research on chromatin, a complex structure of DNA and proteins that makes up the chromosomes that are associated with our genetic traits. Long strands of DNA wrap around proteins called histones, forming bead-like masses called nucleosomes. These nucleosomes, appearing like beads on a string, fold and wrap further into what is called a chromatin fiber. Dr. Singer studied the histone H1, one of the simpler proteins that appears in chromatin. She showed that it plays a role in causing the DNA to tightly coil so that it will fit into the nucleus of cells (Wikipedia 2025).

18.3 Conclusion

Dr. Maxine Singer was a renowned biochemist, scientific enterprise manager, and advocate. She identified triplets of nucleotides as the key unit that codes for protein synthesis. She also showed that some long sequences of DNA can move in location and that this can induce mutations in nearby nucleotide sequences. Dr. Singer was a strong advocate for science and public policy, serving as a leading voice in the development of rules to guide the ethical and safe research of recombinant DNA. She also advocated for science literacy and for equal access for women in science. Dr. Singer received the National Medal of Science in 1992.

References

Asilomar conference on DNA recombinant Molecules. 1975. *Nature* 255:442-4

Berg, Paul and Singer, Maxine. 1993. *Dealing with Genes: The Language of Heredity*. University Science Books.

Berg, Paul and Singer, Maxine. 2003. *George Beadle: An Uncommon FarmerThe Emergence of Genetics in the 20th Century*. Cold Spring Harbor Laboratory Press.

Cave, Abigail. 2024. Maxine Frank Singer (1931-2024), Embryo Project Encyclopedia, https://embryo.asu.edu/pages/maxine-frank-singer-1931-2024. Accessed 12 Aug 2025.

Ferry, Georgina. 02 Aug 2024. Maxine Singer obituary: biologist who shaped genetic engineering and fought discrimination. *Nature* 632:252.

Gellene, Denise. 10 Jul 2024. Maxine Singer, Guiding Force at the Dawn of Biotechnology, Dies at 93. *NY Times*. https://www.nytimes.com/2024/07/10/science/maxine-singer-dead.html. Accessed 12 Aug 2024.

Hazen, Robert and Singer, Maxine. 1997. *Why Aren't Black Holes Black?* Anchor.

Isaacs, Eric and Mulchaey, John. 2024. Maxine Singer, renowned biologist and advocate for STEM inclusion, dies at 93. *Carnegie Science*. https://carnegiescience.edu/news/maxine-singer-renowned-biologist-and-advocate-stem-inclusion-dies-93. Accessed 12 Aug 2025.

Jewish Women's Archive (20 Mar 2009). Maxine Singer, Halvi/Hyman Encyclopedia of Jewish Women. https://jwa.org/encyclopedia/article/singer-maxine. Accessed 12 Aug 2025.
Moyers, Bill. 1988. Bill Moyers Interview of Maxine Singer, National Library of Science. https://profiles.nlm.nih.gov/catalog/nlm:nlmuid-101584644X210-doc. Accessed 12 Aug 2025.
Singer, Maxine and Berg, Paul. 1991. *Genes and Genomes: A Changing Perspective*. University Science Books.
Singer, Maxine. 2018. *Blossoms and the Genes that Make Them*. Oxford University Press.
Singer, Maxine. 1991. Heroines and Role Models. *Science* 253:249.
Speaker, Susan. 7 Mar 2024. Maxine Singer—A Life in Research and Advocacy, National Library of Medicine. https://circulatingnow.nlm.nih.gov/2024/03/07/maxine-singer-a-life-in-research-and advocacy/#:~:text=During%20a%20career%20spanning%20nearly,the%20chemical%20elements%20of%20heredity. Accessed 12 Aug 2025.
Spradling, Allan and Halpern, Marnie. 12 Sep 2024. Maxine F. Singer (1931-2024). *Science* 385(6714):1169.
Wikipedia. 17 Jul 2025. Maxine Singer, https://en.wikipedia.org/wiki/Maxine_Singer. Accessed 12 Aug 2025.

Chapter 19
Eleanor Gibson

For her conceptual insights in developing a theory of perceptual learning; and for achieving adeeper understanding of perceptual development in children and basic processes in reading

Keywords Visual Cliff · Eleanor Gibson · James Gibson · Depth perception · Perceptual learning · Perceptual development · Clark Hull · Richard Walk · Psychology · Psychological experiments

"Have you got a crawling baby? If so, your baby can earn three dollars in an experiment" (Szokolszky 2003). That is the ad used to recruit babies for psychologist Dr. Eleanor Gibson's (Fig. 19.1) famous "visual cliff" experiment, which demonstrated how babies perceive and learn about depth. In her words, "We perceive to learn, as well as learn to perceive" (Baranauckas 2003). She spent her six-decade career studying how we learn, making lasting contributions to the fields of perceptual development and reading. She was awarded the National Medal of Science in 1992.

J. S. Tietjen et al., *Women Laureates 1970 – 2007*, Women in Engineering and Science, https://doi.org/10.1007/978-3-032-23116-1_19

Fig. 19.1 Eleanor Gibson. (Courtesy of the Division of Rare and Manuscript Collections, Cornell University Library)

19.1 Life (1910–2002)

Eleanor "Jackie" Gibson, nee Jack (1910–2002), grew up in Peoria, Illinois, and described her childhood as ordinary (Eppler 1998). Her mother was a housewife, and her father was in the wholesale hardware business. Eleanor's mother and some aunts and cousins attended Smith College, an all-women's school. When Eleanor graduated from high school at 16, she also enrolled at Smith.

Gibson had nothing but praise for her experience at Smith,

> Everyone there was really determined that any girl who had ambitions or potential was going to get a good chance to fulfill it, you know. They helped in every way. (Eppler 1998)

Gibson initially thought that she would major in foreign languages, but the Great Depression had arrived, and she could not afford the required junior year abroad (Gibson and Linn 2001). Having enjoyed her psychology classes, she chose that major instead. In her junior year, she met her future husband, James Gibson, an assistant professor of psychology at Smith, at a college event. She was immediately smitten and enrolled in Professor Gibson's course in advanced experimental psychology.

Eleanor considered taking the class in experimental psychology to be a turning point in her life (Gibson and Linn 2001). Not only did she get to know James Gibson better, but she also discovered that her calling lay in doing psychological experiments. Eleanor graduated from Smith in 1931 but stayed on as a teaching assistant. In 1932, she also began working on a master's thesis at Smith, focusing on verbal memory under the guidance of Professor James Gibson. The two married that summer. After completing her MA, Eleanor was promoted to an instructor position at Smith.

Since Smith did not offer a PhD in psychology, Gibson had to look elsewhere to continue her education. After 2 years as an instructor at Smith, Gibson was accepted to Yale for doctoral studies. She was not given any financial aid, but fortunately, she was awarded a scholarship from Smith that she could use at Yale, which covered tuition with a bit of money left over. She headed to Yale without her husband, who was able to join her later for her second semester while he was on a sabbatical.

Although Yale had admitted women to PhD programs since 1892 (Castellanos-Monfil 2016), Gibson found many contrasts between Yale and Smith. "Smith, a women's college, was a place where women were not only permitted to be scholars, but encouraged, even in the sciences. (Gibson and Linn 2001)," she said.

Not every door was open to her at Yale. Laboratory work was not considered feminine, and within the field of psychology at the time, women were expected to go into developmental and applied psychology (Rodkey 2011). Gibson wanted to work with Professor Yerkes in his chimpanzee lab, studying comparative psychology, but she soon discovered that the professor did not allow women to work in his lab. Gibson ended up with Professor Clark Hull, who was studying learning theory by experimenting with rats. Hull cared more that his students follow his strict research methods than about their gender.

Gibson only studied on campus at Yale for a year, during which she secured Hull as her advisor and took the necessary coursework and preliminary exams for her degree. Back at Smith, Gibson continued on as an instructor while also doing research for her Yale dissertation. Gibson received her PhD from Yale in 1938. The Gibsons were now ready to start a family. Their son, James "Jerry" Gibson, Jr, was born in 1940. Dr. Eleanor Gibson requested maternity leave for a semester, to which the president of Smith agreed, conditioned on approval from the psychology department chair, who so happened to be her husband, Dr. James Gibson. Dr. Eleanor Gibson returned to teaching at Smith after her leave, and both Gibsons remained at Smith until World War II.

In 1941, James Gibson was recruited by the Army to develop visual perception aptitude tests to screen potential pilots. During the next 4 years, Dr. Eleanor Gibson put her career on hold and went with her husband to Army bases in Texas and California. The Gibsons took the opportunity to expand their family, and their daughter Jean was born in 1943. After the four-year absence, Smith welcomed both Gibsons back to the faculty.

After a few years, Dr. James Gibson decided that he needed to leave Smith for an institution where he would have doctoral students to help further his research. James Gibson became a full professor at Cornell, which had a nepotism rule that forbade spouses on the faculty. Dr. Eleanor Gibson was offered a position as a research associate at Cornell's behavioral research farm, working with sheep and goats. This gave her the freedom and time to conduct research without having to teach classes, and she enjoyed working with the animals. However, she left the lab after the baby goats that she was using were given away as Easter presents by the farm manager during spring break.

At the farm, she had become interested in animal rearing and early development, so she moved to a lab where rats were kept and received a grant to compare the effects of rearing rats in the dark versus rearing them in the light. During this time, Dr. Eleanor Gibson collaborated with fellow Cornell psychologist Richard Walk to devise her most famous experiment, known as the "visual cliff," to study the development of depth perception. Drs. Gibson and Walk discovered that, regardless of whether the rats were reared in darkness or light, they would avoid walking over what appeared to be a cliff, indicating that depth perception in rats is an innate ability rather than one learned through visual experience. Experiments using the visual cliff with other animals and human babies have confirmed that depth perception, which is necessary for survival, develops before an animal becomes mobile.

In 1959, Gibson began working with a team funded by the government to study the reading process, an area in which she worked for 12 years. This fit well with her interest in perceptual learning. From her experiments, she concluded that children learn to recognize letters and words through visual perception of the differences between the written letters. They use perceptual learning to distinguish and differentiate patterns as opposed to the prevailing theory of the time, which asserted that reading skills were acquired through stimulus-response, where letters were associated with sounds.

When James Gibson received a fellowship from the National Institutes of Health (NIH), it meant that NIH, instead of Cornell, would pay his salary. This had significant consequences for Dr. Eleanor Gibson because Cornell decided that their nepotism rule no longer applied, and she was offered a faculty position as a full professor. This was in 1966, after 16 years of being a research associate. Perhaps most importantly, she was given a lab of her own. Finally, she did not have to adapt to the research priorities of someone else's lab to continue working. Reflecting on her career before reaching this milestone, she said,

> Couldn't I just set out to work on the research of my choice? No, I couldn't. One needs a lab, and I didn't have one. (Rodkey 2011)

In 1969, Dr. Gibson presented her theory in her book, *Principles of Perceptual Learning and Development*, that perceptual development occurs through learning to distinguish differences in objects, sensory information, and one's environment through experience and practice. The book became a classic in the field. Her book on reading, *The Psychology of Reading,* published in 1975, became another influential classic on learning.

In 1979, James Gibson died. Dr. Eleanor Gibson retired from her faculty position at Cornell that year but remained at her lab until 1987.

Although the Gibsons collaborated on occasion, Dr. Eleanor Gibson pointed out,

> We were never a husband-and-wife team, nor was I a long-suffering wife who gave up a teaching career in order to further her husband's, or to devote herself primarily to tending to her family. (Gibson and Linn 2001)

> We began both our married and our professional lives respecting each other's choices and the need for freedom to think on our own. The conclusion is that a woman can indeed have both a family and a career, if her companion is the right kind of guy. She should not worry about making concessions to his career if she has her own thoughts and ambitions clearly in mind. Her time comes, too, and the rewards for all can be great. (Gibson and Linn 2001)

Dr. Eleanor Gibson's time did indeed come, and she accumulated many honors. In addition to the National Medal of Science in 1992, she received the American Psychological Association (APA) Award for Distinguished Scientific Contributions to Psychology (1968), the G. Stanley Hall Award for Distinguished Contribution to Developmental Psychology (1970), the Guggenheim Fellowship (1972–1973), the Wilbur Cross Medal from Yale University (1973), the Howard C. Warren Medal from the Society of Experimental Psychologists (1977), the Distinguished Scientific Contribution Award from the Society for Research in Child Development (1981), and the American Psychological Foundation Gold Medal Award for Life Achievement in Science (1986), among others. She was elected to the National Academy of Sciences, the National Academy of Education, and the American Academy of Arts and Sciences. She also received 11 honorary doctorates (Gibson n.d.).

When Dr. Eleanor Gibson left Cornell in 1987, she relocated to Middlebury, Vermont, to be closer to her daughter and grandchildren, yet she continued to work and write. She died in 2002 at the age of 92.

19.2 Visual Cliff

Dr. Eleanor Gibson and Richard Walk developed the "visual cliff "experiment to study the development of depth perception. The visual cliff was designed to simulate a precarious drop while actually being safe for the subjects. A glass table, a few feet high, was used with a checkerboard cloth placed on one side, a board in the middle, a few inches thick, and the same pattern of cloth placed on the floor below the uncovered portion of the table. So, one side of the table, the shallow end, was just a few inches lower than the middle, and the other side, the deep end, gave the illusion of being a few feet lower. Figure 19.2 shows the apparatus.

They began by comparing rats that were raised in the dark and therefore had no visual experience with those raised in the light. They placed the rats in the middle of the table and observed if they would move to the shallow side and avoid the "dangers" of the deep end. They found that both sets of rats behaved the same way and avoided crossing over to the deep end. This demonstrated that the rats could perceive the change in depth and avoid the simulated danger of falling off the cliff, regardless of their prior experience. Thus, in rats, depth perception was demonstrated to be an innate ability.

Their next step was to try similar experiments with other animals. They dropped the idea of raising some of the animals in darkness since it was not easy for the scientists to feed the animals and clean cages in the dark. They used kittens, puppies, and baby chicks, observing their behavior at various stages of early

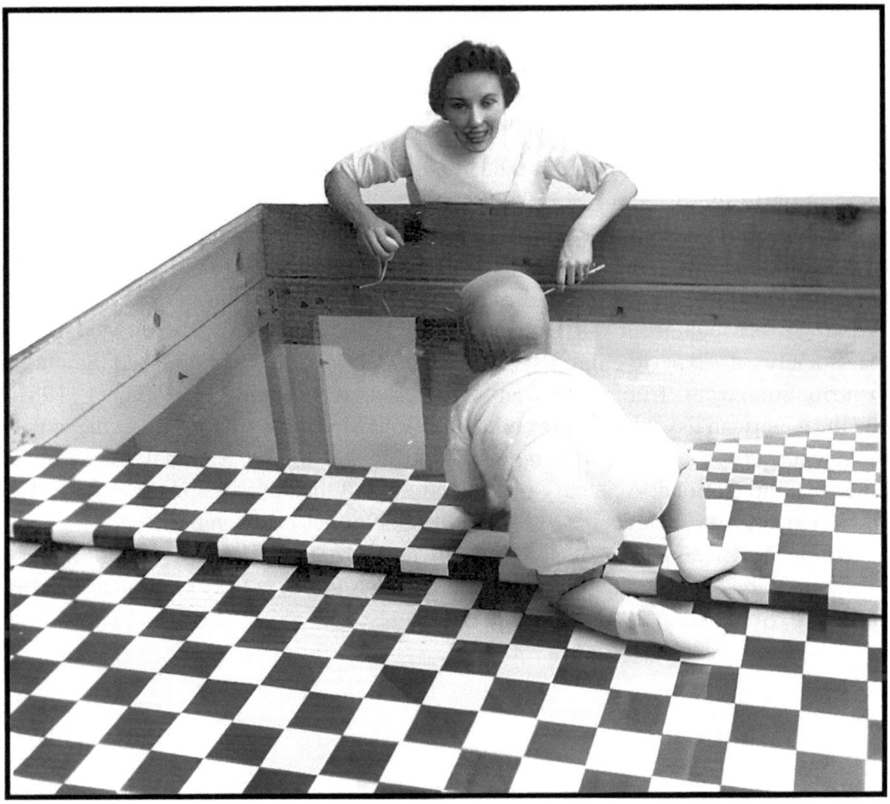

Fig. 19.2 Baby on visual cliff. (From Gibson and Walk (1960). (Copyright 1960 Nature Publishing Group. CC BY 4.0 <https://creativecommons.org/licenses/by/4.0>, via Wikimedia Commons)

development. Finally, it was time to try the experiment with human babies. They put an ad in the paper for crawling babies and paid their parents $3 for their participation. As with other animals, they found that most of the babies avoided the deep end.

Although the experiment is frequently used in discussions of the nature vs. nurture debate, Dr. Gibson went deeper by examining the influence of experience on innate abilities and developing a theory of perceptual learning. Dr. Gibson argued that depth perception was not strictly innate in animals that are not mobile at birth, but rather that these baby animals learn to perceive differences as they explore their environment. She also noted that babies do not develop depth perception from the experience of falling off a cliff, which is fortunate. Instead, they enhance their early ability to perceive depth through learning to differentiate and gather clues from their environment through crawling (Rodkey n.d.). In the case of the "visual cliff" experiment, babies have learned to distinguish between safe and unsafe surfaces through crawling. When they perceive a change in the surface (the illusion of a drop-off), they avoid the deep end. This held even when their mothers called for them from the deep side and tried to entice them with a toy.

19.3 Conclusion

Dr. Eleanor Gibson was one of the few prominent women in psychology during the mid-twentieth century. Her legacy endures in textbooks where her "visual cliff" experiment is still taught to undergraduate psychology majors as they study perceptual development. Dr. Gibson's groundbreaking work on visual perception and the development of reading skills continues to influence how reading is taught today. Her name will forever be associated with developmental psychology and children through the Eleanor J. Gibson Laboratory of Developmental Psychology at Cornell University, where babies and young children continue to be recruited for studies.

References

Baranauckas, Carla. 2003. "Dr. Eleanor J. Gibson, 92, a Pioneer in Perception Studies." *The New York Times*, January 4, 2003, sec. New York. https://www.nytimes.com/2003/01/04/nyregion/dr-eleanor-j-gibson-92-a-pioneer-in-perception-studies.html.

Castellanos-Monfil, Román . 2016. "Portrait of Yale's First Seven Women Ph.D.s Unveiled." Yale News. April 6, 2016. https://news.yale.edu/2016/04/06/portrait-yale-s-first-seven-women-phd-s-unveiled.

Eppler, Marion. 1998. "SCRD Oral History Interview - Eleanor Gibson" Interview by Marion Eppler. Society for Research in Child Development.

Gibson, Eleanor. n.d. "Curriculum Vitae." Society for Research in Child Development. https://www.srcd.org/sites/default/files/file-attachments/gibson_eleanor_cv.pdf. Accessed June 12, 2025.

Gibson, Eleanor J, and Susan Linn. 2001. *Perceiving the Affordances*. Psychology Press.

Rodkey, Elissa. 2011. "The Woman behind the Visual Cliff." *Https://Www.apa.org*, August 2011. https://www.apa.org/monitor/2011/07-08/gibson.

Rodkey, Elissa. n.d. "A Brief History of Eleanor Gibson's Visual Cliff Research." Www.yorku.ca. York University, Toronto. https://www.yorku.ca/rsheese2/1010/egibson.htm. Accessed June 12, 2025.

Szokolszky, Agnes. 2003. "An Interview with Eleanor Gibson." *Ecological Psychology* 15 (4): 271–81. https://doi.org/10.1207/s15326969eco1504_2.

Chapter 20
Salome G. Waelsch

For her lifetime of work on developmental genetics, providing a large body of knowledge on the study of mammalian genetics

Keywords Developmental genetics · Genetics · Mouse genetics · T-complex · T-locus · Embryology · Genetic mutation · Genes · Phenotype · Genotype · Morphogenesis · Tissue differentiation · Chromosomes · Genomics · Developmental disorders

Considered one of the founders of the field of developmental genetics, Dr. Salome Waelsch (Fig. 20.1) overcame gender discrimination throughout much of her career to make significant contributions to the field. Born in what is now Poland, she was educated in Russia and Germany, surviving antisemitism and family tragedy. Her PhD in 1932 was from the University of Freiburg, where she met her first husband. The Spemann lab, where she conducted her PhD research, sparked her interest in genetics and the role of genetics in developmental processes. Emigrating to the USA, she was at first unemployed due to nepotism rules in place at universities but eventually found a role in 1936 at Columbia University—where she laid the theoretical and practical groundwork for the field of developmental genetics through her work with mice. Her first salaried academic appointment occurred in 1955 when the Albert Einstein College of Medicine of Yeshiva University was founded. Dr. Waelsch joined the college in the anatomy department, very shortly thereafter became a full professor, and founded and served as chair of the genetics department there. She continued to work in the laboratory and in the mouse house until she was in her 90s. The recipient of many honors and awards, Dr. Waelsch received the National Medal of Science in 1993.

J. S. Tietjen et al., *Women Laureates 1970 – 2007*, Women in Engineering and Science, https://doi.org/10.1007/978-3-032-23116-1_20

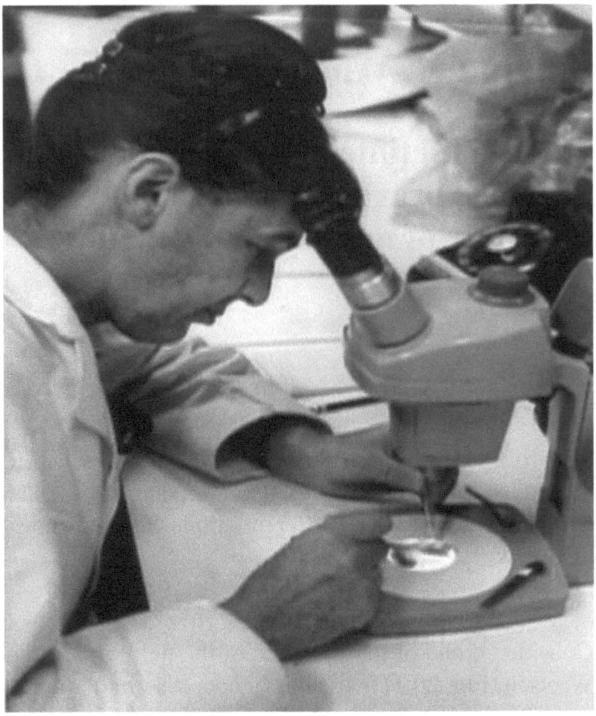

Fig. 20.1 Salome Waelsch at her microscope. (From the Archives of the Albert Einstein College of Medicine)

20.1 Life (1907–2007)

Salome Glücksohn (also Gluecksohn) was born on October 6, 1907, the oldest of three children, in Danzig, West Prussia (today Gdansk, Poland), to Ilya and Nadia Glücksohn. She grew up in Königsberg, East Prussia (today Kaliningrad, Russia), where her father was a grain merchant. As a child, she experienced continuous anti-semitism and tragedy—her father died during the Spanish flu epidemic when she was only 11 years old (1918), and due to hyperinflation, her family lost its wealth (Ogilvie and Harvey 2000; Papaioannou 2019).

Education was very important in her family. Salome thought she would study the classics in college but became entranced with science while an undergraduate student at the University of Königsberg and later at the University of Berlin. Her education was financed by tutoring jobs. Her focuses were zoology and chemistry. She then applied to the lab of Dr. Hans Spemann, an embryologist and later the 1935 Nobel Laureate in Physiology or Medicine, at the University of Freiburg. She was interested in working in his lab because he was a pioneer in experimental embryology; in this field, the experimenter alters the course of development and then observes the changes that result—drawing conclusions from those changes.

Spemann took Salome on as a graduate student in 1928 despite his disapproval of her personally—she was a woman and Jewish. She earned her PhD in 1932 (Genetics Unzipped 2020; Ogilvie and Harvey 2000; Papaioannou 2019).

While at Spemann's lab, Salome was exposed to world-class scientists from throughout the world. However, she was not enthusiastic about her research project: describing the development of limbs in two species of newts. She believed she was given this project to provide support for her male colleagues and that the assignment of the project further demonstrated Spemann's prejudice against women. Nevertheless, she learned skills and discipline for observation and attention to detail that would serve her well during the rest of her long and successful scientific career. In addition, she felt that Spemann's refusal to acknowledge the role of genetics in development was more than unfortunate and sparked a rebellious reaction on her part of furthering her interest in genetics. Fortuitously, the scientist who was assigned to supervise her, Viktor Hamburger, helped spur her interest in genetics and the role of genetics in developmental processes. At the time, scientific thinking did not see a role for genetics in developmental processes. Like Salome, Hamburger later immigrated to the USA. During her time in the Spemann lab, she also met C.H. Waddington, who became her lifelong friend and influenced her scientific thinking (Papaioannou 2019).

After earning her PhD in 1932, Salome briefly worked at the University of Berlin as a research assistant in cell biology before returning to Freiburg and marrying biochemist Rudolph "Rudi" Schoenheimer, who was employed at the University of Freiburg. However, due to nepotism rules at the University of Freiburg, she could not also be hired there. In 1933, when Germany enacted the Law for the Restoration of Professional Civil Service, which led to the firing of non-Aryans, she and her unemployed husband left Germany and came to the USA, where Rudi had been offered a job at Columbia University. Due to nepotism rules at Columbia University, no such job was waiting for Salome; she would be unemployed until 1936 (Papaioannou 2019).

In 1936, Salome's situation changed. She was offered a research associate position in the laboratory of Leslie Clarence Dunn at Columbia University in the zoology department. She would stay there for 17 years. Dunn was supportive of refugees from Nazi Germany, and he believed that chromosomal genes played a role in controlling development. In the early years in Dunn's lab, she worked as a volunteer—at no pay—but later she was compensated, although very poorly. During her time in Dunn's lab, Salome began her work with mouse genetics (Papaioannou 2019).

Her focus was on genetics, inheritance patterns, and the death of an embryo related to the T-locus (a genetic region that regulates embryonic development and is associated with traits including tail length) in mice. She was now working in an environment where genes were deemed to be very important and the science of genetics was expanding. Dunn was examining genes beyond rules of inheritance and mechanisms of transmission of heredity. His lab considered how genes affect adult traits and how those traits developed from the earliest stages of the embryo. In other words, Dunn was interested in how genetics and embryology crossed boundaries—opening up the possibility for the field of developmental genetics (Papaioannou 2019).

In Dunn's laboratory, Salome used naturally occurring genetic mutations to study inheritance. Her primary focus was mutations related to defects of the spine and tail. Her first published paper related to development of embryos and adults of tailless mutants in mice. She laid out the groundwork of what a developmental geneticist would study—mutations that were natural "experiments" that interfered with development. The scientist would examine the course of development and then endeavor to trace back to the original cause. Her paper became a classic of developmental genetics and is regarded as one of the pivotal papers that established this new field. She and Dunn contributed significantly to the understanding of the T-complex on chromosome 17[1] (Papaioannou 2019).

In 1938, Salome became a US citizen. Sometime thereafter, she and Rudi separated. In 1941, he died by suicide after battling depression for years (Papaioannou 2019). She later married Heinrich Waelsch, a Columbia University neurochemist. They had two children, a daughter Naomi (born in 1944) and a son Peter (born in 1947). Heinrich was incredibly supportive of Salome (now Dr. Waelsch) and her career and questioned why she was still a research associate and why she didn't have an academic position (Papaioannou 2019).

Dunn had little incentive to promote Dr. Waelsch at Columbia despite her reputation within scientific circles and the recognitions she received. During the summer of 1948, she was a special lecturer in developmental genetics at the University of Edinburgh and, in 1951, she was a special lecturer at Barnard College (Papaioannou 2019). It would be up to Dr. Waelsch to change her employment situation.

In 1953, Dr. Waelsch applied for and obtained a National Institutes of Health grant and moved out of Dunn's lab to become a research associate in obstetrics and gynecology (genetics) in the College of Physicians and Surgeons at Columbia University. Although still precluded from advancement opportunities at Columbia, this at least provided her the opportunity to have her own lab (Papaioannou 2019).

When the Albert Einstein College of Medicine of Yeshiva University was founded, Dr. Waelsch finally became a professor. Dr. Berta Scharrer, a 1983 National Medal of Science recipient, and her husband Dr. Ernst Scharrer were determined to eliminate the gender discrimination that both Dr. Berta Scharrer and Dr. Waelsch had experienced up to this point in their careers. When the Scharrers co-founded the Department of Anatomy at the Albert Einstein College of Medicine in 1955, both Dr. Waelsch and Dr. Berta Scharrer received their first salaried academic

[1] The T-complex is a large region on mouse chromosome 17 that has multiple linked genes. The T-locus is a specific region within that complex.

appointments. Dr. Waelsch joined as an associate professor of anatomy (genetics). Waelsch taught medical genetics, possibly the first formal course in genetics taught in the USA as part of the medical school curriculum (Papaioannou 2019).

Dr. Waelsch remained at Einstein for the rest of her career. She was promoted to full professor in 1958. In 1963, she helped found the Department of Genetics there and served as chair of the department for 13 years. During the 1960s, she began a collaboration with Nobel Laureate Carl Cori[2] and Robert Erickson examining the genetic regulation of enzyme synthesis. In 1978, she became a professor emerita and maintained a laboratory and mouse house. She continued to work in the laboratory and in the mouse house until she was in her 90s (Papaioannou 2019).

Today, Dr. Waelsch is considered one of the founders of the field of developmental genetics. "She laid the theoretical and practical groundwork for the integration of genetic and developmental studies when the newly burgeoning field of molecular biology turned its attention to development" (Papaioannou 2019). She lived long enough to see the beginning of the applications of technology to genetics that overcame many of the technological limitations that had impacted her life's work (Papaioannou 2019).

Dr. Waelsch received many awards, the bulk of which came later in her life. She was elected to the National Academy of Sciences in 1979 and became an honorary life member of the New York Academy of Sciences. She was elected to the American Academy of Arts and Sciences (1980), as a fellow of the American Association for the Advancement of Science (1988), and in 1995 as a Foreign Member of the Royal Society in London (Papaioannou 2019). She received the National Medal of Science from President Clinton (Fig. 20.2) in 1993 (National Science & Technology Medals Foundation 2025). Her other awards included the Lifetime Achievement Award from the American Cancer Society and the Thomas Hunt Morgan Medal of the Genetics Society of America, which recognized a lifetime contribution to the science of genetics (1999). She received a number of honorary degrees, including one from Columbia University (1995). On the fiftieth anniversary of her graduation, the University of Freiburg awarded her a Golden Diploma in recognition of her lifetime of achievements (Papaioannou 2019).

After her death, the Department of Genetics at the Albert Einstein College of Medicine, jointly with the Spemann Graduate School of Biology and Medicine at the Albert-Ludwigs University of Freiburg, established an annual prize honoring Dr. Waelsch. The purpose of this award is to further the exchange of ideas, build bridges, and open doors for promising scientists at the beginning of their careers (Papaioannou 2019).[3]

---

[2] Carl and his wife Gerty Cori shared the 1947 Nobel Prize in Physiology or Medicine "for their discovery of the course of the catalytic conversion of glycogen." Gerty Cori died in 1957.

[3] Note: She published under four different names: Salome Glücksohn, Salome Gluecksohn-Schoenheimer, Salome Gluecksohn-Waelsch, and Salome G. Waelsch.

Fig. 20.2 Dr. Waelsch receiving her National Medal of Science. (From the Archives of the Albert Einstein College of Medicine)

20.2 Developmental Genetics

Farmers found out centuries ago how to breed plants and animals for their more desirable characteristics. Gregor Mendel in the 1800s, using pea plants, identified recessive and dominant traits by analyzing seven characteristics of those plants. He determined rules of heredity associated with pea plants and said they were due to invisible factors. We now call those invisible factors "genes." Mendel is considered the father of modern genetics (Wikipedia 2025).

Genetics is the study of those genes—a branch of biology that examines how characteristics are passed from one generation to the next. Genetics also looks at relationships between genes, genetic makeup (called genotype), and phenotype (physical, biochemical, and behavioral characteristics, including hair and eye color, blood type, and the presence of disease) (Science of Biogenetics 2023).

Developmental genetics strives to understand how genes control the development of an organism from a single cell to the fully formed animal or plant. Scientists in the field study embryology (the study of an organism from fertilization until birth or hatching), morphogenesis (how the shape and structure of an organism develops), and tissue differentiation (how various types of tissue—connective, epithelial, muscle, and nervous occur) (Science of Biogenetics 2023).

Today, genomics, with its ability to rely on advances in DNA sequencing, has changed the study of developmental genetics from the time that Dr. Waelsch was working in the field. Now, scientists can examine interactions between genes and identify specific genetic factors that are found to impact development processes (Science of Biogenetics 2023).

Developmental geneticists study patterns of gene expression, genomic changes that occur during development (prior to birth), and how those changes affect the growth of cells and their differentiation during development. They use an understanding of the genome to identify specific genetic changes that contribute to the development of various phenotypes. These scientists also study how mutations in genes affect development (Science of Biogenetics 2023). Dr. Waelsch spent many years examining mutations during development.

Developmental genetics combines the fields of genomics (the study of the entire genome—which includes all of an organism's genes), genetics (the study of individual genes and their functions), and embryology. Developmental geneticists can use the knowledge of how an organism develops and how genetic changes influence this process to understand the factors that contribute to normal development and to developmental disorders (Science of Biogenetics 2023).

Applications of developmental genetics include (Science of Biogenetics 2023):

1. Understanding the mechanisms of development which can lead to an understanding of the causes of developmental disorders.
2. Identifying genes and mutations can help uncover the role of specific genetic variations in developmental disorders.
3. Figuring out the genomic basis of phenotype. Developmental genetics provides insights into how genetic variations cause differences in phenotypes—determining how genes and mutations affect the development and functioning of tissues and organs. With this knowledge, new diagnostic tools and therapeutic approaches can be developed for developmental disorders.
4. The genome isn't a static blueprint but can be affected by environmental factors. Understanding how the genetic makeup of an organism can be changed by its environment could have significant impacts on personalized medicine and evolutionary biology.

20.3 Conclusion

Considered one of the founders of the field of developmental genetics, Dr. Salome Waelsch overcame gender discrimination throughout much of her career to make significant contributions to the field. She completed her education in Europe, where her interest in genetics and developmental genetics was sparked. While she was at Columbia University, Dr. Waelsch laid the theoretical and practical groundwork for the field of developmental genetics through her work with mice. She became a full professor, finally, at the Albert Einstein College of Medicine of Yeshiva University and founded and served as chair of the genetics department there. She continued to work in the laboratory and in the mouse house until she was in her 90s. The recipient of many honors and awards, Dr. Waelsch received the National Medal of Science in 1993.

References

Genetics Unzipped. 2020. "A woman and a Jew? Forget it!": The story of Salome-Gluecksohn-Waelsch. https://geneticsunzipped.com/transcripts/2020/9/7/the-story-of-salome-gluecksohn-waelsch. September 10. Accessed September 11, 2025.

National Science & Technology Medals Foundation. Salome G. Waelsch. https://nationalmedals.org/laureate/salome-g-waelsch/. Accessed August 31, 2025.

Ogilvie, Marilyn and Joy Harvey, editors. 2000. The Biographical Dictionary of Women in Science: Pioneering Lives from Ancient Times to the Mid-20th Century. Routledge.

Papaioannou, Virginia E. 2019. Salome Gluecksohn-Waelsch 6 October 1907 – 7 November 2007. Biographical Memoirs of Fellows of the Royal Society. Volume 67. Pp. 153–171.

Science of Biogenetics. 2023. Understanding the role of developmental genetics in shaping organisms – a comprehensive exploration. https://scienceofbiogenetics.com/articles/understanding-the-role-of-developmental-genetics-in-shaping-organisms-a-comprehensive-exploration. December 20. Accessed September 22, 2025.

Wikipedia. Gregor Mendel. https://en.wikipedia.org/wiki/Gregor_Mendel. Accessed September 22, 2025.

Chapter 21
Vera Rubin

For her pioneering research programs in observational cosmology which demonstrated that much of the matter in the universe is dark and for significant contributions to the realization that the universe is more complex and more mysterious than had been imagined

Keywords Dark matter · Astrophysics · Vera Rubin Observatory · Astronomy · Galaxy rotation · Telescopes · Sa galaxies · Sb galaxies · Sc galaxies · George Gamow · Palomar Observatory · Kent Ford · Carnegie Institution · Invisible mass · Cosmology · Observational astronomy

Dr. Vera Rubin (Fig. 21.1) was an astronomer whose observations of galaxies led her to conclude that there is invisible mass (dark matter) in the universe. Gravity acting on the luminous matter alone would not be sufficient to prevent the stars at the outer regions of a galaxy from flying out of it. Although initially met with skepticism, her observations using spectrographic equipment built by Dr. Kent Ford eventually led to acceptance of her theory. Astrophysicists believe that the mass in the universe is about 85% dark matter. Dr. Rubin received a National Medal of Science in 1993.

J. S. Tietjen et al., *Women Laureates 1970 – 2007*, Women in Engineering and Science, https://doi.org/10.1007/978-3-032-23116-1_21

Fig. 21.1 Portrait of Dr. Vera Rubin. (Courtesy of AIP Emilio Segrè Visual Archives, AIP Gallery of Member Society Presidents)

21.1 Life (1928–2016)

Vera Florence Cooper was born in July 1928 in Philadelphia, Pennsylvania, to parents Phillip Cooper and Rose (Appelbaum) Cooper. Her father, originally named Pesach Kobchefski, had emigrated to the USA with his family from Vilnius, Lithuania, when he was 7 or 8 years old and became the only member of his family to go to college, where he earned an engineering degree. Rose had emigrated to the USA from Bessarabia (now part of Moldova and Ukraine). Phillip and Rose met when both were working at Bell Telephone Company. Vera was the younger of their two daughters, the elder being named Ruth (DeVorkin 1995).

From an early age, Vera grew interested in the night sky. After her family moved to Washington, DC, when Vera was 10 years old, she'd watch the night sky stars from her bedroom, memorizing their motions and making maps of their paths in the morning. Her parents encouraged her interest in science and the stars, and she and her father attended an amateur astronomy club once a month starting when she was 14 years old (Soter and Tyson 2000; DeVorkin 1995). The pair also built a telescope from a two-inch lens from Edmund Scientific and a cardboard tube from a linoleum roll. Her mother convinced the local library to allow her access to the books for adults and managed to get a photography store to print the images Vera had taken, which were lines in a dark background—not the error of a young photographer as the store thought, but rather star trails as seen through her telescope (Dawson 2016; Lightman 1989; Stewart 2025).

Vera was inspired by Maria Mitchell, the first female, American astronomer. As Mitchell had taught at Vassar College, Vera decided she would also attend Vassar. When she told her high school physics teacher that she would be going to Vassar, he said, "As long as you stay away from science you should do okay" (Dawson 2016). Although her father was dubious about potential employment as an astronomer and tried to talk his daughter into pursuing mathematics, Vera attended Vassar on scholarship from 1944 to 1948, graduating as the only astronomy major in her class (Dawson 2016; Soter and Tyson 2000; Santisi n.d.).

During summer breaks while attending Vassar, Vera worked in Washington, DC, at the Naval Research Lab (NRL) or at the US Naval Observatory. During the summer of 1947, she spent time with Cornell University physics student, Robert (Bob) Rubin, who was also working in Washington. The two had been introduced by their parents, who knew one another and lived in the same neighborhood. The Rubins married in the Fall of 1947 (Dawson 2016).

Rubin applied to several graduate schools to continue her studies. Princeton University turned her down as they didn't admit women, but Harvard accepted her into their program. Ultimately, Rubin chose to turn down Harvard's offer to go to Cornell University, where her husband was continuing his graduate work. This prompted a note from the Harvard department chair scrawled at the bottom of his acknowledgment of her withdrawal, saying "Damn you women. Every time I get a good one ready, she goes off and gets married" (Lightman 1989).

At Cornell University, Rubin pursued her master's degree in the physics department while her husband completed his PhD. She had been met by a professor of astronomy who told her that there were no jobs in astronomy, and she should go away, but she persevered (DeVorkin 1996). As she neared completion of her MS degree, she opted to present a paper on her research at a meeting of the American Astronomical Society (AAS) in Philadelphia. Unfortunately, the meeting would occur when her firstborn child, David, was only 4 weeks old. Noting this, her department chair suggested that since she wouldn't be able to go and give her talk, he'd give it for her if she added his name to the paper. Rubin declined the offer and announced she would attend the meeting to give her talk. Her parents volunteered to accompany her and their first grandchild to the meeting, driving through heavy snow to get to the venue. The day before her talk was scheduled, Rubin's mother and aunt took her out to buy a dress. Rubin walked into her session, gave her talk about galaxy rotations, and after fielding mostly negative comments and pointed questions, left to return home (DeVorkin 1996; Lightman 1989). After the talk, she was contacted by Dr. George Gamow, a well-known physicist, who asked lots of questions about her work and used her data, with attribution, in talks he gave.

Rubin finished her master's degree in October 1950, but her husband's PhD was not yet completed. They moved to Vera's uncle's tannery in Trumansburg, about

15 miles outside of Ithaca to avoid having to pay Ithaca's high rents and remained there until Bob finished his PhD in June 1951 (DeVorkin 1996).

Upon graduating, Bob Rubin had several offers of employment. He chose a position with the Applied Physics Laboratory (APL), a Navy research lab, because he viewed that as offering Vera the best chance to earn a PhD. At the time, APL was in Silver Spring, very near to Washington, DC. The Rubins moved to an apartment in Washington, DC, with their son, David, and with another child on the way. Although Vera took time off to care for David and prepare for the arrival of her second child, she found herself bored and unhappy staying at home. Within 6 months of moving, she opted to return to school (DeVorkin 1996).

In the Washington, DC, area, Vera had only two reasonable choices for where to pursue a PhD: George Washington University (GWU) or Georgetown University. At the time, GWU did not have an astronomy degree, so she would be required to enroll in the physics department and show proficiency in six areas of physics. That tilted the scales in favor of Georgetown, and Vera started as a student there in February 1952. The astronomy courses at Georgetown were all offered in the evening, so Vera and Bob enlisted help from Vera's parents. Bob would drive Vera to Georgetown since Vera didn't drive, and he would wait while she attended classes. Vera's parents would be at the Rubin apartment caring for the children (DeVorkin 1996).

Georgetown University had a small group of people interested in astronomy. She was given a topic to pursue for her PhD, but after a few days, Rubin realized it was not to her liking. Having had conversations with Dr. Gamow at GWU, she contacted him to ask if she might pursue PhD research with him while officially being a Georgetown student. This was eventually worked out, and Rubin completed her PhD in 1954 (DeVorkin 1996; DeVorkin and Yeager 2007), the same year her third child was born. She had done much of her research work at home after the children had been put to bed.

In 1955, APL moved from Silver Spring to Laurel, Maryland. At that point, Bob decided to leave APL for a two-year appointment at the University of Illinois. Vera had been hired onto the faculty of Georgetown University, supported on grant funding, but she accompanied Bob and worked remotely. They returned to Washington, DC, in 1957, and Bob began his career at the National Bureau of Standards (now the National Institute of Standards and Technology) (DeVorkin and Yeager 2007). In the early 1960s, the fourth Rubin child and only daughter was born.

In 1963, Bob got an NSF senior fellowship for a year and chose to take it at La Jolla, California, so that Vera would have the chance to collaborate with Drs. Margaret and Geoff Burbidge, well-respected astronomers. (Dr. Margaret Burbidge was later a recipient of the National Medal of Science.) Vera valued this connection as the Burbidges were the first astronomers she had a chance to collaborate with on research. It was the first time she felt as if she was a real, working astronomer as she measured spectra they had obtained and contributed to their research (DeVorkin and Yeager 2007).

While in California, Dr. Rubin was invited to come observe at the Palomar Observatory in San Diego. Although women were not permitted to use their large 200-in. telescope, their smaller telescopes were still impressive and useful to Dr. Rubin. However, she discovered while on an observing run that the only bathroom was labeled MEN on its door. When she asked about a ladies' room, she was told the telescope building did not have any. Dr. Rubin took a sheet of paper and drew a figure with a skirt on it and put it on the door, and said, "There you go; now you have a ladies' room" (Carnegie Science 2024; Dawson 2016). When Dr. Rubin returned, the bathroom was marked as a gender-neutral bathroom.

In 1964, Dr. Rubin returned to Washington to find Georgetown University's astronomy effort falling apart. Fatigued from simultaneously caring for four children and teaching and trying to eke out time to observe, she resigned from Georgetown University. She visited a friend at the Carnegie Institution and asked for a job. This was described as "He seemed as surprised as if she'd asked him to marry her" (Carnegie Science 2024). He invited her to attend the campus community lunch with him. At that lunch, Dr. Rubin was invited to talk about her work with the assembled group. She left with a photographic plate from Dr. Kent Ford, an instrumentalist who had just built a very sensitive spectrometer (a device to measure the frequency of light observed). She was asked if she could use the data on the slide to determine stellar velocities. She did, returning the slide with her analysis, and she was then offered a position in the Department of Terrestrial Magnetism at the Carnegie Institution (now part of the Earth and Planets Lab) (Carnegie Science 2024). She was the first woman on the staff of that department. Dr. Rubin explained that she wanted to keep going home at 3 pm but would work after dinner. The management agreed to this but used it to offer her two-thirds of the salary she had been getting at Georgetown. She accepted the job despite this slight (DeVorkin and Yeager 2007).

At Carnegie, Dr. Rubin collaborated with Dr. Ford, using his extraordinarily sensitive spectrometer to observe previously undetected objects, often at Lowell Observatory or Kitt Peak Observatory, both in Arizona. She is shown in Fig. 21.2 at Kitt Peak National Observatory. At that time, working in an observatory could be dangerous as it was always dark and platforms were raised and heavy equipment was moving. For this reason, the Carnegie Institution insured the lives of Drs. Ford and Rubin, with the insurance payout for Dr. Ford set to be four times that for Dr. Rubin (DeVorkin and Yeager 2007).

Dr. Ford's spectrometer, the most sensitive in existence at the time, spread out the light spectrum, making it much more accurate. It was also able to produce results about ten times faster than one could using traditional spectrometers by reducing the exposure time. Using this equipment, Dr. Rubin was able to measure the Doppler shifts of stars as they moved. Moving toward the observer would produce a frequency increase known as a blue shift, and the part of the orbit moving away would produce a frequency drop known as a red shift. By measuring these Doppler shifts and calculating the mass of the star from the data, Dr. Rubin would determine the orbital speed at which an object was moving (Soter and Tyson 2000; Dawson 2016).

At first, Drs. Rubin and Ford focused on the Andromeda galaxy as it is close and thus the stars are bright. Dr. Rubin produced a rotation curve for the galaxy, a graph of the orbital speed of stars in the galaxy as a function of their distance from the center. The expectation was that this curve would fall off as distance increased since gravitational attraction falls off with distance. That means that high-speed orbits far from the orbital center should propel the star out of the orbit and away from the

Fig. 21.2 Dr. Vera Rubin using Kitt Peak National Observatory's 36-inch telescope. (KPNO/NOIRLab/NSF/AURA, CC BY 4.0, via Wikimedia)

center according to Newtonian mechanics, but this was not what they observed. The rotation curves instead were flat, suggesting that the outer stars were moving at the same speed as inner stars. "The surprises came very quickly. By the end of the first night, we were puzzled by the shape of the rotation curve" (Carnegie Science 2024).

Drs. Rubin and Ford published their initial findings in 1970 and noted that the only explanation for what they were seeing was that there was invisible matter in the region of the universe they were observing. By 1980, Drs. Ford and Rubin had orbital speed data on 21 galaxies, and Dr. Rubin noted that "The conclusion is inescapable that non-luminous matter exists beyond the optical galaxy" (Carnegie Science 2024). She is thus credited with having proven the existence of dark matter in the universe.

Dr. Rubin continued to observe the rotational speeds of stars in many galaxies in our universe, ultimately producing rotation curves for more than 75 galaxies. Additionally, in 1992, she discovered a galaxy in which half of the stars orbit in one direction and the other half in the opposite direction. Dr. Rubin suggested that this could be the result of two galaxies merging (Soter and Tyson 2000).

In addition to her research, Dr. Rubin has been known as an advocate for inclusion in science. She challenged conference organizers and hiring committees that overlooked women and mentored many students from underrepresented populations. Frustrated by the lack of recognition for women scientists, she spent much of her time nominating them for prizes and encouraging them to promote themselves (Soter and Tyson 2000; Carnegie Science 2024).

Dr. Rubin was the recipient of many honors and awards. She was elected to the National Academy of Sciences in 1981 and received the National Medal of Science in 1993. In 1994, she received the Dickson Prize in Science from Carnegie Mellon University and the Henry Norris Russell Lectureship from the AAS. In 1996, Dr. Rubin became the second woman to receive the Gold Medal of the Royal Astronomical Society in London, UK. That same year, President Clinton made her a member of the National Science Board. In 2004, Dr. Rubin received the James Craig Watson Medal of the National Academy of Sciences.

In recognition of her discovery of dark matter in the universe, a ridge on Mars has been named in her honor, as has an asteroid and a galaxy. A satellite was named in her honor, and the National Science Foundation's Large Synoptic Survey Telescope was renamed the Vera C. Rubin Observatory by an act of Congress in 2019, making it the first US national observatory named for a woman. Her image graces a quarter that was released in June 2025 as part of the American Women Quarters series, which was started in 2022. The image is shown in Fig. 21.3.

Fig. 21.3 The Dr. Vera Rubin Quarter issued in June 2025. (Courtesy of Carnegie Science)

Dr. Rubin died on December 25, 2016, in Princeton, New Jersey. She was survived by her sons David, Karl, and Allan, all of whom have doctorates in science fields. Her husband died in 2008, and her daughter Judy, an astronomer, passed away in 2014.

21.2 Dark Matter

If you grab the hands of a child and spin them around, a force called the centripetal force is established that runs along the line connecting the two of you. The faster you spin, the more the child is lifted off the ground. If you eventually let go, the child will fly out in a line, not continue to travel in a circular orbit. This behavior is instructive for understanding what happens as planets orbit the sun in our solar system.

In our solar system, what keeps orbits going is a combination of the centripetal force pushing the planet outward and the gravitational force between the sun at the center and the planet pulling the planet in. The centripetal force increases as the orbital speed grows and decreases as the distance between the two objects grows. Doubling the distance halves the force. The gravitational force on the planet increases with the mass of the planet growing and decreases as the distance between the objects increases but more quickly than the centripetal force decreases. Doubling the distance cuts the gravitational force by a factor of four. As a result, the outer planets in our solar system orbit around the sun much slower than the inner planets. Otherwise, the outer planets would have a centripetal force greater than the gravitational force and would fly out of the solar system. From classical mechanics, it was understood that the stars in our universe would similarly follow a pattern of higher-speed orbits near the center of the galaxy and slower rotations as one moved to stars further away.

In 1933, Dr. Fritz Zwicky published data on galaxy motions in the Coma galactic cluster. That cluster contains more than 1000 galaxies. Dr. Zwicky examined the motion of several of the galactic centers about the cluster center and found that some of them were moving so fast at the cluster outer reaches that the centripetal force should be great enough to overcome the gravitational attraction. In other words, the galaxies should fly off rather than remaining part of the cluster. Since Dr. Zwicky had faith in his measurements of the galactic distances, their distance from the cluster center, and the amount of mass present based on the light emitted, he theorized that there had to be some extra mass present that he couldn't see. He called this dark matter and theorized that it was providing the extra gravitational pull necessary to produce the stable orbits observed (Dawson 2016).

Dr. Zwicky's paper was largely ignored as the observations were very laborious, and he had only small amounts of data to support his conclusion. Skeptics assumed the result would not hold up as more data were made available.

More than three decades later, Dr. Vera Rubin, working with Dr. Kent Ford on the new spectrometer he had built, looked at stars in a single galaxy rather than galactic

Fig. 21.4 Three types of spiral galaxies—Sa on the left, Sb in the center, and Sc on the right. (Courtesy of the University of Arizona https://ircamera.as.arizona.edu/NatSci102/NatSci102/text/spiraltypes.htm)

clusters. She measured rotation curves and was surprised that they were flat, indicating a relatively constant speed regardless of location. Remembering Dr. Zwicky's work, Dr. Rubin came to a similar conclusion—that there must be some mass that was not visible.

Again, Dr. Rubin's conclusions were met at first with skepticism, partly because she had modest data and partly because her data focused on bright galaxies. Some astrophysicists suggested that the results would fall apart as less bright galaxies were considered, but that was not the case (DeVorkin and Yeager 2007). Unlike Dr. Zwicky, she and Dr. Ford continued to collect data, and her results remained constant. They looked at spiral galaxies such as shown in Fig. 21.4, starting with roughly twenty type Sa galaxies, which are bright and have a bulge in the center, then twenty type Sb galaxies, in which the spiral arms are more prominent, and finally twenty type Sc galaxies, in which the mass is predominantly in the spiral arms. The rotation curves of all were flat, and the existence of dark matter seemed to be the only explanation.

Dr. Rubin said, "In a spiral galaxy, the ratio of dark-to-light matter is about a factor of ten. That's probably a good number for the ratio of our ignorance-to-knowledge. We are out of kindergarten, but only in about third grade" (Soter and Tyson 2000). She believed that scientists would determine what dark matter is within 10 years of her observations (Vera C. Rubin Observatory (b) 2025). This has not turned out to be the case.

Dark matter is understood to have mass and to be distributed in space, but it does not emit or absorb or reflect light. It does not react with electrical or magnetic fields. In other words, it is invisible by standard methods of observation. It provides a scaffold in the universe on which galaxies and galactic clusters are built. It is thought to account for about 85% of the mass in the universe (Medium 2024, Vera C. Rubin Observatory (a) 2025).

Dr. Rubin continued to study galaxies throughout her career, and her conclusions about the existence of dark matter were confirmed by observations different from hers. Albert Einstein predicted that very massive objects could cause light to bend—a process referred to as gravitational lensing. Observations of gravitational lensing have led to computations of the mass that would be required to cause the light trajectories, and these also lead to a conclusion of dark matter being present. As an example, in the mid-2000s, observation of the Bullet Cluster showed two clusters colliding, and the gravitational lensing showed that most of the mass was far from the glowing area (Vera C. Rubin Observatory (a) 2025).

Late in her life, Dr. Rubin summarized her work and the future by saying:

> We have peered into a new world and have seen that it is more mysterious and more complex than we had imagined. Still more mysteries of the universe remain hidden. Their discovery awaits the adventurous scientists of the future. I like it that way. (Soter and Tyson 2000)

21.3 Conclusion

Dr. Vera Rubin is known as the discoverer of dark matter in the universe. Her careful observations of the orbital speeds of stars showed that there must be some invisible matter or the outer stars would fly out of the galaxy rather than remaining in it. Gravity simply is not enough to explain what she saw. Her conclusions were originally greeted with skepticism but were later verified by different sorts of astronomical measurements. Dr. Rubin was awarded the National Medal of Science in 1993.

References

Carnegie Science. 9 Oct 2024. Vera Rubin: Opening doors to dark matter and women in STEM . Earth & Planets Laboratory. https://carnegiescience.edu/news/vera-rubin-opening-doors-dark-matter-and-women-stem. Accessed 3 Sept 2025.

Dawson, Shay. 2016. Vera Rubin (1928-2016), National Women's History Museum. https://www.womenshistory.org/education-resources/biographies/vera-rubin. Accessed 3 Sept 2025.

DeVorkin, David. 1995. Vera Rubin – Session I, AIP, Niels Bohr Library and Archives. https://repository.aip.org/rubin-vera-1995-september-21-and-may-9-1996-session-i. Accessed 3 Sept 2025.

DeVorkin, David. 1996. Vera Rubin – Session II, AIP, Niels Bohr Library and Archives. https://repository.aip.org/rubin-vera-1995-september-21-and-may-9-1996-session-ii. Accessed 3 Sept 2025.

DeVorkin, David and Yeager, Ashley. 2007. Vera Rubin, AIP, Niels Bohr Library and Archives. https://repository.aip.org/rubin-vera-2007-july-20. Accessed 3 Sept 2025.

Lightman, Alan. 3 April 1989. Vera Rubin, AIP, Niels Bohr Library and Archives. https://repository.aip.org/rubin-vera-1989-april-3. Accessed 3 Sept 2025.

Medium. 2 Dec 2024. What is dark matter and dark energy? The Universe Episodes. https://medium.com/@theuniverseepisodes/what-is-dark-matter-and-dark-energy-fa016e2b5099. Accessed 3 sept 2025.

Santisi, Jennifer. (n.d.) Vera C, Rubin: National Medal of Science, National Science and Technology Medals Foundation. https://nationalmedals.org/laureate/vera-c-rubin/. Accessed 3 Sept 2025.

Soter, Steven and Tyson, Neil (eds). 2000. *Cosmic Horizons: Astronomy at the Cutting Edge*. American Museum of Natural History. New Press. Also available at https://www.org/learn-teach/curriculum-collections/cosmic-horizons-book/vera-rubin-dark-matter. Accessed 3 Sept 2025.

Stewart, Gabrielle. 5 June 2025. New Quarter Honors Vera Rubin, Astronomer Who Revealed the Universe's Hidden Mass, Smithsonian American Women's History Museum. https://womenshistory.si.edu/blog/new-quarter-honors-vera-rubin-astronomer-who-revealed-universes-hidden-mass. Accessed 3 Sept 2025.

Vera C. Rubin Observatory (a). The Nature of Dark Matter, US National Science Foundation, US Department of Energy. https://www.lsst.org/science/dark-matter. Accessed 3 Sept 2025.

Vera C. Rubin Observatory (b). US National Science Foundation, US Department of Energy. https://rubinobservatory.org/about/vera-rubin. Accessed 3 Sept 2025.

Chapter 22
Elizabeth F. Neufeld

For contributions to the understanding of the lysosomal storage diseases, demonstrating the strong linkage between basic and applied scientific investigation

Keywords Lysosomes · Lysosomal storage diseases · GAGs · MPS · Mucopolysaccharidoses · Hunter syndrome · Hurler syndrome · Sanfilippo syndrome · Scheie syndrome · Glycosaminoglycans · Aldurazyme · Enzyme therapy

Dr. Elizabeth F. Neufeld, shown in Fig. 22.1, is a biochemist who discovered the causes of some rare diseases related to defects in the cellular lysosomes, organelles that break down complex molecules to their constituent components for reuse. She identified the causes of these diseases as defective or missing enzymes and developed enzyme therapies for them. Dr. Neufeld received the National Medal of Science in 1994.

J. S. Tietjen et al., *Women Laureates 1970 – 2007*, Women in Engineering and Science, https://doi.org/10.1007/978-3-032-23116-1_22

Fig. 22.1 Dr. Elizabeth F. Neufeld (Santisi 2025)

22.1 Life (1928–)

Elizabeth Fondal was born on September 27, 1928, in Paris to Russian-Jewish immigrant parents, Jacque and Elvire Fondal. Her parents had fled Russia after the Russian Revolution and fled France for the USA in 1940 with their two children when it seemed likely that the Nazis would invade (The Franklin Institute 2025).

After arriving in the USA, Elizabeth and her family lived in New York City. Elizabeth attended Hunter College High School and after graduating, enrolled at Queens College. She graduated in 1948 with her Bachelor of Science degree (Wikipedia 2025).

Immediately following college graduation, Elizabeth went to work as a research assistant at the Jackson Lab, a nonprofit biomedical research institute in Bar Harbor, Maine. There she worked on genetic blood disorders in mice. She left that position to enroll in graduate school at the University of Rochester in Rochester, New York, but left that university in 1951 to work as a research assistant for Drs. Sidney Colowick and Nathan Kaplan at the McCollum-Pratt Institute at Johns Hopkins

University in Baltimore, Maryland. Elizabeth hoped to advance by earning a PhD in biochemistry and asked one of her bosses if he would support her. He answered, "Well you know it's not so good for women; they're not interested in science" (Lasker Foundation Newsletter 2017; Wikipedia 2025). Contrary to this sentiment, Elizabeth has said: "Some people looked at women who wanted a career in science as a little eccentric, but I enjoyed what I was doing, and I decided I would persevere" (Santisi 2025).

In 1951, Elizabeth married Ben Neufeld, a soldier who had grown up in New York City and Baltimore. As a corporal, he was stationed in Germany in the quartermaster's corps at the end of World War II. Although Ben had attended Virginia Military Institute and the University of Maine, he earned his undergraduate degree from New York University. He was stationed near Baltimore when Elizabeth was working at Johns Hopkins University (Mount Sinai Cemetery 2020).

Dissuaded from staying at Johns Hopkins University, Neufeld enrolled in the University of California at Berkeley instead, entering in 1952. There she worked on nucleotides, the building blocks of DNA and RNA, and complex carbohydrates in plants. Ben earned his master's degree at the university and worked at a variety of jobs. Elizabeth graduated with a PhD in 1956 (Mount Sinai Cemetery 2020; Wikipedia 2025).

After earning her PhD, Dr. Neufeld stayed at the University of California at Berkeley as a postdoctoral fellow. While her research work was valued and she was very productive, there was what has been called "a delay in obtaining a faculty position," so Dr. Neufeld opted to leave academia and take a position at the National Institutes of Health (NIH) (Hirschhorn 1983). She continued her work on the biosynthesis of plant cell wall polymers. These polymers generally have uronic acids present, meaning their chemical basis is built upon nucleotide uracil (Maugh 1994; Hirschhorn 1983).

In 1967, someone wandered into Dr. Neufeld's lab and told her about patients with Hurler Syndrome, a disorder in which patients don't metabolize complex carbohydrates properly. Dr. Neufeld thought she knew what might be causing the problem, as she said:

> I had an idea what the problem might be. As it turned out, I was totally wrong, but by the time I had proved that, I was totally hooked. (Maugh 1994)

When Dr. Neufeld first started her research on Hurler syndrome, it was believed that its cause was aberrant regulation producing an overproduction of mucopolysaccharides (MPS, now called glycosaminoglycans, GAGs), a complex series of sugars, that were accumulating in the cells. Dr. Neufeld was able to prove this assumption incorrect. The problem is not overproduction but rather that the degrading process of the carbohydrates into their constituent sugars was not functioning properly (Kresge et al. 2007). This work was first reported in 1968 (Fratantoni (a) 1968).

Continuing her research, Dr. Neufeld ultimately found that the problem in Hurler syndrome could be corrected by introducing an enzyme that was deficient in Hurler patients but normally present. That enzyme, called aldurazyme, is manufactured by

Fig. 22.2 Patient Justin and his infusion team for the first infusion of the enzyme replacement therapy for Hurler syndrome. (From You Tube 2014)

Sanofi and the first infusion of a Hurler patient took place on January 19, 1997, with Dr. Neufeld granted the honor of starting the infusion. The patient and infusion team are shown in Fig. 22.2 (You Tube 2014).

Using similar approaches to those developed for Hurler syndrome, Dr. Neufeld was able to identify causes and missing or defective enzymes for several similar diseases, including Hunter syndrome, Sanfilippo syndrome, and Scheie syndrome. All these disorders are considered lysosome storage diseases, and they are normally fatal to patients in their teens or twenties unless treated (Kresge et al. 2007).

Dr. Neufeld's career at NIH progressed quickly after her discoveries related to rare diseases. In 1973, she became chief of the NIH section of Human Biochemical Genetics. In 1977, she became the first woman from NIH to be elected to the National Academy of Sciences. In 1979, Dr. Neufeld became the chief of the Genetics and Biochemistry branch of the National Institute of Arthritis, Diabetes, and Digestive and Kidney Diseases (NIADDK). From 1981 to 1983, she was the deputy director of NIADDK's division of intramural research, overseeing all the internal research work for the division (Lasker Foundation Newsletter 2017; Wikipedia 2025).

In 1984, the University of California at Los Angeles (UCLA) tried to recruit Dr. Neufeld, but she declined. She was still happy in her work at NIH. Just a few months later, the leadership of NIH changed, and Dr. Neufeld was willing to consider a move to an academic position. She joined UCLA as the chair of the biochemistry department, arriving at a time when the genetics program there was falling apart as faculty left for appointments elsewhere. Dr. Neufeld pioneered an intercampus program like that she had experienced at NIH and ultimately managed to create a department of human genetics in 1995, with the first graduate students recruited in 1998. Dr. Neufeld remained the biochemistry department chair until her retirement in 2004 (You Tube 2014).

Dr. Neufeld served as a member of the Board of Directors of the American Society of Human Genetics from 1977 to 1980. She served on the National MPS Society's Scientific Advisory Board from 1988 to 2020 and was awarded their Visionary Leadership Award in 2022. She was elected President of the American Society for Biochemical and Molecular Biology in 1992. She also served on that board and then as President of the Foundation for Advanced Education in the Sciences (National MPS Society 2022; Lasker Foundation Newsletter 2017).

Dr. Neufeld received many prestigious awards. In 1977, she was elected to membership in the American Philosophical Society and to the American Academy of Arts and Sciences. She received the Canada Gairdner International Award in 1981, the Albert Lasker Award for Clinical Medical Research in 1982, the Elliot Cresson Medal in Life Sciences from the Franklin Institute, the William Allan Award of the Society of Human Genetics in 1982, the Henry Wilkinson Memorial Award of the International Society for Clinical Enzymology in 1983, and the Wolf Prize in 1988. She was named California Scientist of the Year in 1990. She received the National Medal of Science in 1994 (Gairdner 2025; The Franklin Institute 2025; Wikipedia 2025; Lasker Foundation Newsletter 2017; Hirschhorn 1983; Kresge et al. 2007). Upon receiving the National Medal, she said it was

> an enormous honor. I cannot describe how touched I am to have been selected...In a way, I feel I am accepting it on behalf of all the people who have worked with me. (Maugh 1994)

After retiring, Dr. Neufeld and her husband led a quiet life. Ben Neufeld died in 2020 after a few years of declining health. He had always placed his employment as subservient to his wife's and was known for telling long stories. Ben and Elizabeth had two children: Ellis, who is now Executive Vice President of St. Jude Hospital and Physician-in-Chief, and Sophie, who teaches dressage (Mount Sinai Cemetery 2020).

22.2 Lysosomal Storage Diseases

Human cells contain many structures within them, which are called organelles and serve specific purposes. Figure 22.3 shows a typical human cell. Within the cytoplasm of the cell are many lysosomes, generally round structures that break down worn-out or excess cell parts, destroy invading viruses or bacteria, and help self-destruct a cell. Lysosomes contain digestive enzymes and are quite acidic. Thus, they are surrounded by a membrane so that they don't contaminate the cytoplasm (NIH 2025).

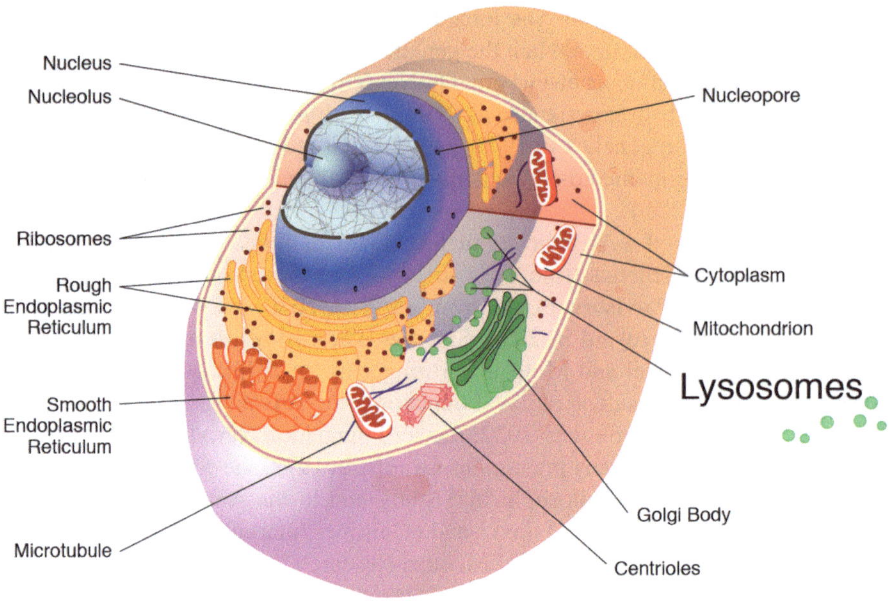

Fig. 22.3 Image of a human cell showing organelles and highlighting the small lysosomes. (From NIH 2025)

The lysosomes in a cell break down molecules into smaller pieces. They break down proteins into amino acids, carbohydrates into sugars, and lipids into simple fatty acids. In the diseases that Dr. Neufeld has studied, the process of breaking down carbohydrates is not functioning normally. Specifically, long, linear chains of sugars known as mucopolysaccharides (MPS) or glycosaminoglycans (GAGs) normally are broken down into component sugars in lysosomes. These complex carbohydrates help build bone, cartilage, tendons, corneas, skin, and connective tissue. Thus, a deficient metabolism of them causes a wide range of symptoms (NIH 2025).

In lysosomal storage diseases, the GAGs are either only partially effective or totally ineffective at the breakdown process. Larger molecules end up stored in the lysosomes, distorting them and distorting the cells they reside in. This can put pressure on nearby tissues and result in distorted features (NIH 2025; Maugh 1994).

Dr. Neufeld's interest in lysosomal storage diseases stemmed from her work on the biosynthesis of cell wall polymers. These, like GAGs, share the presence of uronic acids, so she naturally thought her expertise could be helpful. Fortunately, a new chemical assay technique had been developed as Dr. Neufeld was in the early stages of her work, allowing her to get more accurate measurements.

In her first, transformative experiments, Dr. Neufeld studied GAGs degradation in cultured fibroblasts of people with Hurler and Hunter Syndromes. Fibroblasts are cells that create and maintain a matrix structure in connective tissue. They are often chosen for biological experiments as they are easy to work with and changes in structure are obvious. Dr. Neufeld incorporated a radioactive marker into newly synthesized GAGs and followed it to see how it degraded. It was this experiment that led her to conclude that the problem of these diseases was not overproduction but rather a flaw in the breakdown process (Hirschhorn 1983; Fratantoni (a) 1968).

Dr. Neufeld learned that women who are heterozygous for the X chromosome-linked Hunter syndrome did not show any clinical symptoms. This baffled her because it seemed to be a violation of the generally accepted Lyon hypothesis, which suggested that half the cells should be abnormal for these women. Dr. Neufeld proposed a study of a fibroblast culture made from a combination of Hunter syndrome patient cells and cells from a normal male, but before she could get to this study, her collaborator accidentally mixed cells of Hunter and Hurler patients into a fibroblast. The mixture behaved as though the cells were normal, with a normal breakdown process for GAGs. The results were published in *Science* in 1968 (Fratantoni (b) 1968; Hirschhorn 1983). They showed that the enzyme deficiencies in the two diseases were different, and that a single gene to make each enzyme was adequate to solve the problem.

Over a period of more than a dozen years, Dr. Neufeld pursued this work further, identifying the gene and the specific enzymes that were defective in the various lysosomal storage diseases, and turning her focus to production of those enzymes. She showed that Hurler syndrome could be corrected by a large molecule enzyme derived from the fibroblasts or urine of normal humans. Similarly, she found corrective factors for Hunter syndrome and Sanfilippo syndrome. From this work, Sanofi developed means of making the enzyme treatments at a large enough scale to be therapeutic (Kresge et al. 2007; You Tube 2014). As genetic engineering improves, it is also possible that correcting the errant gene to code for production of the enzyme could be accomplished as an alternative to enzyme therapy.

22.3 Conclusion

Dr. Elizabeth Neufeld determined that a series of rare, fatal diseases that are caused by incomplete breakdown of complex carbohydrates in cells stems from ineffective or missing enzymes. She identified the enzyme key to each of these diseases and, in some cases, found means of creating those enzymes to be delivered to patients as a therapy. Dr. Neufeld was awarded the National Medal of Science in 1994.

References

You Tube. 21 Oct 2014. A Tribute to Dr. Elizabeth Neufeld. https://www.youtube.com/watch?v=fi2_dCxHamw. Accessed 15 Sept 2025.

National MPS Society. 27 Jul 2022. Announcing Dr. Elizabeth Neufeld as Recipient of the National MPS Society Visionary Leadership Award. https://mpssociety.org/resources/announcing-dr-elizabeth-neufeld-as-recipient-of-the-national-mps-society-visionary-leadership-award/. Accessed 15 Sept 2025.

Mount Sinai Cemetery. Mar 2020. Benjamin S. Neufeld (Mar 2020). https://mountsinaiparks.keeper.memorial/BenjaminNeufeld. Accessed 15 Sept 2025.

The Franklin Institute, Elizabeth F. Neufeld. https://fi.edu/en/awards/laureates/elizabeth-f-neufeld. Accessed 15 Sept 2025.

Gairdner. Elizabeth F. Neufeld. https://www.gairdner.org/winner/elizabeth-f-neufeld. Accessed 15 Sept 2025.

Wikipedia. 26 Aug 2025. Elizabeth F. Neufeld. https://en.wikipedia.org/wiki/Elizabeth_F._Neufeld. Accessed 15 Sept 2025.

Fratantoni, Joseph, Hall, Clara, and Neufeld, Elizabeth (a). 1968. The defect in Hurler's and Hunter's syndromes: faulty degradation of polysaccharide *PNAS* 60(2): 699–706.

Fratantoni, Joseph, Hall, Clara, and Neufeld, Elizabeth (b). 1968. Hurler and Hunter syndromes: mutual correction of the defect in cultured fibroblasts. *Science* 162(3853); 570–2.

Hirschhorn, K. 1983. The William Allan Memorial Award. *Am J Genet* 35;1077–80.

Kresge, Nicole, Simoni, Robert and Hill, Robert. May 2007. Lysosomal Storage Disease Factors: The Work of Elizabeth F. Neufeld. *J Biol. Chem* 282(20); e15–e17.

Lasker Foundation Newsletter. Sum 2017. Neufeld's Advice to Young Scientists: Be the Best You Can Be. https://laskerfoundation.org/neufelds-advice-to-young-scientists-be-the-best-you-can-be/. Accessed 15 Sept 2025.

Maugh, Thomas, II. 9 Sept 1994. Biologist at UCLA Wins Medal: Research: National Science Foundation award goes to Elizabeth Neufeld for her work on two genetic diseases. *LA Times*. Available at https://www.latimes.com/archives/la-xpm-1994-09-09-me-36319-story.html. Accessed 15 Sept 2025.

NIH. 15 Sept 2025. Lysosome, National Human Genome Research Institute. https://www.genome.gov/genetics-glossary/Lysosome. Accessed 15 Sept 2025.

Santisi, Jennifer. Elizabeth F. Neufeld, National Science and Technology Medals Foundation. https://nationalmedals.org/laureate/elizabeth-f-neufeld/. Accessed 15 Sept 2025.

Chapter 23
Isabella Karle

For the development and application of a method for determining essentially equal-atom crystal and molecular structures by X-ray analysis, thereby having a profound effect on the practice of organic and biological chemistry

Keywords Crystallography · Electron diffraction · Molecular structure · X-ray diffraction · X-ray crystallography · Three-dimensional molecular structure

Dr. Isabella Karle (Fig. 23.1) revolutionized crystallography, laying the foundation for three-dimensional structures of molecules and the ability to synthesize those molecules. Her techniques, which have now resulted in the identification of thousands of molecular structures, have enabled the development of new pharmaceutical products and other synthesized materials. She received the National Medal of Science in 1995.

J. S. Tietjen et al., *Women Laureates 1970 – 2007*, Women in Engineering and Science, https://doi.org/10.1007/978-3-032-23116-1_23

Fig. 23.1 Isabella Karle holding a construction of a molecule. (Gift of Dr. Karle, courtesy of AIP Emilio Segrè Visual Archives)

23.1 Life (1921–2017)

Isabella Lugowski was born in Detroit in 1921 to parents who had immigrated from Poland. Her father, Zygmunt A. Lugowski, was a house painter, and her mother, Elizabeth Graczyk, was a seamstress. Her mother and a friend decided to go into the restaurant business—Isabella was their accountant, even before she started elementary school. They did not speak English at home; that she learned when she started school. During her high school years, she decided to study chemistry. She got some pushback from her father, who wanted her to be a lawyer, but her parents were very supportive of her going to college (Atomic Heritage Foundation 2015; Hargittai 2015; Proffitt 1999).

She graduated from high school in the middle of the school year and started at Wayne State University. She went to classes and lived at home. That experience was formative as her chemistry professor, assistant professor Dr. Joseph Jasper, at Wayne State was the first person to tell her she should consider going to graduate school to get a master's degree and a PhD—and that became her goal (Atomic Heritage Foundation 2015; Shearer and Shearer 1997).

During her semester at Wayne State, a former high school teacher came to the house and told her there would be a statewide examination for a scholarship to the University of Michigan the following day. The teacher would pick her up and take her to the examination. Isabella was number four out of all the students who took the examination and got a four-year scholarship to go to the University of Michigan. She also got another fellowship and enrolled in the chemistry program at the University of Michigan the following fall (Atomic Heritage Foundation 2015; Shearer and Shearer 1997).

Isabella earned her bachelor's degree with highest honors in 1941 and wanted to continue her studies as a graduate student. But she discovered that all graduate teaching assistantships in the chemistry department were awarded only to men. The American Association of University Women came to her rescue with a fellowship that allowed her to start her graduate work. The following year, she was awarded the Horace H. Rackham fellowship. Isabella stayed at the University of Michigan, receiving her MS in 1942 and marrying fellow graduate student Jerome Karle that same year. She received her PhD in 1944. For her PhD, she worked on electron diffraction by gaseous molecules with Dr. Lawrence Brockway (Shearer and Shearer 1997).

After obtaining their PhDs from the University of Michigan, both Karles worked on the Manhattan Project at the University of Chicago. They were both involved in finding procedures to make the fission fuel plutonium without impurities. Dr. Isabella Karle succeeded in making pure plutonium chloride. When the war was over, they worked for 2 years at the University of Michigan despite the anti-nepotism rules in place at the time. Dr. Karle was the first female member of the chemistry faculty at Michigan (NSTMF 2025).

In 1946, they were both hired at the Naval Research Laboratory (NRL) and even allowed to work together. Dr. Karle characterizes the NRL, "we were immediately thrust into an atmosphere of exciting research and surrounded by exceptional people." Their first work was on electron diffraction (NRL 1994).

Over the course of her distinguished career, Dr. Karle received many awards and honors. Her first award was SWE's 1968 Achievement Award of which she said, "I was rather surprised and pleased that the women engineers were the ones who gave me my first award" (Atomic Heritage Foundation 2015). She was elected to the National Academy of Sciences in 1978. Jerome Karle was the co-recipient of the 1985 Nobel Prize in Chemistry, an award that he shared with Herbert Hauptman. It has been said that the sadness Jerome felt when he learned that Isabella would not share the Nobel Prize never left him (Hargittai 2015).

Although she believed it would have been wonderful to receive the Nobel Prize with Jerome, Dr. Isabella Karle received much recognition for her work. She received the Aminoff Prize from the Royal Swedish Academy of Science in 1988, which was established for pioneers in crystallography. In 1993, she became the first woman to receive the Bower Award and Prize for Achievement in Science from the Franklin Institute in 1993 "for determining three-dimensional structure of molecules with X-ray diffraction" (Hargittai 2015). Karle received the National Medal of Science from President Bill Clinton in 1995 (NSTMF 2025). She also received several honorary degrees, awards from the US Navy, and other honors (Bailey 1994; Shearer and Shearer 1997).

Dr. Karle was active in the profession, serving as President of the American Crystallographic Association. She was a member of the American Physical Society and the American Chemical Society (Bailey 1994).

The Karles retired from the Naval Research Laboratory after more than 60 years of service (Hargittai 2015). Over the course of her career, Dr. Karle had published nearly 300 scientific papers (Shearer and Shearer 1997). She said, "I've had a satisfying and fascinating career" (Shearer and Shearer 1997).

23.2 Contributions

The work the Karles performed at NRL proceeded as had much of their prior work—Jerome undertook the theoretical efforts, and Isabella performed the experiments (Shearer and Shearer 1997). After electron diffraction, they worked on X-ray diffraction and determining the three-dimensional structure of molecules (NRL 1994).

> Our subsequent work concerned X-ray diffraction and the elucidation of the three-dimensional structure of molecules. A key problem existed; in fact, half of the necessary information seemed to be lost in the process of the diffraction procedure. It was called the phase problem. If we could evaluate the phases as well as the intensities, then we could work backwards and determine the structures of these molecules (NRL 1994).
>
> Well, Jerome Karle, with a colleague, Herbert Hauptman, who worked in our group at the time, solved the problem of obtaining the phases theoretically. The results were presented in terms of an infinite set of inequalities, probabilistic functions, and transcendental functions. It was I who bridged the gap between the theory and experiment by translating these mathematical results into practical procedure. . . (NRL 1994).
>
> There was a great amount of trepidation when we gambled our research budget to have IBM do a final refinement on my results. You see, at that time, this was the mid-1950s – IBM was the only company that had a computer that was capable of doing the problem that we needed to be done. . . My results worked and that was the beginning of doing several hundred (I've not counted them) structures over the years, in various areas of science. The successes we've had, in this sort of activity, have spawned numerous advance-study workshops that took place around the world for the next 20-25 years (NRL 1994).
>
> The notes and instructions that I provided for the workshops became teaching materials in various universities and the basis for a number of computer programs. These courses were instrumental in expanding the number of crystal structures solved around the world, from a handful each year to much more than 10,000 a year, at present (NRL 1994).

One of the first successes of solving crystal structures was determining the structure of the venom of a South American frog. The secretions from the frog skin were used by indigenous people as poison on arrow tips. Understanding the mechanism used by the venom to block specific nerve transmission proved to be very useful for the study of nerve transmission for medical purposes. Based on her work, a synthetic form of the venom could be manufactured (Shearer and Shearer 1997).

The National Science and Technology Medals Foundation explains her legacy this way (NSTMF 2025):

> It was at NRL that Karle used a novel x-ray scattering technique to determine the structure of crystals, a technique that is used to study the biological, chemical, metallurgical and physical characteristics, allowing processes to be designed to duplicate the molecules being studied.
>
> Karle's method drastically improved the speed and accuracy of chemical and biomedical analysis and remains the basis of all advanced x-ray crystallography, including computerized programs, used around the world today. The technique has played a major role in the development of new pharmaceutical products and other synthesized materials.

23.3 Conclusion

Dr. Isabella Karle revolutionized X-ray crystallography providing new techniques for the development of new pharmaceutical products and other synthesized materials. Although she did not share the Nobel Prize for this work, the processes that she developed have contributed to the well-being of people around the world.

Author's Note Adapted from Tietjen, Jill S. "Isabella Karle." from the book by Craig, Cecilia, et. al. Women Engineering Legends 1952–1976: Society of Women Engineers' Achievement Award Recipients. Cham, Switzerland: Springer. 2025.

References

Atomic Heritage Foundation. 2015. Voices of the Manhattan Project. Isabella Karle's Oral History Interview. https://ahf.nuclearmuseum.org/voices/oral-histories/isabella-karles-interview-2015/.

Bailey, Martha J. 1994. *American Women in Science: A Biographical Dictionary*. Santa Barbara, California: ABC-CLIO.

Hargittai, Magdolna. 2015. *Women Scientists: Reflections, Challenges, and Breaking Boundaries.* Oxford, England: Oxford University Press.

National Science & Technology Medals Foundation (NSTMF). Isabella L. Karle. https://nationalmedals.org/laureate/isabella-l-karle/. Accessed January 23, 2025.

Office of Naval Research (NRL). 1994. Naval Research Reviews, Two. Volume XLVI. https://apps.dtic.mil/sti/tr/pdf/ADA281435.pdf.

Proffitt, Pamela, ed. 1999. *Notable Women Scientists.* Farmington Hills, Michigan: Gale Group.

Shearer, Benjamin F. and Barbara S. Shearer, Editors. 1997. *Notable Women in the Physical Sciences: A Biographical Dictionary*. Westport Connecticut: Greenwood Press.

Chapter 24
Ruth Patrick

For her algal research, particularly the ecology and paleoecology of diatoms, and for elucidating the importance of biodiversity of aquatic life in ascertaining the natural condition of rivers and the effects of pollution

Keywords Ecology · Paleoecology · Limnology · Patrick Principle · Diatoms · Water pollution · Biodiversity · Rivers · Diatometer · Freshwater biology

Credited with laying the groundwork for modern water pollution control efforts, Ruth Patrick (Fig. 24.1), after whom the biodiversity tenet the Patrick Principle is named, matched the types and numbers of diatoms, single-celled algae, in water to the type and extent of water pollution and invented the diatometer to collect and measure those diatoms. Over her 60-year career, Dr. Patrick advanced the field of limnology, the study of freshwater biology. Along with Rachel Carson, Dr. Patrick is recognized as having ushered in the current concern for the environment and ecology. She has been called the foremost authority of America's river systems (Bailey 1994; Zauzmer 2013). She received the National Medal of Science in 1996.

J. S. Tietjen et al., *Women Laureates 1970 – 2007*, Women in Engineering and Science, https://doi.org/10.1007/978-3-032-23116-1_24

Fig. 24.1 Ruth Patrick. (Courtesy of The Academy of Natural Sciences, Philadelphia)

24.1 Life (1907–2013)

Ruth Patrick was born in Topeka, Kansas, on November 26, 1907, and grew up primarily in Kansas City, Missouri (ANS 2013). Her interest in science was encouraged by her lawyer father, Frank Patrick, who took her and her sister on Sunday walks in the woods where they collected items and put them in a can they carried on the end of a stick (ANS 2013). Her father told her, "Don't cook, don't sew; you can hire people to do that. Read and improve your mind" (Profitt 1999). As her father also said, "Leave the world a better place for having passed through it" (ANS 2013).

Her father gave her a microscope when she was 7 years old, which had a profound effect on her life and career. The walks and the microscope, she later said, would stimulate her lifelong passion for the environment (ANS 2013). She reported, "I forget the whole world when I sit at my microscope" (Lowe 2015). In 2004, she said (Dicke 2013)

> I collected everything: worms, mushrooms, plants, rocks. I remember the feeling I got when my father would roll back the top of his big desk in the library and roll out the microscope. He would make slides with drops of the water samples we had collected, and I would climb up on his knee and peer in. It was miraculous, looking through a window at a whole other world.

Patrick originally took general biology courses at the University of Kansas in Lawrence, Kansas, until her mother insisted that she study elsewhere. She was shipped off to Coker College in South Carolina with her mother hoping she would marry and display social graces instead of studying botany! (Lowe 2015) Regardless of her mother's wishes, Patrick studied botany, receiving her BS degree from Coker College in 1929 (Profitt 1999). Her father feared that her science at Coker would not be an adequate education. Thus, her summers as an undergraduate student were spent taking courses at the Woods Hole Biological Laboratory and the Cold Spring Harbor Biological Laboratory (Lowe 2015). She knew that graduate school was going to be necessary for the science career she envisioned, and she selected the University of Virginia, where she could continue to study diatoms. She earned both her MS (1931) (thesis title: *A Study of the Diatoms of Charlottesville and Vicinity*) and PhD in botany (1934) (dissertation title: *A Taxonomic and Distributional Study of Some Diatoms from Siam and the Federated Malay States*) there (Profitt 1999).

Originally hired as a "volunteer" (without pay—as she was told that women scientists at the time were not paid and discouraged from wearing lipstick) in 1933 at the Academy of Natural Sciences in Philadelphia, Pennsylvania, Dr. Patrick's initial efforts were in microscopy to work with their collection of diatoms, considered one of the best collections in the world. Diatoms are microscopic, single-celled algae that are at the bottom of the food chain. They are an important part of the food supply of freshwater ecosystems. She continued without pay until 1945, while supporting herself through part-time teaching at the Pennsylvania School of Horticulture and making chick embryo slides for Temple University (Profitt 1999: Shearer and Shearer 1996; Dicke 2013).

She progressed through several positions at the Academy, including curator of the Leidy Microscopical Society (1937–1947) and associate curator of the Academy's microscopy department (1937–1947). In 1947, she became the curator and chairwoman of the Academy's limnology department, which she founded, today called the Patrick Center for Environmental Research.[1] In 1973, Dr. Patrick was named the Francis Boyer Research Chair of limnology at the Academy. From 1973 to 1976, she served as chairwoman of the Academy's board, the first woman to hold that position. Concurrently, she taught at the University of Pennsylvania. Her courses included limnology, pollution biology, and phycology. Her research included taxonomy; ecology and the physiology of diatoms; the biodynamic cycle of rivers; and forces that determine the diversity of ecosystems (ANS 2013; Bailey 1994; Bott 2013; Profitt 1999; Shearer and Shearer 1996).

[1] In 2011, the Academy of Natural Sciences became affiliated with Drexel University (ANS 2011).

In the late 1940s, Dr. Patrick gave a paper at a scientific conference on her research on the diatoms of the Poconos. An oil company executive in the audience was so impressed with the ability of diatoms to predict the health of a body of water that he provided funds to support her research. With these funds, in 1948, Dr. Patrick undertook a survey of the then severely polluted Conestoga Creek in Pennsylvania; the first study of its kind. The Creek contained many types of pollution, including fertilizer runoff, sewage, and waste products (some of them toxic) from industries in the area. She matched the types and numbers of diatoms in the water to the type and extent of pollution. This procedure is today used universally but was groundbreaking at the time. She said, "By studying the conditions in which diatoms lived, I discovered that the presence of different species often pointed to different types of water" (Bott and Sweeney 2013; Dicke 2013; National Science and Technology Medals Foundation 2025; Profitt 1999).

To aid in the effort, she invented the Catherwood diatometer, a clear acrylic device that holds glass microscope slides and is equipped with flotation capability. The diatometer collects the diatoms from bodies of water; they attach to the slides and grow there. Her research showed that healthier bodies of water should contain many species of organisms from different groups. The belief that biodiversity (the number and kinds of species) is the key indicator of water health is today known as the Patrick Principle in her honor. The Patrick Principle is the foundation of all current environmental assessments (Bott and Sweeney 2013; Dicke 2013; National Science and Technology Medals Foundation 2025; Profitt 1999).

> My great aim has been to be able to diagnose the presence of pollution and develop means of cleaning things up. (Zauzmer 2013)

An advisor to presidents, Dr. Patrick was actively involved in the drafting of the federal Clean Water Act, passed in 1972, and made frequent appearances on Capitol Hill as an environmental expert. Called the foremost authority on America's river systems, Dr. Patrick estimated at one point that she had waded into 850 different rivers around the globe, including the Amazon River. She established baseline conditions on the Savannah River before the opening of the Savannah River Nuclear Power Plant and then monitored river conditions after it opened. At the request of the Army Corps of Engineers, she evaluated the effect of channelization on rivers throughout the USA. After the Three Mile Island accident in 1979, she was asked to assess radionuclide contamination of the Susquehanna River (Bott and Sweeney 2013).

In 1967, she was instrumental in the founding of the Stroud Water Research Center on the banks of White Clay Creek in Pennsylvania. The first project of the Center was devoted to understanding the biological structure and ecological functioning of a riffle-pool sequence in this creek. The interdisciplinary team approach examined chemistry, microbial ecology, entomology, hydrology, and fish, among other topics and produced a new paradigm for flowing water systems called "The River Continuum"—one of the most highly cited papers in the field of stream ecology. The Center continues in operation today, adhering to Dr. Patrick's original vision: To advance knowledge and stewardship of freshwater systems through global research, education, and watershed restoration (Bott and Sweeney 2013: Stroud 2025).

Dr. Patrick taught limnology and botany at the University of Pennsylvania for more than 35 years, during which she always had time for her students and was beloved. She wrote hundreds of technical articles and published multiple books. She collaborated with Charles Reimer on the two-volume *Diatoms of the United States*. Before these volumes were published, scientists used European literature for the study of diatoms. Because of her work, the Academy of Natural Science in Philadelphia became the mecca for diatomists. She also wrote *Groundwater Contamination in the United States* (1983, 1987), *Power So Great*, *Colors of Tomorrow*, and the five-book series *Rivers of the United States*. She became known as "The River Doctor" (ANS 2013; Bailey 1994; Dicke 2013; Lowe 2015; Profitt 1999; Shearer and Shearer 1996).

Dr. Patrick was the first woman to serve on the board of directors of DuPont and its first environmental activist. She also served on the board of directors of Pennsylvania Power and Light and advised Presidents Lyndon B. Johnson (on water pollution) and Ronald Reagan (on acid rain), as well as several Pennsylvania governors on water quality issues. She served on water pollution and water quality panels for the National Academy of Sciences and the Department of the Interior, as well as other federal advisory groups. These included the National Academy of Sciences Committee on Science and Public Policy, the General Advisory Committee of the Environmental Protection Agency, the Advisory Council of the Renewable Resources Foundation, and the Smithsonian Institution Council. Her professional affiliations included the Phycological Society of America (president), American Society of Naturalists (president), Botanical Society of America, American Society of Limnology and Oceanography, American Institute of Biological Sciences, Ecological Society of America, and the American Society of Plant Taxonomists. She also served on the board of directors of the World Wildlife Fund (ANS 2013; Bott and Sweeney 2013; Legacy.com 2013).

Dr. Patrick was a great believer in scientists, government, and industry working together to solve problems. She said (Dicke 2013; Profitt 1999; Shearer and Shearer 1996):

> My great theme in life is that academia, government, and industry have got to work closely on all the big problems of the world. Unless academics and industry get together there will not be many bright young people trained in the future. We have to develop an atmosphere where the industrialist trusts the scientist and the scientist trusts the industrialist. You've got to trust people.

Elected to the National Academy of Sciences in 1970 (the twelfth woman to receive this form of recognition), Dr. Patrick received the National Medal of Science in 1996 from President Bill Clinton. A Fellow of the American Academy of Arts and Sciences (elected 1976) and the recipient of over 25 honorary degrees, she has been inducted into the National Women's Hall of Fame (Dicke 2013; NSTMF 2025; Ogilvie and Harvey 2000). In 2008, Coker College named her the "Alumna of the Century" (ANS 2013).

Her many other awards include the 1975 John and Alice Tyler Ecology Award, Pennsylvania Award for Excellence in Science and Technology, the Gold Medal from the Royal Zoological Society of Belgium, the Heinz Awards' Chairman's Medal, and the Benjamin Franklin Award for Outstanding Scientific Achievement from the American Philosophical Society. In 1996, she was inducted into the South Carolina Hall of Science and Technology. A public television profile of her, produced by PBS Channel WHYY in Philadelphia, was nominated for a Mid-Atlantic Emmy in 2005. Buildings and locations named for her include the Ruth Patrick Science Education Center at the University of South Carolina (ANS 2013). In 1998, the Board of the Association of Sciences for Limnology and Oceanography initiated the Ruth Patrick Award to honor outstanding research by a scientist in the application of basic aquatic science principles to the identification, analysis, and/or solution of important environmental problems (ASLO 2025).

Dr. Patrick was married twice. Her first husband was Charles Hodge IV, with whom she had a son, Charles Hodge V. Her husband died in 1985. Her second husband was Lewis H. Van Dusen, Jr., who died in 2004. When she married Hodge, an entomologist, in 1931, her father asked him if she could keep her maiden name. Ruth Patrick says, "When I got married my father said to Charles": "I always wanted to be a scientist. Would you mind if Ruth kept the name of Patrick in her research? I'd like to see it go on" (ANS 2013; Lowe 2015).

24.2 Diatoms

Diatoms (Fig. 24.2) are single-celled algae that live in oceans, lakes, and rivers. They are transparent and their walls are made of opaline silica. Their cell walls often feature intricate patterns. Diatoms collect energy from the sun and convert it into chemical energy through photosynthesis. They are responsible for 20–30% of the air we breathe as they remove carbon dioxide from the atmosphere and release oxygen. They are also an important component of the food chain in bodies of water, serving as a food source for everything from zooplankton to whales, including aquatic insects and fish (Spaulding 2021).

As Dr. Patrick discovered, diatoms are very sensitive to their environment and thus are excellent indicators of water condition—and water pollution. Any species will have a specific range of salinity and acidity levels that it tolerates. Other characteristics of the water for which each species has ranges and tolerances for include water level, flow regime, nutrient concentration, and suspended sediment (Spaulding 2021).

Dr. Patrick established the Diatom New Taxon File—also called the "New Species File" in the 1950s. Her goal was to have a 5″ × 8″ card for every diatom name that had been published after the publication of the Mill's Index—which covered 1816–1932. The original card index was digitized by The Academy of Natural Sciences in 2014–2016 (Potapova et al. 2024). There are estimated to be anywhere from 20,000 to two million species of diatoms. More are found each year (Spaulding 2021).

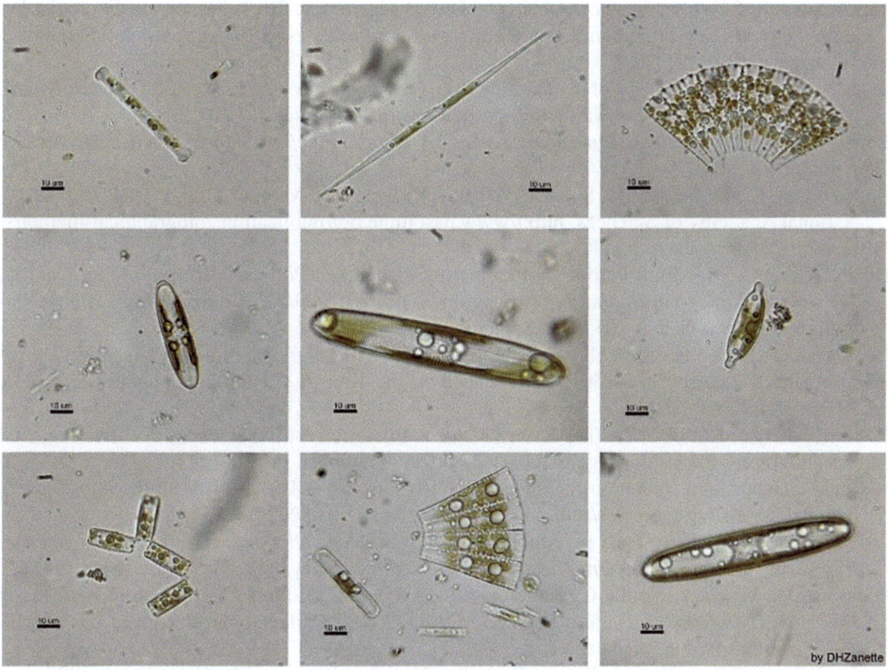

Fig. 24.2 Diatoms. (Courtesy of Wikipedia)

24.3 Conclusion

"The River Doctor," Ruth Patrick established that diatoms could indicate the health of bodies of water and laid the groundwork for modern water pollution control efforts. A beloved teacher and researcher, Dr. Patrick is recognized, along with Rachel Carson, with ushering in the current concern for the environment and ecology.

Author's note Adapted from the chapter by Jill S. Tietjen titled "Pioneering Women in Water Quality," from the volume in the Springer Women in Engineering and Science series *Women in Water Quality: Investigations by Prominent Female Engineers*, Deborah Jean O'Bannon, editor, 2020. Cham, Switzerland. Springer.

References

The Academy of Natural Sciences (ANS) of Drexel University. 2011. The Academy of Natural Sciences and Drexel University Announce a Historic Affiliation. http://www.ansp.org/about/drexel-affiliation/. Accessed July 10, 2025.

The Academy of Natural Sciences (ANS) of Drexel University, 2013. Pioneering Ecologist Dr. Ruth Patrick Dies https://ansp.org/about/press-room/releases/2013/ruth-patrick/. September 13. Accessed July 10, 2025.

Association for the Sciences of Limnology and Oceanography (ASLO). Ruth Patrick Award. https://www.aslo.org/aslo-awards/ruth-patrick-award/. Accessed July 10, 2025.

Bailey, Martha J., *American Women in Science: A Biographical Dictionary*, Denver, Colorado: ABC-CLIO, 1994.

Bott, Thomas L. and Bernard W. Sweeney. 2013. A Biographical Memoir, Ruth Patrick 1907-2013. National Academy of Sciences. http://www.nasonline.org/publications/biographical-memoirs/memoir-pdfs/patrick-ruth.pdf.

Dicke, William. 2013. Ruth Patrick, A Pioneer in Science and Pollution Control is Dead at 105. The New York Times. http://www.nytimes.com/2013/09/24/us/ruth-patrick-a-pioneer-in-pollution-control-dies-at-105.html. September 24.

Legacy.com. Ruth Patrick obituary. https://www.legacy.com/us/obituaries/kansascity/name/ruth-patrick-obituary?id=4456157. Published in the Kansas City Star September 29, 2013. Accessed July 10, 2025.

Lowe, Rex. 2015. Ruth Patrick: The River Doctor. Association for the Sciences of Limnology and Oceanography. https://doi.org/10.1002/lob.10070. November. Pp. 108–111.

National Science and Technology Medals Foundation (NSTMF). Ruth Patrick. https://www.nationalmedals.org/laureates/ruth-patrick. Accessed July 10, 2025.

Ogilvie, Marilyn and Joy Harvey, editors. *The Biographical Dictionary of Women in Science: Pioneering Lives from Ancient Times to the Mid-20th Century*, New York: Routledge, 2000.

Potapova, M., Veselá, J., Smith, C., Minerovic, A., Aycock, L. (Editors) 2024. Diatom New Taxon File at the Academy of Natural Sciences (DNTF-ANS), Philadelphia. Retrieved on July 10, 2025, from http://dntf.ansp.org.

Profitt, Pamela, Editor, *Notable Women Scientists*, Detroit, Michigan: The Gale Group, 1999.

Shearer, Benjamin H. Shearer and Barbara S. Shearer, Editors, *Notable Women in the Life Sciences: A Biographical Dictionary*, Westport, Connecticut: Greenwood Press, 1996.

Spaulding et al. 2021. Diatoms.org: supporting taxonomists, connecting communities. Diatom Research 36(4): 291–304. https://doi.org/10.1080/0269249X.2021.2006790.

Stroud Water Research Center. https://stroudcenter.org. Accessed July 10, 2025.

Zauzmer, Julie. 2013. Ruth Patrick, ecology pioneer, dies at 105. https://www.washingtonpost.com/national/health-science/ruth-patrick-ecology-pioneer-dies-at-105/2013/09/23/2bcde762-245e-11e3-b75d-5b7f66349852_story.html?utm_term=.5d9dd8392428. September 23.

Chapter 25
Stephanie Kwolek

For her contributions to the discovery, development, and liquid crystal processing of aramid fibers, which allowed for the development of lifesaving products used around the world

Keywords Chemist · Kevlar® · Aramid fibers · Polymer · Polymer chemistry · DuPont · Lycra® · Nomex® · Rayon · Liquid crystal processing · Spandex · Liquid crystalline polymer · Polyamides · Condensation polymers · Intractable polymers · Nylon · Low-temperature polymerization

Stephanie Kwolek (Fig. 25.1), the fourth woman inducted into the National Inventor's Hall of Fame (1995), is best known for her invention of Kevlar®, the lightweight yet very strong polymer used in bulletproof vests and many other products. Kwolek spent 40 years with DuPont, during which time she obtained 16 patents for a variety of groundbreaking materials and devised new processes in polymer chemistry. In 1996, Kwolek became the second individual woman to receive the National Medal of Technology.

J. S. Tietjen et al., *Women Laureates 1970 – 2007*, Women in Engineering and Science, https://doi.org/10.1007/978-3-032-23116-1_25

Fig. 25.1 Stephanie Kwolek. (Courtesy of Science History Institute—Wikimedia Commons)

25.1 Life and Career (1923–2014)

The oldest daughter of John and Nellie Kwolek, Stephanie Kwolek was born in New Kensington, Pennsylvania. Her father loved the natural world, and she spent many hours with him exploring the woods and fields near their home. She created scrapbooks that she filled with leaves, seeds, grasses, wildflowers, and their descriptions. Her mother was at first a homemaker but had to enter the workforce after her husband died when Stephanie was 10 years old. Stephanie inherited an interest in fabric and sewing from her mother (Proffitt 1999; Science History Institute Museum and Library 2025). Her family's circumstances led to her saying, "I realized early in life that it was a matter of sink or swim" (Ferguson 1986).

Kwolek showed an early interest in science—and in fashion design. Maybe it is not surprising that much of her scientific research thus centered on fibers. Intending to pursue a career as a doctor, Kwolek graduated with a BS in chemistry from Margaret Morrison Carnegie College (then the women's college of Carnegie Mellon, today part of Carnegie Mellon University) in 1946. She accepted a position as a chemist in the Rayon Department with DuPont, in Buffalo, New York, planning to save the money she needed to attend medical school. One of the reasons she chose DuPont was that the starting salaries for men and women were equal (Ferguson 1986). But she became so interested in the research in which she was involved that she decided medical school was no longer in her future.

Four years after she began work at DuPont, the Pioneering Research Laboratory where she worked was transferred to Wilmington, Delaware (Artwork Archive 2025; Ferguson 1986; Proffitt 1999).

> I became so interested in the work that I was doing that I stayed on. It was very challenging. It differed from day to day. It was very exciting because you never knew what you might come up with. It was a constant learning process. (Proffitt 1999)

"There was tremendous interest in new fibers, and new ways to make fibers," said Kwolek, who would spend her entire career conducting polymer research (Busch-Vishniac et al. 2024; Rogin et al. 2024). Polymers are large molecules whose building blocks are smaller molecules chained together (Busch-Vishniac et al. 2024).

One of her early projects was the preparation of polymers that formed Lycra® spandex fibers, the stretchy material used in athletic wear. Another notable product to emerge from her work is Nomex®, which is fire-resistant and now commonly used in protective gear worn by firefighters. In the course of looking for a strong, lightweight fiber that could be used in radial tires, Kwolek discovered what became Kevlar® (Carnegie Mellon University 2014; Proffitt 1999).

Kwolek described the environment for women during the years of her career:

> It certainly was difficult at times because the opportunities were not open to women from the 1940s through probably the 1970s. Things finally got better in the 1980s for women but the opportunities for advancement were limited. (Proffitt 1999)

After her retirement, Kwolek worked as a consultant in polymer chemistry. She also served on the Liquid Crystalline Polymer Committee of the National Research Council of the National Academy of Sciences. Kwolek actively mentored other women scientists and participated in outreach to schools to encourage children to pursue scientific careers. Her paper, written with Paul W. Morgan, *The Nylon Rope Trick*, and published in the *Journal of Chemical Education* in April 1959, describes a science experiment that can be performed in the classroom. The two polymers used to make nylon are dissolved in two solvents, one of which is water. The two solutions are poured, one on top of the other, into a beaker. A film forms at the interface that can be pulled out of the center as a rope, resembling scarves that are pulled out of a magician's hat. It is used in classrooms across the country. It was even exhibited at the Smithsonian Institution and at the DuPont exhibit at the 1964 World's Fair in New York City (Carnegie Mellon University 2014; Ferguson 1986; Proffitt 1999; Science History Institute Museum and Library 2025).

Among her many awards, Kwolek received the National Medal of Technology in 1996, only the second individual woman to do so. She was the fourth woman inducted into the National Inventors Hall of Fame (1995). She was the first woman inducted into the Plastics Hall of Fame (1997), she received the Perkin Medal from the Society of Chemical Industry's American section, and she was elected to the National Academy of Engineering in 2001. Kwolek received the Fred O. Conley Award in Plastics Engineering and Technology from the Society of Plastics Engineers and the American Innovator Award from the U.S. Patent and Trademark Office. She was the first woman to receive DuPont's Lavoisier Medal for outstanding technical achievement (1995). Kwolek received a number of honorary doctorates and has been inducted into the Hall of Fame of Delaware Women and the

National Women's Hall of Fame (Artwork Archive 2025; Carnegie Mellon University 2014).

Kwolek was a member of the American Chemical Society, the American Institute of Chemists, and the Franklin Institute of Philadelphia. She was also a member of two honorary societies—Sigma Xi and Phi Kappa Phi (Carnegie Mellon University 2014).

In her retirement, she enjoyed gardening, music, traveling, and reading (Proffitt 1999). On the day Kwolek died in 2014, *The New York Times* reported that DuPont had just announced the one-millionth vest made with Kevlar® technology had been sold (Warren 2025).

She had said,

> I was fortunate enough to do something that would be a benefit to mankind. It's been an extremely satisfying discovery. I don't think there's anything like saving someone's life to bring you satisfaction and happiness. (Rogin et al. 2024)

25.2 Kevlar®

Kwolek worked extensively with polyamides which are polymers in which the building blocks are linked by amide groups. Amide groups (CO-NH) may be produced by an amine (NH_2) and a carboxyl (CO_2H) or by polymerization of amino acids or amino-acid derivatives. Polyamides include proteins and peptides as well as synthetic polymers (Britannica 2026). Aromatic polyamides (shortened to aramids) are synthetic fibers in which there are rings of six carbon atoms and the rings are connected by amides (Wikipedia—Aramid 1996).

Condensation polymers occur not by chain-growth processes but by reactions, often with the loss of a byproduct (water). Dacron (polyester), Nylon 66 (polyamide), Mylar (polyester), Lexan (polycarbonate), Perlon (polyamide), Kevlar® (polyamide), and Nomex® (polyamide) are all condensation polymers (Reusch 2026). Many of these polymers presented problems in that either they would not melt when subjected to the heat required for melt polymerization or they degraded when subjected to the heat. Thus, they were termed "intractable polymers" as in "difficult to deal with or control." The polymer chemistry research and development that Kwolek conducted also contributed to the development of polyimide film which is classified as a high-performance plastic (Wikipedia—Polyimide 1996).

She characterizes her major contributions as follows (Ferguson 1986):

> The first would be the discovery of liquid crystalline solutions of synthetic aromatic polyamides and of fibers from these solutions. These fibers have exceptionally high tensile strength and stiffness combined with light weight. This discovery resulted in the development of "Kevlar" aramid fiber.
>
> Number two would be the discovery and development with others in the laboratory of low-temperature polymerization processes for preparing aromatic polyamides and other condensation polymers. These polymers either did not melt, or else degraded before they melted. Until this time, the only known method for preparing these polymers was by melt polymerization. When we discovered the various low-temperature methods of polymeriza-

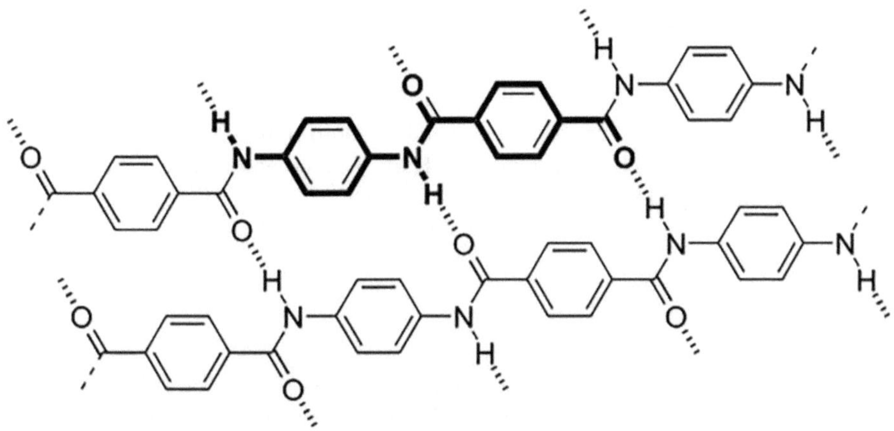

Fig. 25.2 Kevlar's molecular structure. (Courtesy of Wikipedia)

> tion, we opened up the entire field of intractable polymers. This work served as the basis for the future large-scale preparation of "Lycra" spandex fiber, "Kapton" polyimide film, and "Nomex" and "Kevlar" aramid fibers. Now we were able to make hundreds of polymers that could not be made before.

In 1964, Kwolek and her colleagues began looking for a fiber that could be used to replace the steel used in radial tires. She unexpectedly discovered that polyamides that have rod-like molecules will form liquid crystalline solutions under certain conditions. In most polymers, the molecules flow in a disorderly fashion. However, in liquid crystals, the molecules are orderly—they all line up in one direction, which is why the fiber that results is so strong. The polymer creation process involves "spinning" a solution in a machine called a spinneret. Most of the liquid crystal solutions that had been spun up to this point had been clear. The solution that Kwolek asked the machine technician to spin, however, was cloudy. He did not want to spin it, because he thought it would clog the machine (Profitt 1999).

But he finally did spin the solution. The resulting product was so strong that Kwolek didn't want to report her findings. She thought there had been a mistake. But there wasn't a mistake—she had invented Kevlar®. Figure 25.2 shows the molecular structure of Kevlar®.

Five times stronger than steel (20 times stronger under water), Kevlar® is resistant to wear, corrosion, fatigue, and flame and is non-conductive. It has more than 200 applications, including protective clothing for firefighters and police officers, ropes and cables, plastic reinforcement, industrial fabrics, and materials in the aerospace and aircraft fields (Carnegie Mellon University 2014; Profitt 1999).

When Stephanie Kwolek was inducted into the National Inventors Hall of Fame, she was escorted to the stage by a policeman whose Kevlar® bulletproof vest had saved his life.

In 1987, the International Association of Chiefs of Police (IACP) and DuPont began a partnership honoring officers wearing Kevlar® body armor who have been saved from death or serious injury. It is called the IACP/DuPont™ Kevlar® Survivors' Club®. To date, 3100 officers have been honored (DuPont 2025).

25.3 Conclusion

Polymer chemist Stephanie Kwolek is best known for her invention of Kevlar®, a polymer five times stronger than steel, used in more than 200 products. She has been inducted into the National Inventors Hall of Fame and received the National Medal of Technology in 1996.

References

Artworkarchive.com. The Delaware Women's Hall of Fame - Stephanie Louise Kwolek. https://www.artworkarchive.com/profile/owaa/artwork/stephanie-louisekwolek. Accessed November 12, 2025.

Britannica. Polyamide. https://www.britannica.com/science/polyamide. Accessed January 11, 2026.

Busch-Vishniac, Ilene, Busch, Lauren, and Tietjen, Jill S. 2024. Women in the National Inventors Hall of Fame: The First 50 Years. Springer.

Ferguson, Raymond C. 1986. Oral History of Stephanie Louise Kwolek. The Beckman Center for the History of Chemistry. May 4. https://digital.sciencehistory.org/works/d217qq72t. Accessed November 12, 2025.

Carnegie Mellon University. 2014. Obituary: Carnegie Mellon Alumna and Hall of Fame Inventor Stephanie Kwolek Dies at 90. https://www.cmu.edu/news/stories/archives/2014/june/june20_obituarykwolek.html. June 20. Accessed November 13, 2025.

DuPont. IACP/DUPONT™ KEVLAR® SURVIVORS' CLUB®. https://www.dupont.com/military-law-enforcement-and-emergency-response/the-kevlar-survivors-club.html. Accessed November 18, 2025.

Proffitt, Pamela, Editor, *Notable Women Scientists*, Detroit, Michigan: The Gale Group, 1999.

Reusch, William. Condensation Polymers. Virtual Textbook of Organic Chemistry. https://chem.libretexts.org/Bookshelves/Organic_Chemistry/Supplemental_Modules_(Organic_Chemistry)/Polymers/Condensation_Polymers. Accessed January 11, 2026.

Rogin, Al, Mufson, Claire, and Lana Green. 2024. The life and achievements of chemist Stephanie Kwolek, inventor of Kevlar. PBS News Weekend. https://www.pbs.org/newshour/show/the-life-and-achievements-of-chemist-stephanie-kwolek-inventor-of-kevlar. March 17. Accessed November 13, 2025.

Science History Institute Museum and Library. Stephanie Kwolek. https://www.sciencehistory.org/education/scientific-biographies/stephanie-l-kwolek/. Accessed November 12, 2025.

Warren, Robert. National Science & Technology Medals Foundation. Stephanie Kwolek. https://nationalmedals.org/laureate/stephanie-kwolek/. Accessed November 12, 2025.

Wikipedia. Aramid. (1996) https://en.wikipedia.org/wiki/Aramid. Accessed January 11, 2026.

Wikipedia. Polyimide. (1996) https://en.wikipedia.org/wiki/Polyimide. Accessed January 11, 2026.

Chapter 26
Darleane C. Hoffman

For her discovery of primordial plutonium in nature and the symmetric spontaneous fission of heavy nuclei; for pioneering studies of elements 104, 105, and 106, and for her outstanding service to education of students in nuclear chemistry and as director of the Seaborg Institute for Transactinium Science of the University of California

Keywords Darleane Hoffman · Nellie Naylor · Marie Curie · Donald Martin · Roderick Spence · Nuclear chemistry · Spontaneous fission · Seaborgium · Rutherfordium · Dubnium · Plutonium-244 · Fermium · Heavy elements · Oak Ridge National Laboratory · Los Alamos National Laboratory · Lawrence Berkeley National Laboratory · Lawrence Livermore National Laboratory · Transuranium · Transactinium · Seaborg Institute · Glenn Seaborg · Radiochemistry

Career or family? In the 1950s and 1960s, women often felt that they could not have both, but nuclear chemist Dr. Darleane C. Hoffman (Fig. 26.1) carefully plotted her career and family choices to achieve success in both. Dr. Hoffman became a world authority on nuclear fission while sustaining a long marriage and raising two children. Her accomplishments include being the first to discover that plutonium existed in nature, changing the understanding of spontaneous fission, and confirming the discovery of element 106 (seaborgium). Dr. Hoffman won the National Medal of Science in 1997.

J. S. Tietjen et al., *Women Laureates 1970 – 2007*, Women in Engineering and Science, https://doi.org/10.1007/978-3-032-23116-1_26

Fig. 26.1 Darleane C. Hoffman. (US Department of Energy)

26.1 Life (1926–2025)

Darleane Hoffman, née Christian, was born in 1926 in the small town of Terril, Iowa. Her father, Carl, was a school superintendent and math teacher, and her mother, Elverna, was a housewife who also wrote and directed plays. The family, which included a younger brother, lived in Terril until Darleane was in third grade. They then relocated to Coon Rapids, Iowa, and then, at the start of her 10th grade year, moved to West Union, Iowa (Domush 2012).

Darleane was a good student and had time to participate in many extracurricular activities. Her mother, who had studied oratory and music in college, encouraged Darleane in these pursuits. Darleane did not take to oratory, but liked singing and playing the piano. She also enjoyed swimming and drawing and took art lessons. Science instruction at her high school was lacking. There were no chemistry courses, and physics instruction was impacted when World War II broke out. The male physics teacher left to join the service, so an unqualified substitute teacher took over, leaving Darleane with a dislike of the subject. Math was her favorite subject, and she took all the math courses she could, but when the time came for college, she decided to major in applied art, with the idea of becoming a fashion designer (Domush 2012).

After graduating from high school as a co-valedictorian, Darleane enrolled at Iowa State College (now Iowa State University) in Ames, Iowa. Her time as an applied art major was brief. Applied art was part of the school of home economics, which required its students to take a chemistry class for non-science majors. She found the professor, Nellie Naylor, who taught both theory and practical applications of chemistry, to be inspiring. Hoffman came away thinking that "chemistry was the most interesting, most logical, and most useful thing" she had ever studied, and changed her major accordingly. Her home economics advisor was not supportive, asking, "Do you think that is a suitable profession for a woman?" Hoffman did. After all, Professor Naylor was a woman. Hoffman was concerned, though, that she would be expected to give up a career if she married, as was common at the time. Professor Naylor had never married, but Darleane wanted both a career and a family. Hoffman had read a biography of Marie Curie and took inspiration from Curie's life as a scientist, wife, and mother (Domush 2012).

During the summer of 1947, Hoffman was hired by Dr. Donald Martin to work at Iowa State's Institute for Atomic Research, splitting sheets of mica to make radiation detectors. Again, Hoffman thought of Marie Curie and her work detecting radioactivity and was excited to follow in the famed scientist's footsteps. She was also happy that the job paid well and kept her from having to return home to a summer job as a bank teller (Domush 2012).

Hoffman graduated from Iowa State in 1948 with her bachelor's degree, and not feeling ready to leave Iowa, stayed at Iowa State for graduate school. She had already begun research in nuclear chemistry that she would use for her dissertation. She and Dr. Martin used the new synchrotron, a particle accelerator, at Iowa State to discover the existence of some isotopes. An isotope is a form of an element with the same number of protons but a different number of neutrons in the nucleus. In their experiments, the synchrotron accelerated electrons to near the speed of light, causing the electrons to emit high-energy X-ray beams, or gamma rays. These beams were used to bombard heavy elements, such as cobalt, platinum, and iridium, and Hoffman and Martin studied the resulting nuclear reaction to uncover isotopes of the elements (University of California, Berkeley 2012).

Hoffman completed her PhD in just 3 years. During her time in graduate school, two significant events occurred in her life. Her father died, which led her mother to join her in Ames a year later. She also met her future husband, Marvin M. Hoffman, a physics PhD student, in a statistics class. Marvin worked on the synchrotron and was authorized to run it at night. The two came to spend many late nights together running experiments on the synchrotron (University of California, Berkeley 2012).

Darleane found the atmosphere at Iowa State very supportive. There was only one other female doctoral student in chemistry, but the male students were good friends and colleagues, and the professors did not discriminate. Ever mindful of marriage's potential to derail her plans, she vowed not to marry until after she finished her PhD. So, 6 days after she received her PhD, she married Marvin in 1951 (Domush 2012).

After graduation, Dr. Hoffman took a job at Oak Ridge National Laboratory (Oak Ridge) in Tennessee, while Marvin stayed behind in Iowa for a year to finish his studies. The situation turned out to be beneficial when Darleane discovered that she had made a typo in her dissertation, and Marvin stole the dissertation from the library and made the correction, returning the original before anyone noticed (Domush 2012).

When Marvin graduated, Oak Ridge was willing to offer him a job, but nothing there interested him. As a result, he took a position at Los Alamos National

Laboratory (LANL or Los Alamos) in New Mexico, which assured him that there would be a place for his spouse in nuclear chemistry or radiochemistry. The two set off for New Mexico in December of 1952 without knowing precisely what Darleane's position would be, but with a vague sense that it was to be in the radiochemistry group. When she called up the personnel office at LANL to announce her arrival and readiness to start work in the test division's radiochemistry group, she was told that the group did not employ women. Having never encountered gender discrimination before, she was shocked. Fortunately, the situation was straightened out when Darleane attended a welcoming party for new hires with her husband. There she met the head of the radiochemistry group, Roderick Spence, who said, "Well, where have you been? I've been looking for you." From this encounter, she gained a distrust of personnel departments (Domush 2012).

Even though Dr. Hoffman got on the payroll, there was another obstacle to overcome before she could be given work to do, and that was obtaining a government security clearance. She had received a security clearance while at Oak Ridge, and it should have been an easy matter to transfer the clearance to Los Alamos since they are both Department of Energy labs, but bureaucracy got in the way. Los Alamos's personnel department had been put in charge of this process (Domush 2012).

This was during the Cold War arms race, and the radiochemistry group was analyzing debris from nuclear tests. In 1952, the USA tested what it called the "Ivy Mike," or "Mike" for short, a hydrogen bomb, which at the time was the most powerful bomb ever built. The test was performed on Elugelab Island, part of the Marshall Islands in the Pacific Ocean. The explosion destroyed the island and left a crater 6300 ft. in diameter and 160 ft. deep in its place (Fabry 2015).

Debris was extracted from the crater and sent to Los Alamos. Unfortunately, while Dr. Hoffman was still awaiting her security clearance and therefore unable to participate, the group analyzed the debris and discovered two new elements, einsteinium and fermium. This was a huge disappointment and missed opportunity for Dr. Hoffman (Domush 2012).

After 2 months without results from the personnel department regarding the security clearance, a push from Dr. Spence and a call to the FBI cleared things up within 2 days, and Dr. Hoffman was finally able to get to work. She began analyzing debris from the "Mike" test and discovered a new isotope of plutonium, ^{244}Pu (Domush 2012).

Dr. Hoffman enjoyed the radiochemistry group. She did not have any problem collaborating with the men, and there were several women in the department, apparently unbeknownst to the personnel department. She rose to group head and eventually to division leader, overseeing the radiochemistry group. She was the first female technical division leader at Los Alamos and commented, "Nobody seemed to resent having a woman division leader to work for" (Domush 2012).

The Hoffmans initially lived in a small apartment in Los Alamos. The Lab controlled housing and allocated it on a points system, so it was tough to get a house unless you had children. Dr. Hoffman wanted to be established in her career before starting a family. Her first child, Maureane, was born in 1957, six and a half years after the Hoffmans married. This was so unusual at the time that their families began suggesting that the couple adopt, assuming they were unable to have children of their own. Dr. Hoffman worked up until 2 days before her daughter was born. She hired a nanny to take care of Maureane and returned to work immediately. Two years later, the Hoffmans added a son, Daryl, to the family (Domush 2012).

In 1964, Dr. Hoffman received a National Science Foundation Senior Postdoctoral Fellowship to work at the Institute for Atomic Energy in Kjeller, Norway, so she took a sabbatical from Los Alamos, and the entire family, including Dr. Hoffman's mother, moved to Norway for a year. She found her experience there to be "liberating." Women were treated equally and had more freedom to be out on their own without negative societal judgment, which she found eye-opening (Darleane C. Hoffman et al. 2000).

During her time at Los Alamos, Dr. Hoffman primarily researched heavy elements, which have high densities and atomic weights. Having discovered the existence of ^{244}Pu, she set about trying to prove that the isotope existed naturally and was not just man-made through nuclear reactions. She succeeded in doing this through some complicated chemistry that Nobel prize-winning chemist, Glenn Seaborg, called an "experimental tour de force" (University of California, Berkeley 2012; Hoffman et al. 1971).

Another significant discovery came in 1971, when Dr. Hoffman was working with ^{257}Fm, an isotope of the heavy element fermium, which was isolated from the nuclear testing debris at the Nevada Test Site (now known as the Nevada National Security Site). Dr. Hoffman's team was studying spontaneous fission, where the nucleus splits in two without the use of external stimulation. It was thought that the split always resulted in two unequal, or asymmetric, nuclei. However, Dr. Hoffman's experiments "showed a spectacular symmetric fission distribution" (Mainz and Strom 2019, pp. 311–318).

Although her finding was met with skepticism, it was later validated at the Lawrence Livermore National Laboratory (LLNL) and was a significant advancement in the understanding of spontaneous fission (University of California, Berkeley 2012).

Dr. Hoffman continued working with elements that have an atomic number greater than 92, called transuranium elements. These elements have extremely short half-lives, so Dr. Hoffman's team had to develop new techniques to isolate the elements. Most notably, she studied the properties of lawrencium (103), rutherfordium (104), and dubnium (105).

In 1978, Dr. Hoffman received a Guggenheim Fellowship to continue studying nuclear fission at Lawrence Berkeley National Laboratory (LBNL), where she worked with Dr. Glenn Seaborg. When she returned to Los Alamos the following year, it was as the Chemistry and Nuclear Chemistry division leader, a position she held for 5 years until 1984, when she was lured back to LBNL. Seaborg was nearing retirement and looking for a successor. He was legendary. By the end of his career, Seaborg had taken part in the discovery of ten elements and had been awarded the 1951 Nobel Prize in Chemistry (Wikipedia Contributors 2019). Seaborg had remained so impressed by Dr. Hoffman that he kept a picture of her in his office for inspiration (Chapman 2021).

Dr. Hoffman found leaving Los Alamos, where she had been for most of the previous 31 years, difficult, but with the move, she became a tenured professor of chemistry at the University of California, Berkeley, and the Group Leader, Heavy Element Nuclear and Radiochemistry, at LBNL. Her husband had become a business consultant, so the couple did not have to worry about finding a position for him. Dr. Hoffman had never wanted to teach, but by this time, her feelings had changed. "I felt rather strongly about educating the next generation of students in my discipline" (Domush 2012).

She also noted that roughly 33% of the undergraduates and 50% of the chemistry graduate students were women. Yet, there was only one other female professor in Berkeley's chemistry department when she arrived (Domush 2012).

In addition to teaching, Dr. Hoffman continued her research. Her group confirmed the discovery of element 106, which led to the USA earning the right to name the element over a Soviet team that had also claimed the discovery. Through Dr. Hoffman's leadership, the element was named seaborgium after Glenn Seaborg (Darleane C. Hoffman et al. 2000). She also developed the "one atom at a time" technique to study the chemical properties of extremely short-lived, heavy elements by performing rapid chemical separations on individual atoms and evaluating the results over a statistically significant number of these individual experiments (Hoffman and Lee 1999).

In 1991, Dr. Hoffman retired from active teaching to co-found the Glenn T. Seaborg Institute for Transactinium Science at LLNL and become its first director. Transactinium refers to elements with atomic numbers above Actinium (89). Seaborg Institutes were subsequently established at LBNL, LANL, Oak Ridge, and the Idaho National Laboratory. Per its charter, the Institute studies "transactinium elements with special emphasis on education and training of the future generation of scientists in heavy-element research" (Domush 2012; Darleane C. Hoffman et al. 2000).

Dr. Hoffman retired from her position as director of the Seaborg Institute in 1996, but stayed on as a research advisor until 2007 (Domush 2012).

Dr. Hoffman had been a strong advocate for the environment. While at Los Alamos, she made many trips to the Nevada Test Site and raised concerns over the proper storage of nuclear waste (Mainz and Strom 2019, pp. 311–318).

In addition to the National Medal of Science in 1997, Dr. Hoffman received the American Chemical Society (ACS) Award for Nuclear Chemistry in 1983 (the first female recipient), the ACS Garvan-Olin Medal in 1990, the ACS Priestley Medal in 2000 (the second woman recipient), the Sigma Xi Proctor Prize for Scientific Achievement in 2003, Iowa State University's John V. Atanasoff Research and Discovery Award in 2007, and the Hevesy Medal Award in 2010 (Zanish-Belcher 2021).

Most recently, in 2023, Dr. Hoffman was a co-recipient of the Enrico Fermi Presidential Award from the Biden-Harris Administration for her,

> scientific discoveries advancing the field of nuclear and radiochemistry, for distinguished service to the Department of Energy's missions in national security and nuclear waste management, and for sustained leadership in radiochemistry research and education (The White House 2023)

Dr. Hoffman credited her success to the help she had around the house and in caring for her children, the mentorship of Dr. Glenn Seaborg, and especially the support of her husband Marvin (Domush 2012). She died in 2025 at the age of 98.

26.2 Finding ^{244}Pu in Nature

In 1941, Dr. Glenn Seaborg and others created the first man-made plutonium by starting with uranium and inducing a chain of nuclear reactions by bombarding the uranium with neutrons in a particle accelerator. Uranium is a naturally occurring element that can be found in the Earth's crust, but for years, it was thought that transuranium elements, such as plutonium, existed only in the labs of nuclear scientists and not in nature (Poston, Sr. 1998).

This began to change when the plutonium isotope, ^{244}Pu, was found in the debris from the "Mike" hydrogen bomb test. Studies showed that ^{244}Pu's half-life was an extremely long 80 million years, leading scientists to speculate that plutonium was created during the birth of the solar system and that some might still exist. But where? That is the question that Dr. Hoffman and her colleague at Los Alamos, Francine Lawrence, sought to answer. They needed to locate ancient ore deposits that were enriched with enough plutonium at the Earth's formation 4.5 billion years ago that trace amounts of ^{244}Pu might still be detectable (Darleane C. Hoffman et al. 2000).

Their search led to the Molybdenum Corporation of America's Mountain Pass Mine in Southern California (*The Santa Fe New Mexican* 1971). Molybdenum was mining Precambrian bastnasite, a mineral that contains rare earth elements, thought to be about 1.4 billion years old (Wikipedia Contributors 2024). This bastnasite contained approximately 500,000 times more cerium dioxide, an oxide of the rare earth element cerium, than the average amount found in the Earth's crust. Because the chemistry of plutonium is similar to that of cerium (element 58), Dr. Hoffman hypothesized that where there is a lot of cerium, there might still be a trace of plutonium. Molybdenum Corporation used an acid solvent to extract cerium from the mined bastnasite for their commercial purposes and provided the resulting solvent, for which they had no use, to Dr. Hoffman for her experiments. The next step was to separate the plutonium from the solvent by washing it in a solution to remove other elements, such as iron, cerium, and thorium. The washing was performed multiple times to obtain the purest sample possible. Dr. Hoffman still had to prove that what remained contained ^{244}Pu. So, the resulting sample was sent to the Knolls Atomic Power Laboratory in New York, which had one of the world's most sensitive mass spectrometers. This instrument separates particles according to mass. Jack Mewherter and Frank Rourke performed the mass-spectrometric analysis at Knolls and, after 2 years of experimentation, delivered the news that Dr. Hoffman proved the existence of naturally occurring plutonium (Darleane C. Hoffman et al. 2000).

Figure 26.2 illustrates the process described above.

This discovery was of particular interest to cosmologists. It supported the hypothesis that heavy elements were made by a supernova collapsing in on itself at the time our solar system was formed, rather than much earlier in the galaxy's history, as previously thought (*The Santa Fe New Mexican* 1971). Thus, Dr. Hoffman added to the understanding of the formation of our solar system.

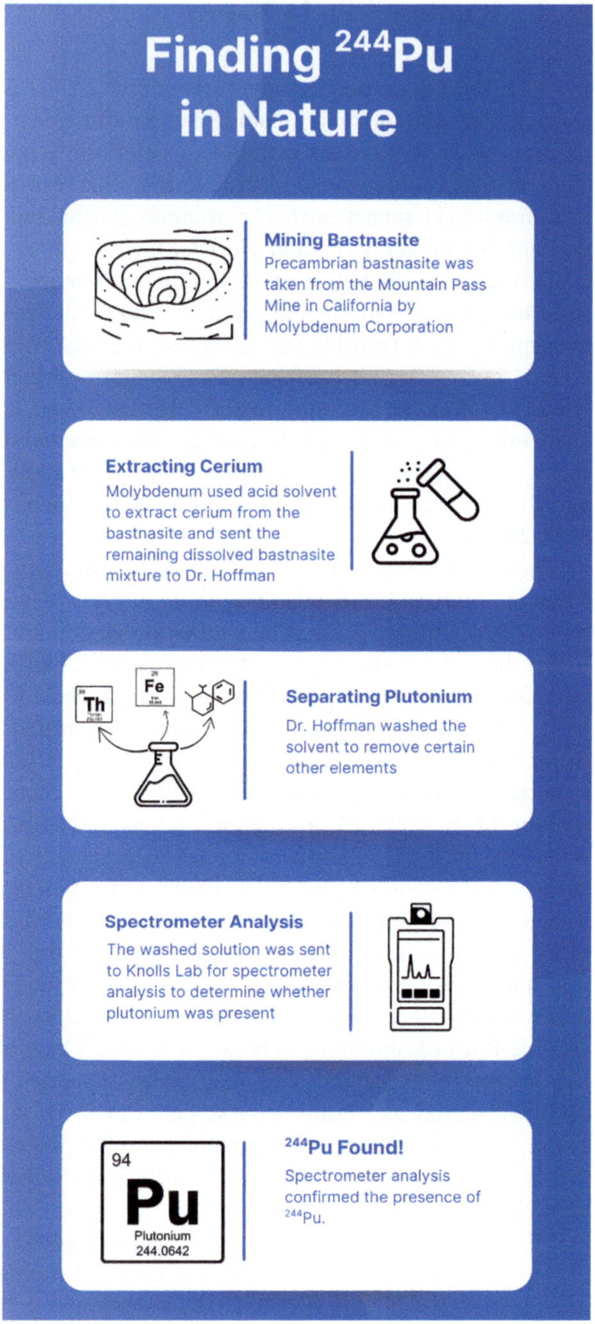

Fig. 26.2 Finding ^{244}Pu in nature. (Courtesy of Lauren Busch)

26.3 Conclusion

Dr. Darleane Hoffman was a pioneer in the field of heavy elements, who furthered the understanding of the creation of our solar system and nuclear fission. She developed techniques for confirming the existence of heavy elements and studying their chemical properties. Dr. Hoffman balanced a career and family, achieving the highest levels of recognition in chemistry. Her dedication to both scientific research and education left an indelible mark on the field of nuclear chemistry and the generations of scientists who followed her.

References

Chapman, Kit. 2021. "The Women Written out of Nuclear Science." Lady Science. December 9, 2021. https://www.ladyscience.com/features/the-women-written-out-of-nuclear-science-2021.

Domush, Hilary L. 2012. "Darleane C. Hoffman, Interview by Hilary L. Domush." Philadelphia: Chemical Heritage Foundation.

Fabry, Merrill. 2015. "What the First H-Bomb Test Looked Like." Time. November 2, 2015. https://time.com/4096424/ivy-mike-history/.

Hoffman, D. C., F. O. Lawrence, J. L. Mewherter, and F. M. Rourke. 1971. "Detection of Plutonium-244 in Nature." *Nature* 234 (5325): 132–34. https://doi.org/10.1038/234132a0.

Hoffman, Darleane C, Albert Ghiorso, and Glenn T Seaborg. 2000. *The Transuranium People : The inside Story*. London: Icp.

Hoffman, Darleane C, and Diana M Lee. 1999. "Chemistry of the Heaviest Elements - One Atom at a Time." *Journal of Chemical Education* 76 (3): 331–47. https://doi.org/10.2172/860292.

Mainz, Vera V, and E. Thomas Strom. 2019. *The Posthumous Nobel Prize in Chemistry: Ladies in Waiting for the Nobel Prize (Volume 2)*. American Chemical Society.

Poston, Sr., John W. 1998. "Do Transuranic Elements such as Plutonium Ever Occur Naturally?" Scientific American. March 23, 1998. https://www.scientificamerican.com/article/do-transuranic-elements-s/.

The Santa Fe New Mexican. 1971. "Rare Plutonium Isotope Reported," September 9, 1971.

The White House. 2023. "Biden-Harris Administration Announces Recipients of the Enrico Fermi Award." March 28, 2023. https://bidenwhitehouse.archives.gov/ostp/news-updates/2023/03/28/biden-harris-administration-announces-recipients-of-the-enrico-fermi-award/.

University of California, Berkeley. 2012. "Adventures in the Nature of Matter." *Catalyst, University of California, Berkeley* 6 (2): 12–15.

Wikipedia Contributors. 2019. "Glenn T. Seaborg." Wikipedia. Wikimedia Foundation. June 19, 2019. https://en.wikipedia.org/wiki/Glenn_T._Seaborg.

Wikipedia Contributors. 2024. "Mountain Pass Rare Earth Mine." Wikipedia. Wikimedia Foundation. November 25, 2024. https://en.wikipedia.org/wiki/Mountain_Pass_Rare_Earth_Mine.

Zanish-Belcher, Tanya. 2021. "Hoffman, Darleane C." Iowa State University Biographical Dictionary. July 30, 2021. https://isubios.pubpub.org/pub/ls5ryoo3/release/1.

Chapter 27
Janet D. Rowley

For revolutionizing cancer research, diagnosis, and treatment through her discovery of chromosomal translocations in cancer and her pioneering work on the relationship of prior treatment to recurring chromosome abnormalities, for epitomizing the bench to bedside philosophy in her application of basic discoveries to clinical medicine, and for her leadership nationally and internationally in the oncology and biomedical communities

Keywords Janet Rowley · Cytogenetics · Chromosomal translocation · Breakpoint regions · Acute myeloid leukemia · Chronic myeloid leukemia · Promyelocytic leukemia · Acute lymphoblastic leukemia · Philadelphia chromosome · Gleevec · Donald Rowley · Jerome Lejeune · David Hungerford · Peter Nowell · Argonne Cancer Research Hospital · Hutchins College

Dubbed "the matriarch of modern cancer genetics" (Dreifus 2011), Dr. Janet Rowley (Fig. 27.1) reshaped the understanding of cancer by proving it to be a fundamentally genetic disease. In the 1970s, her observations of chromosomal abnormalities in leukemia patients led her to discover that pairs of chromosomes exchanged genetic material. She established that these abnormalities, known as chromosomal translocations, caused cancer. Although initially met with skepticism, Dr. Rowley's groundbreaking findings ushered in a new era of cancer research, targeted treatment, and medications. Dr. Rowley was considered a pioneer in translational research, the process of taking scientific research from the lab through to the treatment of human disease (The University of Chicago Medicine Comprehensive Cancer Center 2021). For her research, contributions to cancer treatment, and impact on the medical and scientific community, Dr. Rowley was awarded the National Medal of Science in 1998.

J. S. Tietjen et al., *Women Laureates 1970 – 2007*, Women in Engineering and Science, https://doi.org/10.1007/978-3-032-23116-1_27

Fig. 27.1 Dr. Janet Rowley. (Photo credit: Jason Smith)

27.1 Life (1925–2013)

Janet Rowley, nee Davison, was born in New York City in 1925 but raised in Chicago from the age of two. She was the only child of Hurford Davison, a department store manager with an MBA from Harvard, and Ethel Ballantyne Davison, a high school English teacher. Hurford and Ethel met as undergraduates at the University of Chicago.

As a child, Janet liked things that were orderly. Her father was an avid stamp collector, and Janet took up the hobby too. She loved studying the stamps and their patterns, and comparing them to the pictures in stamp catalogs. This skill in precise pattern recognition served her well in her scientific career, enabling her to detect differences in chromosomes (Maestrejuan 2005).

Janet went to Chicago public schools through the 8th grade. She enjoyed Latin because she thought it was logical, and for a time, thought she would be a Latin teacher (Maestrejuan 2005). To avoid being taught by her mother or her mother's friends, she moved to a Catholic girls' school at the beginning of high school. There, she became more interested in science, which her parents encouraged, since they felt it offered greater career potential than Latin. After her sophomore year, she applied to the University of Chicago's Hutchins College, which was a four-year program for gifted high school students, covering the last 2 years of high school and the first two of college (Maestrejuan 2005). At this experimental "lab" school, students were taught to learn independently and read primary texts (Collins 1998). It also taught her to "stick to my convictions if I really thought that I was correct, even when others disagree (The University of Chicago Medicine Comprehensive Cancer Center 2021)." At Hutchins College, Janet took several biology courses and noticed that many of her male lab partners planned to go on to medical school. The idea of caring for people instead of working in a lab appealed to her, and she decided to pursue a pre-med track. She graduated in 1944 with a Bachelor of Philosophy (PhB), which was a degree unique to the University of Chicago. She spent the summer finishing off the necessary pre-med courses and hoped to enter the University of Chicago's medical school that fall, but their quota of three women in a class of 65 students had been filled. She was given a scholarship to take another year of undergraduate courses while waiting for a spot in medical school. She entered medical school in June of 1945 at the age of 20. Courses in her first year of medical school were counted toward a Bachelor of Science, which she received in 1946 (Maestrejuan 2005).

Janet met her husband, Donald Rowley, during medical school, where he was also a student. The two married the day after Janet's graduation in 1948. She worked as a research assistant in a lab for a year and a half while she waited for Donald to finish medical school in 1950. Afterwards, she and Donald both had internships with the US Public Health Service.

Dr. Janet Rowley had not put a lot of thought into planning her career. She intended to work part-time while prioritizing her husband's career and raising a family. In 1951, the couple moved to Bethesda, Maryland, so that Donald could conduct pathology research at the National Institutes of Health (NIH) (Maestrejuan 2005). Of this move, she said, "He went and I went along. This is really the story of my life. I just sort of go along and then see what happens" (Maestrejuan 2005).

Their first child, Donald, was born in 1952. Three more sons followed: David, Robert, and Roger (The University of Chicago Medicine Comprehensive Cancer Center 2021).

In 1953, Dr. Rowley went to work two afternoons a week as an attending physician at the Montgomery County, Maryland, Infant Welfare and Prenatal Clinic. Some of the stay-at-home moms she knew provided babysitting (Maestrejuan 2005).

In 1954, the Rowley family moved back to Chicago so Donald could take an academic position at the University of Chicago. Again, Dr. Janet Rowley found part-time work at a clinic. Unfortunately, finding part-time help at home was not as easy. Eventually, she hired full-time help while she worked 2 days a week doing physicals at a clinic at Cooke County Hospital that cared for children with developmental disabilities. In the late 1950s, Dr. Rowley began working 3 days a week at a clinic at the University of Illinois School of Medicine in Chicago that treated patients with epilepsy. Here she continued to perform physicals, but she began wanting a greater intellectual challenge (Maestrejuan 2005).

A turning point in Dr. Rowley's career came in 1961. At the University of Illinois clinic, she saw many patients with Down syndrome, and just the year before, Jerome Lejeune discovered the cause of Down syndrome to be an extra copy of the 21st chromosome. This was the first identification of chromosomal abnormality in humans. When Donald decided to take a sabbatical year in Oxford, England, Dr. Rowley applied for an NIH Fellowship to study cytogenetics. Oxford University happened to be one of the few places to learn cytogenetics, the study of chromosomes (Maestrejuan 2005). At Oxford, Dr. Rowley worked in Laszlo Lajtha's lab researching chromosome duplication (Dreifus 2011).

Upon her return to Chicago in 1962, Dr. Rowley decided to continue with her research. She arranged a part-time position at the Argonne Cancer Research Hospital of the University of Chicago, saying to the director, Leon Jacobson, "All I need is a microscope and a darkroom. And by the way, will you pay me? I must earn enough for a babysitter (Dreifus 2011)."

At Argonne, Dr. Rowley sometimes helped Jacobson, a hematologist, look at slides from his leukemia patients. They saw chromosomal abnormalities, but technological limitations at the time prevented them from identifying specific chromosomes.

In 1970, Donald Rowley took another sabbatical year in Oxford, and the family went along. Just before their move, a new technique emerged that stained genetic material with special dyes. The bands on the chromosomes show up in contrast under a fluorescent microscope. Careful examination could identify subtle distinctions in the material and distinguish different chromosomes. At Oxford, Dr. Janet Rowley borrowed a fluorescent microscope at night and on weekends and realized she could use this new banding technique to study the chromosomes of leukemia patients (Dreifus 2011). Back in Chicago the following year, she did just that.

Dr. Rowley stained chromosomes from patients with acute myeloid leukemia (AML). She took photographs of the stained chromosomes, cut them up, and laid them out on her dining table to study individual chromosomes. She arranged and rearranged the pictures to look for patterns or abnormalities. Her children jokingly referred to the photograph fragments as their mother's paper dolls, and Dr. Rowley, in turn, begged them not to sneeze on her careful arrangements (Fox 2013). In 1972, while looking at the photographs for one of the patients, she noticed that chromosomes 8 and 21 were broken, with a piece of chromosome 8 now on chromosome 21, and vice versa. This phenomenon is known as chromosomal translocation, and this is its first documented example in human cancer cells (Dreifus 2011). Dr. Rowley confirmed the results in other AML patients. She then discovered that chromosomes 9 and 22 swapped segments in patients with chronic myeloid leukemia (CML).

With these exciting discoveries, Dr. Rowley was working full-time but was classified and paid as a part-time employee. She officially put her career front and center, becoming a full-time employee of the Department of Medicine and Argonne Cancer Research Hospital in 1975. She had been made an associate professor in 1969, a title she held until 1977, when she was promoted to full professor.

As with many paradigm-shifting results, it took some time for her work to be accepted. With Dr. Rowley's discovery of a third swap between chromosomes 15 and 17 with promyelocytic leukemia in 1977, the link between chromosomal translocation and cancer was undeniable (Fox 2013). Dr. Rowley's research had shattered the long-held assumption that genetic changes were a consequence of cancer. By proving they were, in fact, the cause, she established cancer as a fundamentally genetic disease and ushered in a revolutionary new era for cancer research and treatment. (Note that 5–10% of cancers are from inherited genetic defects, but the 90–95% of cancers that result from environmental and lifestyle factors cause genetic defects, as well (Anand and Kunnumakara 2008)).

In this new era of cancer genetics, researchers sought chromosomal translocations that cause other cancers. Dr. Rowley also continued her research and identified additional chromosomal translocations associated with leukemia and lymphoma. In 1984, the University of Chicago recognized her contributions by appointing her the Blum-Riese Distinguished Service Professor (The University of Chicago Medicine Comprehensive Cancer Center 2021).

Dr. Rowley subsequently adapted and used advanced molecular techniques, such as fluorescence in situ hybridization (FISH), to identify the genes underlying these translocations. This included mapping and cloning the rearranged gene found in most infant leukemias, enabling her team to better understand how its disruption drives uncontrolled cell growth. This work pinpointed the exact locations where chromosomes break and rejoin, known as breakpoint regions, leading to more accurate diagnostic tests for these cancers (The University of Chicago Medicine 1998).

Among her most significant later findings was the discovery that high doses of a topoisomerase II inhibitor, a cancer drug, could induce specific translocations involving the *MLL* gene on chromosome 11q23. This demonstrated that the treatment could cause secondary leukemias in some patients (Gruber Foundation n.d.).

Dr. Rowley was devoted to helping others succeed in science. She mentored nearly 100 students and fellows and was very generous with the use of her lab (Gollin and Reshmi 2014). She encouraged researchers to take risks, saying, "Do something different if it looks interesting… I didn't do anything noteworthy until I was 50" (Gollin and Reshmi 2014).

Dr. Rowley also had an impact on public policy. Appointed by President Jimmy Carter, she served on the National Cancer Advisory Board from 1979 to 1984, where she pushed for funding for innovative cancer research. Later, she was a member of the President's Council on Bioethics, established by President George W. Bush, from 2002 to 2009. She was a strong advocate for stem cell research, opposing the restrictions on federal funding for embryonic stem cell studies, and stood next to President Barack Obama when he lifted the federal moratorium in 2009 (University of Chicago Medicine Comprehensive Cancer Center 2021).

In addition to the National Medal of Science in 1998, Dr. Rowley received numerous other prestigious awards, including the Lasker Award (co-recipient) in 1998, the American Philosophical Society's Benjamin Franklin Medal in 2003, the Presidential Medal of Freedom in 2009, the Lifetime Achievement Award from the American Association for Cancer Research in 2010, and the Gruber Genetics Prize in 2012. She was elected to several honor societies, including the National Academy of Sciences, the Institute of Medicine, the American Philosophical Society, and the American Academy of Arts & Sciences (University of Chicago Medicine Comprehensive Cancer Center 2021). She has been inducted into the National Women's Hall of Fame.

Dr. Rowley remained active researching, publishing, and teaching until shortly before her death. She had over 500 publications by the end of her career. She also found time for hobbies, such as gardening and biking, and could be seen biking to her lab well into her 80s (The University of Chicago Medicine Comprehensive Cancer Center 2021).

In March 2013, Donald Rowley, her husband of 65 years, died. In November 2013, Dr. Rowley received her last award, the Partners in Discovery Award from the Women's Board of the University of Chicago Cancer Research Foundation. She was suffering from ovarian cancer, but she made it to the ceremony because it was important to her to advocate for funding for scientific research. She died that December from the disease and donated her cancer tissue to science (Gollin and Reshmi 2014).

27.2 Chromosomal Translocation

One of the first two chromosomal translocations that Dr. Rowley discovered was between chromosomes 9 and 22. An abnormality in chromosome 22 had been found in patients with CML by David Hungerford and Peter Nowell in 1960. This altered chromosome was named the Philadelphia chromosome. However, Hungerford and Nowell mistakenly thought that the abnormality was caused by a deletion of genetic material from the chromosome. It was Dr. Rowley who realized that pieces of chromosomes 9 and 22 were being swapped.

Dr. Rowley employed a technique developed in Sweden by Torbjörn Caspersson, staining chromosomes with quinacrine mustard. Each of the 23 pairs of chromosomes in the human body could be identified by the unique visible pattern of light and dark regions, or bands, that it produces (Sarath 2025).

Now that Dr. Rowley could distinguish between chromosomes, she set out to look for abnormalities. Through painstaking examination of the chromosomal bands and patterns, she saw that when the Philadelphia chromosome was present, an abnormal chromosome 9 was also present. She realized that pieces of the chromosomes had been traded. Figure 27.2 shows a representation of the translocation. It highlights the ABL gene on chromosome 9 and the BCR gene on chromosome 22, the regions that break off, and the modified chromosomes. (9q34.12 and 22q11.21 refer to exact locations on the chromosomes.)

Figure 27.3 shows a photo of a normal chromosome 9 on the left adjacent to its modified version on the right, along with a normal chromosome 22 on the left adjacent to the Philadelphia chromosome on the right. The arrows point to the region that has been swapped.

The resulting Philadelphia chromosome contains both the ABL gene from chromosome 9 and the BCR gene from chromosome 22, forming the BCR-ABL fusion gene. This gene produces a protein that continuously signals cells to divide, leading to an overproduction of cells. Additionally, the protein inhibits these abnormal cells from dying. The result of this uncontrolled cell growth is cancer. The BCR-ABL gene is associated with CML, but it is also found in some patients with acute lymphoblastic leukemia (ALL) and occasionally in patients with AML (Ayatollahi et al. 2018).

One of the first significant treatments to result from Dr. Rowley's work was the development of the drug imatinib, brand name Gleevec, which blocks the protein

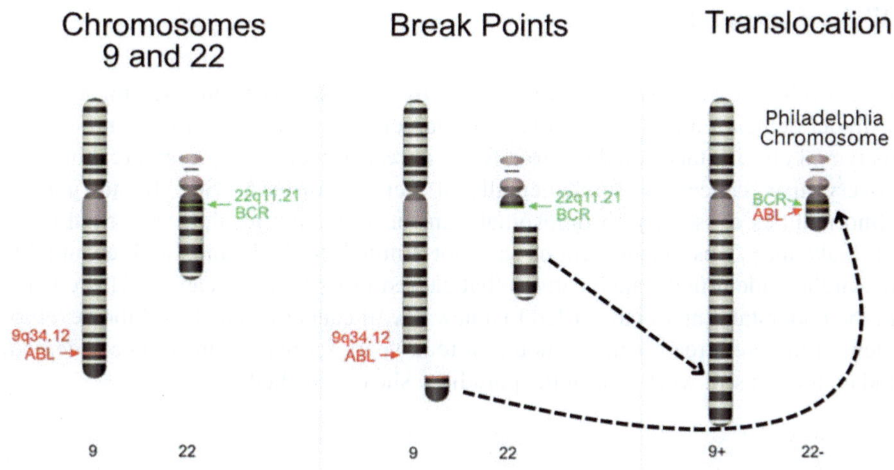

Fig. 27.2 Translocation of chromosomes 9 and 22. (Philadelphia_chromosome.jpg: A Obeidat in ar.wikipedia Chromosome_9.svg: Mysid Chromosome_22.svg: Master Uegly (talk), derivative work of *Chr-22.jpeg: NIH.gov derivative work: Master Uegly, CC BY-SA 3.0 via Wikimedia Commons (further modified to show break points))

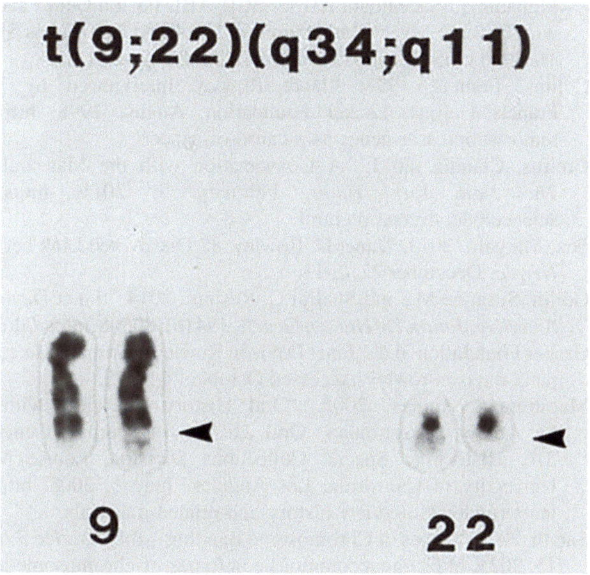

Fig. 27.3 Chromosomes 9 and 22 before and after translocation. (The Armed Forces Institute of Pathology (AFIP), Public domain, via Wikimedia Commons)

produced by the translocation of chromosomes 9 and 22 and thereby prevents continuous cell division and cancer cell death inhibition. Prior to the development of Gleevec, patients with CML had a life expectancy of just 3–5 years, but with this treatment, 90% of patients are considered cured (Fox 2013).

27.3 Conclusion

Dr. Janet Rowley's early career was shaped by her desire to balance raising a family with her dedication to helping others through clinical medicine. However, her curiosity and circumstances led her to a focus in research and the groundbreaking discovery that cancer is fundamentally a genetic disease. She found genetic abnormalities caused by chromosomal translocations that resulted in certain types of leukemia. This phenomenon was not limited to leukemia, and eventually, researchers identified translocations that caused other types of cancer. This change in the understanding of cancer led to a new age of cancer research and the development of targeted treatments. Although it took years for her findings to be accepted, today doctors still work within the paradigm she established.

References

Anand, Preetha, and Ajaikumar B. Kunnumakara. 2008. "Cancer Is a Preventable Disease That Requires Major Lifestyle Changes." *Pharmaceutical Research* 25 (9): 2097–2116. https://doi.org/10.1007/s11095-008-9661-9.

Ayatollahi, Hossein, Mohammad Keramati, Mohammad Kooshyar, Sepideh Shakeri, and Mohammad Sadeghian. 2018. "BCR-ABL Fusion Genes and Laboratory Findings in Patients with Chronic Myeloid Leukemia in Northeast Iran Abbas Shirdel (MD) 2." *Caspian J Intern Med* 9 (1): 65–70. https://doi.org/10.22088/cjim.9.1.65.

Collins, Francis. 1998. "Janet Rowley Interviewed by Francis Collins" Interview by Francis Collins. Lasker Foundation. August 1998. https://laskerfoundation.org/winners/tumor-suppressor-genes-as-a-cause-of-cancer/.

Dreifus, Claudia. 2011. "A Conversation with the Matriarch of Modern Cancer Genetics." *The New York Times*, February 7, 2011. https://www.nytimes.com/2011/02/08/science/08conversation.html.

Fox, Margalit. 2013. "Janet D. Rowley, 88 Discovered Link between Cancer, Genes." *The Miami Herald*, December 22, 2013.

Gollin, Susanne M., and Shalini C. Reshmi. 2014. "Janet Davison Rowley, M.D. (1925–2013)." *American Journal of Human Genetics* 94 (6): 805–8. https://doi.org/10.1016/j.ajhg.2014.05.008.

Gruber Foundation. n.d. "Janet Davison Rowley." Gruber.Yale.edu. https://gruber.yale.edu/person/janet-davison-rowley. Accessed October 24, 2025.

Maestrejuan, Andrea. 2005. "Oral History Interview with Dr. Janet Rowley" Interview by Andrea Maestrejuan. Oral History of Human Genetics Collection (Ms. Coll. No. 316. History & Special Collections Division, Louise M. Darling Biomedical Library, University of California, Los Angeles. June 6, 2005. https://ohhgp.semel.ucla.edu/people/janet-rowley/#interview-history-and-related-materials.

Sarath. 2025. "Types of Chromosome Banding—the Barcode Explained." Gene Commons. August 15, 2025. https://genecommons.com/types-of-chromosome-banding-the-barcode-explained/.

The University of Chicago Medicine. 1998. "Janet Rowley, MD, Awarded National Medal of Science." Uchicagomedicine.org. UChicago Medicine. December 8, 1998. https://www.uchicagomedicine.org/forefront/news/janet-rowley-md-awarded-national-medal-of-science.

The University of Chicago Medicine Comprehensive Cancer Center. 2021. "Janet Rowley, Cancer Genetics Pioneer, 1925-2013." Cancer History Project. December 2021. https://cancerhistoryproject.com/people/janet-rowley-cancer-genetics-pioneer-1925-2013.

Chapter 28
Cathleen Synge Morawetz

For her many pioneering contributions to the theory of partial differential equations and wave propagation that resulted in applications in aerodynamics, acoustics, and optics. Her research accomplishments are matched by her leadership and inspiration, judgment and vision, and knowledge and generosity to colleagues and collaborators

Keywords Cathleen Morawetz · Cathleen Synge Morawetz · John Synge · Cecilia Krieger · Herbert Morawetz · Harold Gad · Richard Courant · Harold Gad · Kurt Otto Friedrichs · C. C. Lin · Courant Institute · Morawetz's inequalities · Morawetz's estimates · Aerodynamics · Acoustics · Relativity · Partial differential equations · Wave scattering · Transonic flow · Airfoil · Supercritical airfoil

Dr. Cathleen Synge Morawetz (Fig. 28.1) was a Canadian-American mathematician who grew up in Canada and Ireland. Although her father was a well-known mathematician, Dr. Morawetz did not set out to study mathematics; however, she was encouraged by mentors and the availability of academic scholarships to pursue this path. She came to the USA for graduate school and stayed, forging a distinguished career at the Courant Institute at New York University (NYU). Her work in partial differential equations and wave scattering not only led to her accomplishments in aeronautics and mathematics but also provided a foundation for advances in a wide range of disciplines, including acoustics, relativity theory, and optics. Dr. Morawetz was awarded the National Medal of Science in 1998.

J. S. Tietjen et al., *Women Laureates 1970 – 2007*, Women in Engineering and Science, https://doi.org/10.1007/978-3-032-23116-1_28

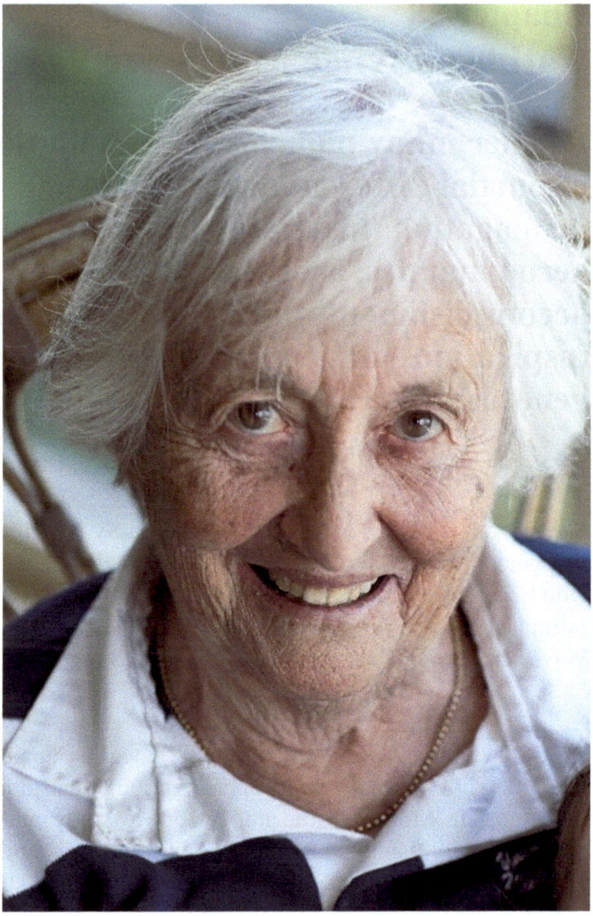

Fig. 28.1 Cathleen Morawetz. (Wikiijohn, CC BY-SA 4.0, via Wikimedia Commons)

28.1 Life (1923–2017)

Cathleen Morawetz, née Synge, was born in Toronto, Canada, in 1923. Her parents were Irish and met as students at Trinity College in Dublin. Her father, John L. Synge, was a prominent mathematician (Morrow and Perl 1998). Her mother, Elizabeth, initially studied math at Trinity but was convinced by her brother to switch to history, where he thought there was a better future (Albers et al. 1990; Morrow and Perl 1998). John and Elizabeth married while still in college, and Elizabeth dropped out to support her husband through school. Cathleen believed that her mother regretted not having received a degree, and therefore, encouraged her daughter to pursue a career (Chang 2017; Morrow and Perl 1998).

From 1920 to 1925, John was a lecturer at the University of Toronto, where Elizabeth was born. In 1925, John was offered a teaching position back at Trinity College, so the family went to Ireland (Roberts 2012). Cathleen was the middle of three daughters. She was considered precocious and stubborn. When her older sister started school at age five, Cathleen insisted on going too. Her mother managed to enroll her in kindergarten at a Protestant private school, where she learned to read at the age of three (Albers et al. 1990).

The family returned to Canada in 1930, where John became a professor at the University of Toronto. Cathleen's primary academic interest growing up was in history (Roberts 2012). Ironically, she had not been good at arithmetic, and her father, who was not all that interested in her life, had not encouraged her. It was not until her last year of high school that she began concentrating in mathematics at the suggestion of one of her teachers, who encouraged a small group of top students to apply for college scholarships in math and science. She did win a scholarship, although not the top prize, an outcome that greatly disappointed her. Cathleen was expected to live at home during college, so she enrolled at the University of Toronto (Albers et al. 1990).

At the University of Toronto, Cathleen began studying math, physics, and chemistry. However, her studies were interrupted by World War II. By the end of her junior year, many of her classmates had gone off to war, including her boyfriend. She initially planned to join the Navy but changed her mind when she found out that she would have to attend boot camp, while all the men were immediately commissioned as officers (Albers et al. 1990). Instead, her father arranged a position for her at the Inspection Board of the United Kingdom and Canada near Quebec City. She enjoyed being away from home for the first time and appreciated the break from the pressure of school and living in fear of losing her scholarship (Albers et al. 1990).

After a year, Cathleen returned to the University of Toronto to finish her degree. She majored in math, studying applied rather than pure mathematics. Cathleen's parents had left Toronto for the USA, so she lived in the dorms her senior year. That was also the year she met her future husband, chemist Herbert Morawetz. Herbert was a Jewish refugee from Czechoslovakia who had managed his family's textile business until the Nazi occupation in 1939, at which time he was able to escape to Canada. Deciding to pursue a college education, he enrolled at the University of Toronto at the age of 24 (Bohning 1986). He received an undergraduate degree in chemical engineering in 1943 and his master's degree in 1945 (Wang 2018). When they met, Herbert was working for the Bakelite Company in Toronto, a job he had secured through family connections after being rejected everywhere else he had applied due to antisemitism (Albers et al. 1990).

When Cathleen graduated from college in 1945, she considered going to India to teach and get another taste of adventure, but Cecilia Krieger, a prominent mathematics professor and family friend, talked her out of it. Krieger encouraged her to apply for a scholarship from the Canadian Association of University Women for graduate school. Cathleen wanted to go to Caltech to study engineering, but the school informed her that it did not accept women (Albers et al. 1990). Instead, she went to the Massachusetts Institute of Technology (MIT) in Cambridge, Massachusetts. Cathleen and Herbert got engaged the summer before she entered MIT. To be closer to her, Herbert arranged a transfer to Bakelite's facility in Bound Brook, New Jersey.

During her first semester at MIT, Cathleen decided engineering was not for her, and she pursued a master's degree in applied mathematics instead. She also married Herbert, and the two commuted between Cambridge and Bound Brook. This was an unusual arrangement at the time, but Herbert had promised that he would not interfere with her career, and he kept his word (Albers et al. 1990).

After graduating from MIT in 1946, Cathleen Morawetz moved to New Jersey to be with her husband. She applied for a job at Bell Labs in New Jersey, but they offered her a position in the pool of women with bachelor's degrees despite her master's from MIT. She refused. That winter, her father attended a meeting of the American Mathematical Society (AMS), where he met Richard Courant, an influential professor at the Institute at NYU. (The Institute later became known as the Courant Institute.) Courant's daughter had recently married, and the two fathers lamented the fact that their intellectual daughters were not finding opportunities to pursue their careers (Albers et al. 1990). Courant purportedly said, "You can't do anything for my daughter, but perhaps I can do something for yours (Albers et al. 1990)." After an interview and reference check, Morawetz was hired to solder connections on a machine that Harold Gad was building to solve linear equations (Albers et al. 1990). Unfortunately, Gad objected to having a woman do the soldering, so he found a man for the job instead. Courant did not revoke Morawetz's employment offer, but he wasn't sure how to make use of her. She ended up editing Courant and Kurt Otto Friedrichs' book on shock waves (Albers et al. 1990). Simultaneously, Morawetz began work on a PhD and started a family. Morawetz's daughter, Pegeen, was born in 1947, followed by John in 1949, Lida Jeck in 1952, and Nancy in 1954 (Albers et al. 1990).

After Pegeen was born, Morawetz passed her oral exams and began working on her PhD thesis, with Friedrichs as her advisor, but she had trouble finding a topic that excited her. After a couple of years, she was concerned about being able to complete her degree. By this time, Herbert was working on his PhD in chemistry at the Polytechnic Institute of Brooklyn. They had two children and were living in a small apartment in Brooklyn. Although she struggled, Morawetz managed to finish her thesis on imploding shock waves in 1950 (Chang 2017). (Her degree was conferred in 1951.). Herbert completed his PhD in 1950 and was offered a postdoc at Harvard Medical School. The family moved to Massachusetts for a year, where Dr. Cathleen Morawetz was a research associate at MIT under Chia-Chiao (C.C.) Lin (Morrow and Perl 1998).

Lin's approach helped Dr. Morawetz find success. He gave her problems to work on, but if she did not make progress after 5 weeks, he had her tackle something different. Dr. Morawetz's research with Lin led to her first two published papers (Albers et al. 1990).

After a year in Massachusetts, Dr. Morawetz returned to NYU's Courant Institute as a research associate, and Herbert took a position at the Polytechnic Institute of Brooklyn, which later became part of NYU.

Dr. Morawetz's research flourished in the 1950s, focusing on transonic flow and shock waves. When an aircraft travels close to the speed of sound, some regions of airflow around it are subsonic while others are supersonic. This phenomenon, known as transonic flow, creates shock waves on the aircraft's wings, resulting in increased drag and instability (Wikipedia Contributors 2021a). At the time, aeronautical engineers hoped to design airplane wings that would eliminate shock waves entirely. Using mixed partial differential equations, which help describe patterns, Dr. Morawetz proved this to be a mathematical impossibility, redirecting efforts to minimize shock waves and their effects instead (Morrow and Perl 1998; Chang 2017).

In the 1960s and 1970s, Dr. Morawetz continued her work on partial differential equations and transonic flow and also expanded into scattering theory, particularly the scattering of waves by obstacles. Using mixed partial differential equations, she proved that energy decays around an obstacle, causing the waves to scatter. Her techniques for calculating the maximum amount of wave energy near an object at a given time became known as "Morawetz's inequalities" and "Morawetz's estimates" (Chang 2017). These methods continue to influence the fields of aerodynamics, fluid dynamics, general relativity, and other areas where wave propagation is relevant (New York University 2017). Her research on partial differential equations continued through the 1980s and 1990s (Morawetz 1997).

Meanwhile, Dr. Morawetz rose from a research associate in 1951 to a full professor in 1965 (Albers et al. 1990). With four kids at home, she chose to work part-time and receive an 80% salary, but she put in full-time hours regardless and was annoyed that it took 14 years to become a full professor (Morawetz 2002). Of her part-time and pay arrangement, she said, "I felt much better that way. No one could question, as they had at MIT, whether I was really earning what I was being paid (Albers et al. 1990)."

She faced questions about choosing to have both a career and a family. When asked whether she worried about her children while she was at work, she responded, "No, I'm much more likely to worry about a theorem when I'm with my children (Chang 2017)."

She was also fond of joking that she became a mathematician because she was no good at housework (Roberts 2012).

Although Dr. Morawetz was considered an outstanding mentor and collaborated with many young mathematicians (Gamba et al. 2018), she did not enjoy classroom teaching.

> I used to be a very nervous and unhappy teacher, and I spent an awful lot of time preparing my lectures. If you taught in the Institute, you taught only graduate courses, so a full load was two graduate courses. I just found that a big burden. (Albers et al. 1990)

To reduce her teaching load, she took on administrative duties instead. In 1978, Dr. Morawetz became Associate Director of the Courant Institute at NYU. For 3 years, beginning in 1981, she served in dual capacities as Deputy Director of the Courant Institute and Chairman of NYU's Department of Mathematics. From 1984 to 1988, she was Director of the Institute (Morawetz 2002). In 1993, she became Professor Emerita.

Dr. Morawetz also served on the boards of several societies and institutions. She was on the board of trustees of Princeton University, the AMS (serving as president for one term), and the Alfred P. Sloan Foundation. She was a member of the Advisory Committee for the National Science Foundation for the Mathematical Sciences, the

Mathematical Advisory Committee to the National Bureau of Standards, the New York City Mayor's Commission on Science and Technology, and the National Research Council's Board on Mathematical Sciences. In industry, she served on the board of NCR Corporation for 13 years (Morawetz 2002).

In addition to the National Medal of Science, Dr. Morawetz was recognized with numerous other honors, including the Guggenheim Fellowship in 1966 and again in 1978 (New York University 2017). In 2004, she became the first woman to receive the AMS's Leroy P. Steele Prize for Lifetime Achievement, and in 2006, she was the first woman to be awarded the AMS's George David Birkhoff Prize in Applied Mathematics (Gamba et al. 2018). She also received ten honorary degrees (New York University 2017).

Dr. Morawetz died in August 2017 at the age of 94. Her husband died less than 3 months later at the age of 102.

28.2 Far-Reaching Implications

Dr. Morawetz's work was so influential because it provided a framework that others used and expanded upon to solve problems in a wide range of areas. In aeronautics, her mathematical insights led to the development of the supercritical airfoil by NASA engineer Richard Whitcomb (Gamba et al. 2018), which is standard on almost all commercial and military jets. An airfoil is the cross-sectional shape of a wing or blade designed to control how air flows around it, reduce drag, and produce lift. A supercritical airfoil is designed specifically for airplanes flying near the speed of sound to reduce the formation of shock waves.

With a conventional airfoil, a region of supersonic airflow forms, beginning at speeds of approximately 0.85 times the speed of sound. Figure 28.2 illustrates the results of transonic flow for a conventional airfoil versus a supercritical airfoil. The teardrop shape of a conventional airfoil causes strong shock waves, flow separation, increased drag, and instability. A supercritical airfoil has a flattened upper surface and a more curved lower surface, which delays transonic flow to higher speeds and creates weaker shock waves, making flight more efficient (Wikipedia Contributors 2021b).

In acoustics, Dr. Morawetz studied how sound waves disperse when encountering an obstacle and proved that if a sound wave hits a star-shaped object, the sound will disperse and lose energy over time, rather than being trapped and increasing in intensity. In geometry, a star-shaped obstacle has at least one internal point from which the entire shape is visible. This is relevant in the design of concert halls and other buildings to ensure that sound does not get trapped and echo (Gamba et al. 2018).

Dr. Morawetz's work on wave scattering was applied to the design and use of sonar systems. These systems emit electromagnetic waves and analyze the signals that scatter back from objects to determine their location, speed, and other properties. Dr. Morawetz's work helped to describe and predict this scattering behavior mathematically (Jackson 1999).

Later researchers, such as Sergiu Klainerman, Elena Giorgi, and Jeremie Szeftel, expanded on her work to prove the stability of black holes (Klainerman 2025). They

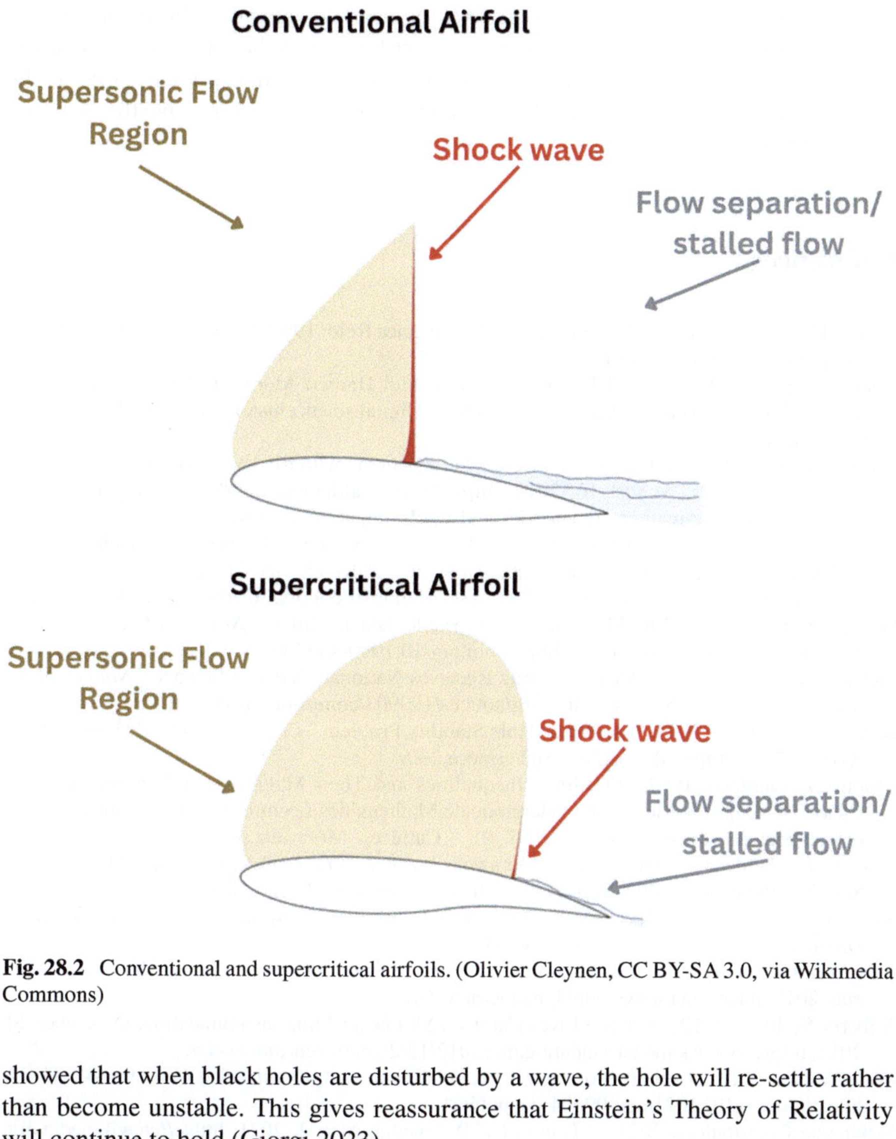

Fig. 28.2 Conventional and supercritical airfoils. (Olivier Cleynen, CC BY-SA 3.0, via Wikimedia Commons)

showed that when black holes are disturbed by a wave, the hole will re-settle rather than become unstable. This gives reassurance that Einstein's Theory of Relativity will continue to hold (Giorgi 2023).

28.3 Conclusion

Mathematician Dr. Cathleen Morawetz led a distinguished career, making fundamental and lasting contributions in and beyond mathematics. Her work in partial differential equations, particularly Morawetz's inequalities and estimates, has been applied by scientists across disciplines to better understand the behavior of waves.

Her influence, either direct or indirect, extends from the oceans to the skies, through the design of airplanes, sonar systems, concert halls, and the study of black holes. Dr. Morawetz was a leader and mentor, becoming the first woman to helm the Courant Institute at NYU and the first female mathematician to receive the National Medal of Science.

References

Albers, Donald J, Gerald L Alexanderson, and Constance Reid. 1990. *More Mathematical People*. Houghton Mifflin Harcourt P.

Bohning, James J. 1986. "Oral History Interview with Herbert Morawetz." Sciencehistory.org. Science History Institute. April 1, 1986. https://digital.sciencehistory.org/works/q811kk726#tab=ohTranscript.

Chang, Kenneth. 2017. "Cathleen Morawetz, Mathematician with Real-World Impact, Dies at 94." The Seattle Times. August 16, 2017. https://www.seattletimes.com/seattle-news/obituaries/cathleen-morawetz-mathematician-with-real-world-impact-dies-at-94/.

Gamba, Irene, Christina Sormani, Terence Tao, Leslie Greengard, Tonatiuh Sánchez-Vizuet, and Kevin R. Payne. 2018. "The Mathematics of Cathleen Synge Morawetz." *Notices of the American Mathematical Society* 65 (07): 764–78. https://doi.org/10.1090/noti1706.

Giorgi, Elena. 2023. "The Mathematics of Stable Black Holes." *Notices of the American Mathematical Society* 70 (04): 1. https://doi.org/10.1090/noti2662.

Jackson, Allyn. 1999. "Cathleen Morawetz Receives National Medal of Science." *Notices of the AMS* 46 (3): 352. https://www.ams.org/notices/199903/comm-morawetz.pdf.

Klainerman, Sergiu . 2025. "The Black Hole Stability Problem." *Comptes Rendus Mécanique* 353 (G1): 555–81. https://doi.org/10.5802/crmeca.290.

Morawetz, Cathleen. 1997. "Quadratic Inequalities and Their Magical Help." Presented at the Charles R. Deprima Memorial Undergraduate Mathematics Lecture, January 14. https://www.pma.caltech.edu/documents/2635/1997_01_-_Cathleen_Morawetz.pdf.

Morawetz, Cathleen Synge. 2002. "Morawetz — CV 2002." Celebratio.org. Mathematical Sciences Publishers. 2002. https://celebratio.org/Morawetz_CS/article/675/.

Morrow, Charlene, and Teri Perl. 1998. *Notable Women in Mathematics: A Biographical Dictionary*. Westport, Conn.: Greenwood Press.

New York University. 2017. "In Memoriam: Professor Emerita Cathleen Synge Morawetz." Nyu.edu. 2017. https://math.nyu.edu/dynamic/news/7/.

Roberts, Siobhan. 2012. "Science Lives Cathleen Morawetz." Simons Foundation. December 20, 2012. https://www.simonsfoundation.org/2012/12/20/cathleen-morawetz/.

Wang, Linda. 2018. "Herbert Morawetz." *C&EN Chemical & Engineering News* 96 (4): 32–32. https://doi.org/10.1021/cen-09604-peopobits4.

Wikipedia Contributors. 2021a. "Transonic." Wikipedia. May 2, 2021. https://en.wikipedia.org/wiki/Transonic.

Wikipedia Contributors. 2021b. "Supercritical Airfoil." Wikipedia. October 5, 2021. https://en.wikipedia.org/wiki/Supercritical_airfoil.

Chapter 29
Susan Solomon

For key scientific insights in explaining the cause of the Antarctic Ozone hole and for advancing the understanding of the global ozone layer; for changing the direction of ozone research through her findings; and for exemplary service to worldwide public policy decisions and to the American public

Keywords Ozone layer · Ozone hole · Ozone depletion · Susan Solomon · National Oceanic and Atmospheric Administration (NOAA) Aeronomy Lab · John Noxon · Art Schmeltekopf · Antarctica · CFCs · Climate change · Montreal Protocol · Chlorofluorocarbons

Dr. Susan Solomon (Fig. 29.1) is an atmospheric scientist who traveled to the "end of the world" to solve the mystery of whether chlorofluorocarbons (CFCs) were causing the hole in the ozone layer of Antarctica. She was the first woman to lead an expedition to Antarctica and stood on rooftops in extreme cold to collect scientific data that proved the negative environmental impact of human-made CFCs. This breakthrough led to an environmental success story with the quick adoption of the highly effective Montreal Protocol, an agreement to reduce the use of CFCs ratified by every country in the United Nations. For over 40 years, Dr. Solomon's career has focused on addressing climate change and enhancing the health of the planet and its inhabitants. Dr. Solomon received the National Medal of Science in 1999.

J. S. Tietjen et al., *Women Laureates 1970 – 2007*, Women in Engineering and Science, https://doi.org/10.1007/978-3-032-23116-1_29

Fig. 29.1 Susan Solomon. (Bengt Nyman from Vaxholm, Sweden, CC BY 2.0, via Wikimedia Commons)

29.1 Life (1956–)

Susan Solomon was born in Chicago in 1956. She grew up in a middle-class neighborhood in the city with her father, Leonard, an insurance salesman, her mother, Alice, a fourth-grade teacher, and one brother. No one in the family had much interest in science, but when Susan was about 9 years old, she saw the *Undersea World of Jacques Cousteau* on TV, and she became hooked. She was captivated by the program, taken with the beauty of the ocean creatures (Kellogg 1997).

However, Susan did not become a marine biologist. When she began high school and took a biology course, she found it frustrating. She felt the discipline was not quantitative enough and that the experimenter could too easily influence results. She found that chemistry suited her better. Applying quantitative methods to obtain results provided greater certainty, and she found chemistry both elegant and fascinating (Kellogg 1997).

Art was also an interest of Susan's, and she was unsure whether she should study it in college rather than science. Her success at the international-level science fair in high school helped steer her to a career in science. She won the City of Chicago's Science Fair project, "Using Light to Determine Percentage of Oxygen." She went to the International Science Fair, where she placed third. This experience built her confidence and also reinforced what she liked about chemistry, saying, "I was able to build a quantitative framework that actually helped me understand something about the physical and natural world" (Kellogg 1997).

After high school, Solomon decided to attend the Illinois Institute of Technology in Chicago and majored in chemistry. By her senior year, she knew she wanted to pursue a PhD, but she did not want her work confined to experiments conducted in a test tube in a laboratory. Fortunately, she found atmospheric chemistry where she would not be so confined, saying, "Engage in chemistry that took place on a planet, rather than in a test tube" (Kellogg 1997).

Solomon graduated from Illinois Institute of Technology in 1977 (Solomon 2023). She received a graduate fellowship from the University Corporation for Atmospheric Research (UCAR), which funded 2 years of studies and provided a job at the National Center for Atmospheric Research (NCAR) in Boulder, Colorado, during the summers (Kellogg 1997). She chose the University of California, Berkeley, for graduate school. She received her master's degree in chemistry from UC Berkeley in 1979 (Solomon 2023). She went straight into the PhD program at Berkeley, but she was not particularly interested in conducting research there, as its focus was not in atmospheric chemistry (Kellogg 1997). She was allowed to do her PhD research at NCAR and received her PhD from UC Berkeley in 1981 (Solomon 2023).

After graduation, Dr. Solomon wanted to stay in Boulder, and she was hired by the National Oceanic and Atmospheric Administration (NOAA) Aeronomy Lab there. Just 5 years later, she was standing at the South Pole. In May 1985, scientists Joseph Farman, Brian Gardiner, and Jonathan Shanklin from the British Antarctic Survey published their observations on ozone depletion over Antarctica (Filus 2019). Their data showed that the ozone layer had decreased by about half from the 1970s to 1984 (Smith 2024). Although many scientists initially doubted this shocking result, Dr. Solomon reviewed the paper and data and believed that the assertion was correct. What was not known at this stage was the cause of the depletion. Dr. Solomon posited that hydrogen chloride (HCl) and chlorine nitrate ($ClONO_2$) interact at the surfaces of clouds that form in the stratosphere (the second layer of Earth's atmosphere) over Antarctica, producing a highly reactive form of chlorine that damages the ozone layer (Kellogg 1997).

A team from the University of Wyoming was headed to Antarctica in August 1986, and NOAA wanted to send someone along to take measurements to confirm the existence of the ozone hole and collect data to test Dr. Solomon's theory. John Noxon and Arthur Louis (Art) Schmeltekopf at Aeronomy Lab had developed an instrument to measure NO_2 (nitrous oxide) and ozone. Noxon had died in 1985, and Schmeltekopf was committed to another project, so Dr. Solomon volunteered to go on the Antarctic expedition, even though at that time she was more of a theoretician than a field scientist (Kellogg 1997). The expedition team consisted of 18 researchers from the University of Wyoming, NOAA, State University of New York at Stony Brook, and the National Aeronautics and Space Administration (NASA) (Solomon 1986; Welch 2006). Dr. Solomon, the only woman on the trip, was selected to serve as the expedition leader.

When asked in an interview as to why she thought she was selected as the leader, Dr. Solomon credited her willingness and ability to speak to the media in layperson's terms. She also felt that, as a theoretician, she was viewed as less personally invested in NOAA's experiment and, therefore, able to allocate logistical resources objectively between the four experiments being conducted (Kellogg 1997).

The expedition was a great success. Dr. Solomon's group measured NO_2, ozone, and ClO_2 (chlorine dioxide). The data and subsequent analysis showed that the chlorine dioxide level was 100 times higher than expected, providing the first direct evidence that chlorine chemistry and, in turn, CFCs were the cause of the Antarctic ozone hole (Kellogg 1997). The findings garnered considerable public attention and concern. CFCs are a man-made chemical consisting of chlorine, fluorine, and carbon, which is stable, non-flammable, and nontoxic. At the time, CFCs were used in refrigerators, freezers, air conditioners, and as propellants in spray cans, among other applications. If their use was not put in check, the ozone layer would thin all over the world, not just Antarctica (Smith 2024).

Public outcry over the depletion of the ozone layer, which protects Earth from harmful ultraviolet (UV) radiation from the Sun, prompted action on multiple fronts. The demand for spray cans that used CFCs as propellants dropped significantly, prompting the industry to search for substitutes. In astonishingly short order, and less than a year after the 1986 Antarctic expedition, the Montreal Protocol on Substances That Deplete the Ozone Layer, an international treaty designed to protect the ozone layer by phasing out CFCs, was adopted in September 1987 (United Nations Environment Programme n.d.) and was subsequently ratified by every United Nations member, the first treaty to be universally ratified in the history of the United Nations, a still very rare feat (Wikipedia Contributors 2019). As of 2024, emissions of CFCs and related substances have decreased by more than 99% (Drollette 2024). Because CFCs stay in the atmosphere for decades, recovery of the ozone layer has been slow but steady. The ozone layer is expected to make a full recovery by 2040 over most of the world, but not until 2045 over the Arctic and 2066 over Antarctica ("Ozone Layer Recovery Is on track, Helping Avoid Global Warming by 0.5°C" 2023).

Dr. Solomon returned to Antarctica in 1987 to collect additional data and later conducted investigations on ozone depletion in the Arctic and elsewhere (Science History Institute n.d.). Her group has also researched how sulfur emitted in volcanic eruptions affects ozone depletion (Schaffer 2019). In addition to ozone depletion, Dr. Solomon has studied ozone in the lower atmosphere. While ozone in the stratosphere creates a protective layer between the Sun and Earth, in the lower atmosphere, ozone is a secondary pollutant formed through chemical reactions between sunlight and pollutants, such as vehicle and industrial plant emissions, which trap heat and cause smog (Scientific Asia n.d.).

While still at NOAA, Dr. Solomon served on the Intergovernmental Panel on Climate Change (IPCC). In 1994, she co-authored a report on radiative forcing, which is the difference between the energy that Earth absorbs from the Sun and the energy that it releases back into space (NOAA 2021). The report introduced improvements in quantifying the contributions of various greenhouse gases to radiative forcing, furthering understanding of human factors in climate change and providing a scientific foundation for international climate policies (Houghton et al. 1996).

From 2002 to 2008, Dr. Solomon co-chaired the IPCC's working group that produced the report "Climate Change 2007: The Physical Science Basis" (Solomon et al. 2007). The report stated that "warming of the climate system is unequivocal" and that it was greater than 90% likely to be a result of an increase in greenhouse gases caused by human activity (Solomon et al. 2007). The IPCC has noted that climate change will lead to a scarcity of resources, such as water, potentially sparking global conflicts and war. In 2007, the IPCC shared the Nobel Peace Prize with former US Vice President Al Gore Jr.

> for their efforts to build up and disseminate greater knowledge about man-made climate change, and to lay the foundations for the measures that are needed to counteract such change (The Nobel Prize n.d.).

In recognition of her contributions, Dr. Solomon attended the Nobel ceremony and shared the stage with Gore and other IPCC members (Colorado Public Radio Staff 2010).

In 2011, Dr. Solomon left NOAA for a position at the Massachusetts Institute of Technology (MIT), where she serves as Chair of the Program in Atmospheres, Oceans and Climate (PAOC) and the Lee and Geraldine Martin Professor of Environmental Studies in the Department of Chemistry. Additional areas of research over the years have included ocean-atmosphere interactions, the behavior of the ice-age stratosphere, the influences of volcanoes and wildfires on atmospheric chemistry and dynamics, possible changes to the width of the Earth's tropics, and documenting significant changes in climate in both the stratosphere and troposphere (MIT Earth, Atmospheric, and Planetary Sciences n.d.).

In a full-circle moment in 2016, Dr. Solomon used balloon and satellite ozone data, along with other measurements and scientific models, to show that the ozone layer over Antarctica was recovering (Solomon et al. 2016). "We can now be confident that the things we've done have put the planet on a path to heal," she said (University Corporation for Atmospheric Research 2016).

Dr. Solomon's scientific contributions were recognized early and often. In 1992, at just the age of 36, she was elected to the National Academy of Sciences. In 1994, both a snow saddle (a glaciological feature with a saddle-shaped depression) and a glacier in Antarctica were named in her honor. In addition to the National Medal of Science, she received the highest honors from the American Meteorological Society (Carl-Gustaf Rossby Research Medal in 2000), the Geochemical Society (the Goldschmidt Medal in 2006), and the American Geophysical Union (William Bowie Medal in 2007). She also received Japan's Asahi Glass Foundation Blue Planet Prize in 2004 and the Grande Médaille of the French Academy of Sciences in 2008. Also in 2008, she was elected to the American Philosophical Society and named one of the year's 100 most influential people in *Time* magazine. Dr. Solomon has been inducted into the Colorado Women's Hall of Fame and the National Women's Hall of Fame. She shared one of the highest honors in the Earth Sciences, the Vetlesen Prize (the first female recipient) in 2012. In 2018, Dr. Solomon was awarded the Swedish Academy of Science's Crafoord Prize, and in 2019, she was inducted into the inaugural class of the Government Hall of Fame (Wikipedia Contributors 2021).

Dr. Solomon is the sole author of *The Coldest March: Scott's Fatal Antarctic Expedition*, and *Solvable: How We Healed the Earth, and How We Can Do It Again.*

Although Dr. Solomon has given many interviews over the years, she has said little about her personal life. Dr. Solomon has a love of nature and enjoys bird photography, particularly capturing images of penguins. She also likes making jewelry from glass and shells that she collects on the beach (Pewsey 2022). She has been married to Barry Sidwell since 1988 (Kellogg 1997).

29.2 The Ozone Layer

Single oxygen atoms are unstable, so they primarily exist in pairs, as O_2. In the stratosphere, when O_2 molecules are struck by high-intensity UV radiation (UVC) from the Sun, they split apart. The resulting unstable atoms each collide with and bond to O_2 molecules to form O_3, a molecule with three oxygen atoms, known as ozone. This process is shown in Fig. 29.2.

When lower-energy UV radiation (known as UVB) from the Sun hits O_3, a single oxygen atom breaks off, converting it back to a free oxygen atom, O, and an O_2 molecule. Thus, ozone is continuously produced and removed from the atmosphere, a process called the Chapman Cycle. In the absence of pollutants, a relatively stable layer of ozone is maintained.

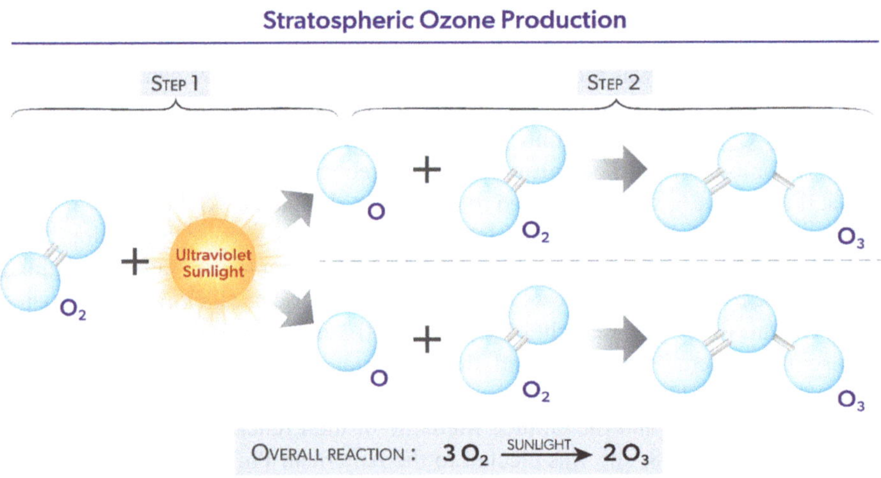

Fig. 29.2 Ozone production. (Courtesy of NOAA)

The ozone layer is often called Earth's sunscreen because it prevents harmful UVB and UVC radiation from reaching the Earth's atmosphere. This protection is essential for human life and the entire ecosystem. UV radiation can cause skin cancer, cataracts, and other eye diseases and weakened immune systems in people and animals. It can also inhibit plant growth and crop yields, increasing the possibility of famine. For marine life, UV rays can penetrate water and impact the health of phytoplankton, the foundation of the aquatic food web, disrupting the entire food chain (IERE Team 2025).

Contrary to popular belief, the ozone hole is not actually a hole. The term refers to a significant thinning of the ozone layer. Scientists observed that the ozone hole was bigger over Antarctica during the Southern Hemisphere's spring. The prevailing theory held that CFCs released chlorine atoms in the stratosphere, but scientists did not believe this was the primary mechanism for the massive, localized depletion observed over Antarctica.

Dr. Solomon hypothesized that in the stratosphere, UV radiation from the Sun weakens the CFC carbon–chlorine bond, causing a chlorine atom to break free. These chlorine atoms react with other gases, primarily forming inactive HCl and $ClONO_2$, which act as "storage molecules" by locking up the chlorine atoms released during the breakdown of CFCs.

Dr. Solomon suggested that HCl and $ClONO_2$ accumulate on the surface of polar stratospheric clouds that form in the winter due to the extremely cold conditions unique to the stratosphere over Antarctica. When temperatures warm in the spring, sunlight triggers a reaction with HCl and $ClONO_2$ molecules that forms Cl_2. Figure 29.3 depicts the ozone-destruction cycle as theorized by Dr. Solomon. UV radiation breaks the Cl_2 molecular bond, releasing free chlorine atoms. The free chlorine atoms then react with an ozone molecule, resulting in an oxygen atom from the ozone bonding with the chlorine atom, yielding chlorine monoxide (ClO) and an oxygen molecule (O_2). The cycle continues when the ClO reacts with another O atom in the stratosphere, forming O_2 and regenerating a Cl atom, which can attack another ozone molecule. Thus, a chain reaction occurs where a single chlorine atom from a CFC can destroy thousands of ozone molecules (Fahey et al. 2022).

In Antarctica, she sought to prove her theory by measuring the levels of reactive chlorine compounds in the atmosphere. Some measurements were taken at night, with Dr. Solomon standing on a rooftop in subzero temperatures and high winds to adjust the mirrors for Noxon and Schmeltekopf's instrument to use the moonlight. Based on the data she obtained, these high ClO levels corresponded precisely to areas with the most severe ozone depletion. This ruled out alternative theories of ozone destruction, such as natural phenomena, because they did not predict significant levels of chlorine. Only the breakdown of CFCs could produce massive amounts of ClO. Thus, Dr. Solomon provided empirical evidence that confirmed the role of CFCs in damaging the ozone layer (Kellogg 1997).

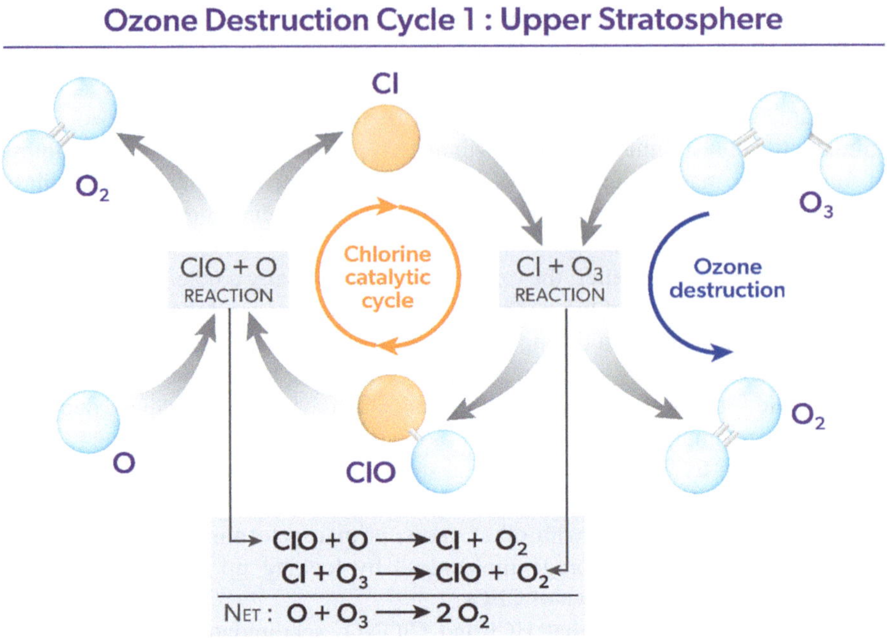

Fig. 29.3 Ozone-destruction cycle. (Courtesy of NOAA)

29.3 Conclusion

Dr. Susan Solomon has played a critical role in protecting the ozone layer. Her ability and passion for applied science, along with her communication and leadership skills, have kept her at the forefront of the fight against climate change. She is an effective steward of the planet, whose work led to the rapid and universal adoption of the Montreal Protocol, a rare triumph of international scientific and political cooperation. Dr. Solomon serves as a reminder that humanity can solve seemingly unsolvable problems and work together to heal the planet.

References

Colorado Public Radio Staff. 2010. "Boulder Scientist on Nobel Winning Panel." Colorado Public Radio. April 5, 2010. https://www.cpr.org/2010/04/05/boulder-scientist-on-nobel-winning-panel/.

Drollette Jr., Dan. 2024. "Interview with Susan Solomon: The Healing of the Ozone Hole, and What Else We Can Learn from Atmospheric Near-Misses." Bulletin of the Atomic Scientists. November 2024. https://thebulletin.org/premium/2023-05/interview-with-susan-solomon-the-healing-of-the-ozone-hole-and-what-else-we-can-learn-from-atmospheric-near-misses/.

Fahey, David W., Paul A. Newman, John A. Pyle, and Bonfils Safari. 2022. "Scientific Assessment of Ozone Depletion 2022: Twenty Questions and Answers about the Ozone Layer." Csl.

noaa.gov. NOAA Chemical Sciences Laboratory. 2022. https://csl.noaa.gov/assessments/ozone/2022/twentyquestions/#section-i.
Filus, Michael. 2019. "How One Nature Paper in 1985 Led to the Discovery of the Ozone Hole." Springer Nature Research Communities. Springer Nature. July 13, 2019. https://communities.springernature.com/posts/how-one-nature-paper-in-1985-led-to-the-discovery-of-the-ozone-hole.
Houghton, John Theodore, Filho, Callander, Harris, Kattenberg, and Maskell. 1996. *Climate Change 1995: The Science of Climate Change.* Cambridge: Cambridge University Press for the Intergovernmental Panel on Climate Change.
IERE Team. 2025. "Why Do We Need the Ozone Layer? - the Institute for Environmental Research and Education." The Institute for Environmental Research and Education. June 30, 2025. https://iere.org/why-do-we-need-the-ozone-layer/.
Kellogg, Dale. 1997. "Interview with Susan Solomon" Interview by Dale Kellogg. American Meteorological Society University Corporation for Atmospheric Research, Tape Recorded Interview Project. September 7, 1997. https://voices.nmfs.noaa.gov/sites/default/files/2024-12/Solomon_Susan_09-05-1997.pdf.
MIT Earth, Atmospheric, and Planetary Sciences. n.d. "Susan Solomon." MIT. https://eaps.mit.edu/people/faculty/susan-solomon/. Accessed November 4, 2025.
NOAA. 2021. "Climate Forcing." Climate.gov. September 22, 2021. https://www.climate.gov/maps-data/climate-data-primer/predicting-climate/climate-forcing.
Pewsey, Emma. 2022. "In Situ with Susan Solomon." Chemistry World. Royal Society of Chemistry. April 29, 2022. https://www.chemistryworld.com/culture/susan-solomon-we-had-to-literally-stand-out-on-the-roof/4015443.article.
Schaffer, Amanda. 2019. "The Climate Optimist." MIT Technology Review. February 27, 2019. https://www.technologyreview.com/2019/02/27/137093/the-climate-optimist/.
Science History Institute. n.d. "Susan Solomon." https://www.sciencehistory.org/education/scientific-biographies/susan-solomon/. Accessed November 3, 2025.
Scientific Asia. n.d. "Which Type of Pollution Includes CFCs and Smog?" https://scientificasia.net/which-type-of-pollution-includes-cfcs-and-smog/. Accessed November 4, 2025.
Smith, Chris. 2024. "Susan Solomon: Chemistry of the Ozone Hole." Thenakedscientists.com. July 16, 2024. https://www.thenakedscientists.com/articles/interviews/susan-solomon-chemistry-ozone-hole.
Solomon, Susan. 1986. "The National Ozone Expedition, 1986." *Antarctic Journal of the United States* 22 (5).
Solomon, Susan. 2023. "Curriculum Vitae." *MIT Earth, Atmospheric, and Planetary Sciences.* MIT. https://www.ssolomongroup.mit.edu/_files/ugd/eee3f7_d2687ee-a99794b53b868afecf37559e4.pdf.
Solomon, Susan, Diane J. Ivy, Doug Kinnison, Michael J. Mills, Ryan R. Neely, and Anja Schmidt. 2016. "Emergence of Healing in the Antarctic Ozone Layer." *Science* 353 (6296): 269–74. https://doi.org/10.1126/science.aae0061.
Solomon, Susan, Dahe Qin, Martin Manning, Melinda Marquis, Kristen Averyt, Melinda M.B. Tignor, Henry LeRoy Miller, Jr., and Zhenlin Chen, eds. 2007. "AR4 Climate Change 2007: The Physical Science Basis — IPCC." *Ipcc.ch.* IPCC. https://www.ipcc.ch/report/ar4/wg1/.
The Nobel Prize. n.d. "Intergovernmental Panel on Climate Change – Facts." https://www.nobelprize.org/prizes/peace/2007/ipcc/facts/. Accessed November 4, 2025.
United Nations Environment Programme. 2023. "Ozone Layer Recovery Is on Track, Helping Avoid Global Warming by 0.5°C." 2023. https://www.unep.org/news-and-stories/press-release/ozone-layer-recovery-track-helping-avoid-global-warming-05degc.
United Nations Environment Programme. n.d. "About Montreal Protocol." UN Environment Programme. United Nations Environment Programme. https://www.unep.org/ozonaction/who-we-are/about-montreal-protocol. Accessed November 3, 2025.

University Corporation for Atmospheric Research. 2016. "Scientists Observe First Signs of Healing in the Antarctic Ozone Layer." NCAR News. June 30, 2016. https://news.ucar.edu/121673/scientists-observe-first-signs-healing-antarctic-ozone-layer.

Welch, Charles. 2006. "Twenty Year Anniversary of the Discovery of the Cause of the Antarctic Ozone Hole." Theozonehole.org. September 15, 2006. https://www.theozonehole.org/twenty-year.htm.

Wikipedia Contributors. 2019. "Montreal Protocol." Wikipedia. Wikimedia Foundation. June 13, 2019. https://en.wikipedia.org/wiki/Montreal_Protocol.

Wikipedia Contributors. 2021. "Susan Solomon." Wikipedia. October 11, 2021. https://en.wikipedia.org/wiki/Susan_Solomon.

Chapter 30
Lynn Margulis

For her outstanding contributions to understanding of the development, structure, and evolution of living things, for inspiring new research in the biological, climatological, geological, and planetary sciences, and for her extraordinary abilities as a teacher and communicator of science to the public

Keywords Evolution · Biology · Neo-Darwinism · Serial endosymbiosis theory · Genetics · Symbiogenesis · Evolutionary biologist · Symbiosis · Natural selection · Random mutations · Prokaryote · Eukaryote · Organelle · Mitochondria · Chloroplasts · Bacteria · Plastid

Enthralled by genetics from her late teens when she graduated from college, evolutionary biologist Lynn Margulis, shown in Fig. 30.1, developed the "serial endosymbiosis theory" (SET), which demonstrates that symbiotic mergers among bacteria led to plant and animal cells, not the random mutations espoused by neo-Darwinism. Regarded as heresy at the time of its publication, SET changed humankind's understanding of the evolution of cells. SET is today a central tenet of modern biology and taught in high school. Margulis spent her career in academia, wrote books—both textbooks and popular science books—and hundreds of scientific papers. A member of the National Academy of Sciences, Margulis received numerous awards, including the National Medal of Science from President Bill Clinton in 1999.

J. S. Tietjen et al., *Women Laureates 1970 – 2007*, Women in Engineering and Science, https://doi.org/10.1007/978-3-032-23116-1_30

Fig. 30.1 Lynn Margulis. (Courtesy of Bob O'Connor)

30.1 Life and Career (1938–2011)

Lynn Petra Alexander was born in 1938 in Chicago, Illinois, the eldest of four daughters of Morris and Leona Wise Alexander. Her father was an attorney who became a businessman, and her mother was a travel agent (Poundstone 1999; Wikipedia—Margulis 2025a). A teenage prodigy, Lynn entered the University of Chicago in the spring of 1954 when she was just 15 (Douglass Funeral Service 2025; Margulis 1998). While there, she met astronomer Carl Sagan, who was a graduate student. They married in June of 1957, when she was 19, almost immediately after her graduation with an AB in liberal arts. The marriage produced two sons (Dorion in 1959 and Jeremy in 1960) (Poundstone 1999).

After a summer at the University of Colorado, where she studied ecology and plant physiology, Lynn started graduate studies at the University of Wisconsin in the Fall of 1957 (Poundstone 1999). She began as a teaching fellow in zoology and studied cell biology and genetics in two separate departments. At Wisconsin, she studied under geneticist James F. Crow. She says,

> I adored Crow's general genetics course; it changed my life. I knew when I left the University of Chicago I wanted to study genetics, but after Crow's class I knew I wanted to study only genetics. (Margulis 1998)

One of her professors at the University of Wisconsin was Hans Ris of whom she said, "Hans Ris was a fine teacher—the best of my whole career. He was a scientist's scientist; he made no concessions at all" (Goldscheider 2009). Ris had examined chloroplasts—the bodies within plant cells that contain green chlorophyll—under an electron microscope. He thought they looked like cells of blue-green bacteria. Lynn decided that this was evidence that chloroplasts were bacteria that were living symbiotically inside higher plant cells (Poundstone 1999). Her conviction and subsequent research would revolutionize biology (Poundstone 1999).

Lynn earned her MS in zoology and genetics in 1960 and entered the University of California, Berkeley, later that year. Her objective was to study all aspects of evolution, paleontology, and genetics as she was interested in the evolutionary history of cells (Margulis 1998). At the time she said,

> Although I was twenty-two years old and already the mother of two persistently active boys, my enthusiasm for pursuing cell genetics and evolution overwhelmed any thoughts of becoming a full-time housewife. More than husbands, I had always wanted children. (Margulis 1998)

Her marriage to Carl Sagan was very difficult for her—she described it later like "a torture chamber shared with children." She and her two sons moved out of a shared domicile when Jeremy started walking (Poundstone 1999).

At Berkeley, Lynn continued her study of chloroplasts. She found that the bacteria in the chloroplasts retained their own DNA. She also came to believe that mitochondria (responsible for oxygen respiration) were also descended from formerly free-living bacteria. The theory that would result from her research and conviction is called "serial endosymbiosis theory (SET)" (Poundstone 1999).

When Carl Sagan was offered a position at Harvard, which he accepted beginning in February 1963, he convinced Lynn and the boys to move East with him. She originally said no but decided that she did not want to continue living in California and did not want to return to the Midwest. She saw it as a move of convenience and relocated to Massachusetts, determined to stay only a short time with Carl. They moved to Watertown, Massachusetts, in 1963. At this point, she had not finished her PhD thesis (Poundstone 1999). From 1963 to 1966, she was first a research associate, then a lecturer, and later the biology coordinator for the Peace Corps/Columbia Project (one of the original programs established by the Peace Corps in 1961), all at Brandeis University (Library of Congress 2024; Peace Corps 2025). Nights and weekends, she worked on her thesis. Lynn told Carl they were getting a divorce. The divorce was final in 1964 (Poundstone 1999).

Lynn completed her PhD in biology in 1965. She also continued to work on SET and wrote a paper, *On the Origin of Mitosing Cells*, that was rejected by 15 publications before its 1966 acceptance for publication in the *Journal of Theoretical Biology.* The article appeared in print in 1967 under her then name of Lynn Sagan. The paper was accepted; SET itself would take many years before acceptance—but now is even taught in high school (Margulis 1998).

She married Thomas N. Margulis in 1967 and had two children—Jennifer and Zachary—before their 1980 divorce (Douglass Funeral Service 2025).

Dr. Margulis is also recognized for her contributions to the Gaia Hypothesis, which posits that the Earth is a self-regulating system that maintains conditions to sustain life (Douglass Funeral Service 2025).

In 1966, Dr. Margulis joined Boston University, where she remained for 22 years, receiving regular promotions and achieving Full Professor status in 1977 and University Professor appointment in 1986. In 1988, she moved to the University of Massachusetts Amherst as Distinguished University Professor of Botany. In 1993, she was named Distinguished Professor of Biology and, in 1997, she moved to the Department of Geosciences as Distinguished Professor of Geosciences (Library of Congress 2024).

Dr. Margulis wrote numerous books—textbooks and those for the general public. She wrote a number of books with her son Dorion Sagan, including *Microcosmos*, translated into many languages (Archibald 2012). She said that she and Dorion "Make real science accessible to readers" (Goldscheider 2009). Her textbook, *Symbiosis and Cell Evolution*, has been called "An indisputable classic in the literature of science" (Douglass Funeral Service 2025).

In 1975, Dr. Margulis was named a fellow of the American Association for the Advancement of Science, "for her contributions to cell biology, in particular for her studies on the origin of eukaryotic cells" (Margulis 2011). She was elected to membership in the National Academy of Sciences in 1983 (National Academy of Sciences 2025). She received the National Medal of Science from President Bill Clinton in 1999 (National Science and Technology Metals Foundation 2025). That year, she also received the William Procter Prize for Scientific Achievement from Sigma Xi, which recognizes "outstanding contribution to scientific research and has demonstrated an ability to communicate the significance of this research to scientists in other disciplines" (Sigma Xi 2025).

In 2002, she was named by *Discover* magazine as 1 of the 50 most important women in science (Svitil 2002). In 2008, she received the Darwin-Wallace Medal from the Linnean Society of London. This medal is "awarded to a person or group who have made major advances in evolutionary science" and from 1908 to 2008 was only awarded every 50 years. Her many other awards include numerous honorary doctorates from institutions both in the USA and abroad (Margulis 2011).

She served as President of Sigma Xi, the Scientific Research Society from 2005 to 2006 (Library of Congress 2024). She was the co-founder of two international societies, Evolutionary Protistology (ISEP) and Symbiosis (ISS). Dr. Margulis served on committees for the National Academy of Sciences, NASA, the National Science Resource Center, and the American Association for the Advancement of Science (NASA 2025).

Dr. Margulis continued to work and teach (and bicycle to work) until she died of a stroke in 2011. Her papers are archived permanently at the Library of Congress (Library of Congress 2024).

30.2 Serial Endosymbiosis Theory

Dr. Margulis defines symbiosis as the system in which members of different species live in physical contact. Humans have bacterial and animal symbionts in our digestive system and on our eyelashes. Clover plants have little balls on their roots, which are nitrogen-fixing bacteria. Trees have mushrooms entwined in their roots. Dogs have symbiotic worms in their digestive system. The world is filled with symbiosis (Margulis 1998).

The evolutionary concept that she developed, "Serial Endosymbiosis Theory (SET)" (sometimes used interchangeably with symbiogenesis), posits that evolutionary adaptations arise primarily from the merging of two separate organisms (single-celled bacteria) to form a single new organism, not from random mutations (Goldscheider 2009). Those bacteria had either been eaten or had infected the host cells in which they now resided. This action didn't kill either organism—both cell lines survived. In her 1967 article, Dr. Margulis proposed that specific organelles (subunits of cells that perform specific functions much like organs in the body) of eukaryotes (nucleus-containing cells)—mitochondria, plastids (photosynthesis and food storage), and the flagella apparatus (locomotion)—were endosymbiotic. They resulted from bacteria (prokaryotes) that were now symbiotic with the host (Archibald 2012). Different types of bacteria merged to form the cells that today comprise plants and animals and fungi (Margulis 1998). The DNA from the original bacteria can still be found today in mitochondria and plastids within eukaryotes.

Prokaryotes (Fig. 30.2), such as bacteria, are single-celled organisms that have no nucleus or other internal bodies (Ashworth 2024).

Eukaryotes (Figs. 30.3 and 30.4) include amoebas and almost all of the cells in a human body and are more complicated than prokaryotes. Eukaryotes have a nucleus as well as organelles—ribosomes, mitochondria, and for plants, chloroplasts (Ashworth 2024). An organelle is a subunit within a cell that has a specialized function (much like organs in a body). Functions include energy production, processing of proteins, locomotion, and others (Wikipedia—Organelle 2025b). Mitochondria are an organelle that oxidizes glucose to produce energy (Wikipedia—Organelle 2025b).

Today, mitochondrial and plastid genomes have around 200 of the original more than 1500 genes believed to have been present millions of years ago when they were free living. Many of those missing genes reside instead in the host's nuclear genome (Archibald 2012).

One of the many important consequences of SET was that it maintained that the merging of cell lines, not just the accumulation of random mutations, was one of the drivers of evolution (Poundstone 1999). This was contrary to the theory of neo-Darwinism (natural selection combines with genetics leading to specific evolution), which was dominant at the time.

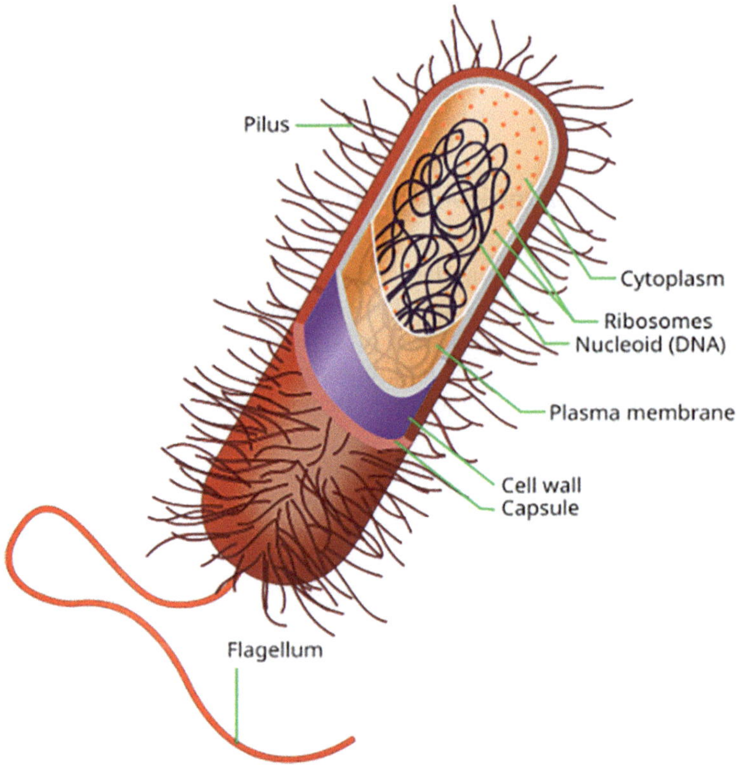

Fig. 30.2 Prokaryote cell. (Courtesy of Wikimedia)

SET demonstrates how symbiosis generates innovation in evolution. Dr. Margulis' contribution to evolution is considered on par with Darwin's theory of natural selection (Douglass Funeral Service 2025). She further elucidated SET in her 1970 book *Origin of Eukaryotic Cells: Evidence and Research Implications for a Theory of the Origin and Evolution of Microbial, Plant and Animal Cells on the Precambrian Earth* (Margulis 2011). Rejected by its original publisher, Academic Press, it found a more receptive audience at the Yale University Press, which published the book a year after its original rejection (Margulis 1998). In this book, she uses data from many scientific disciplines—geology, paleontology, ecology, cell biology, microbiology, genetics, and biochemistry. This interdisciplinary thinking permeates much of her work (Archibald 2012).

> "The evolution of eukaryotic cells was the single most important event in the history of the organic world," said Ernst Mayr, the leading evolutionary biologist of the last century. "Margulis's contribution to our understanding the symbiotic factors was of enormous importance." (Teresi 2011)

By 1981, when her book *Symbiosis in Cell Evolution* was published, SET was on its way to becoming textbook fact, aided by the development of new access to

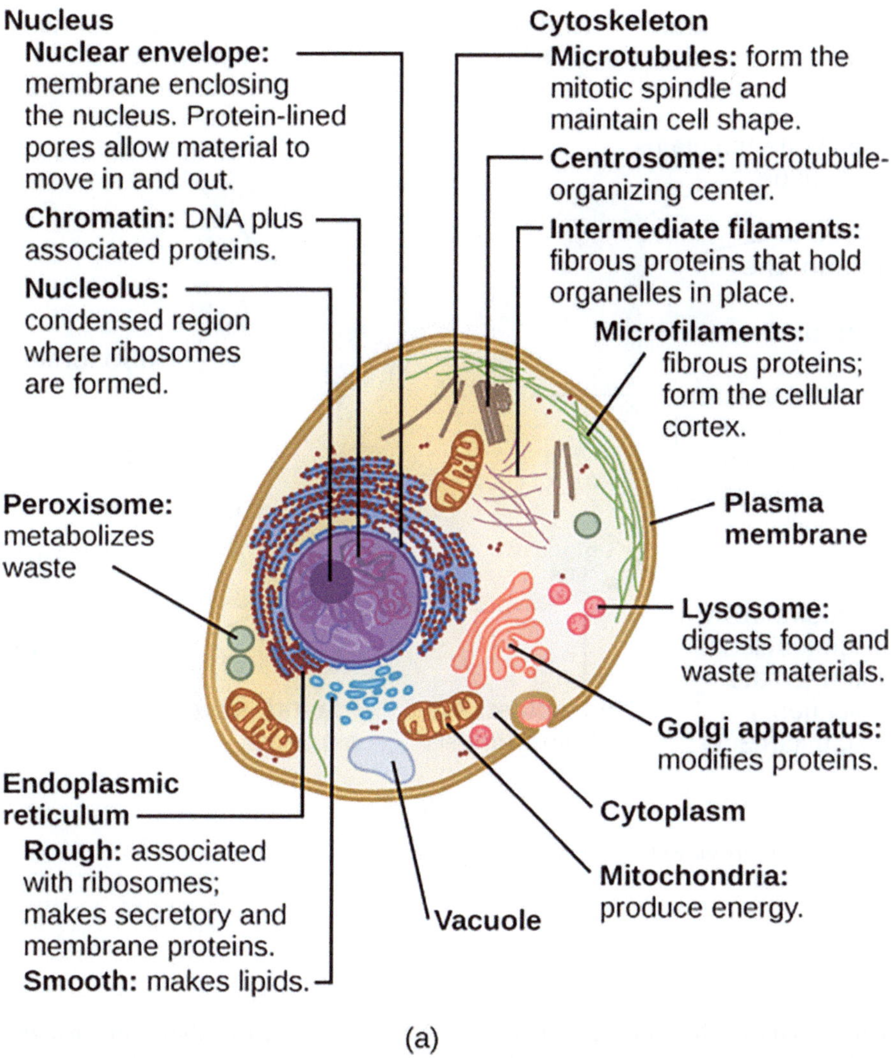

Fig. 30.3 Eukaryotic cell—animal (Molnar and Gair 2015)

molecular data. Later, sequencing of genomes would provide further evidence that eukaryotic cells contain bacteria DNA, validating SET (Archibald 2012; James and Fitzsimmons 2025). So far, SET has been proven correct for mitochondria and plastids (Archibald 2012). In 1998, Dr. Margulis said,

> My students and I still work on the central idea [microbial symbionts and membrane-bounded organelles]: the origin of cells with nuclei is exactly equal to the evolutionary integration of symbiotic bacterial communities. (Margulis 1998)

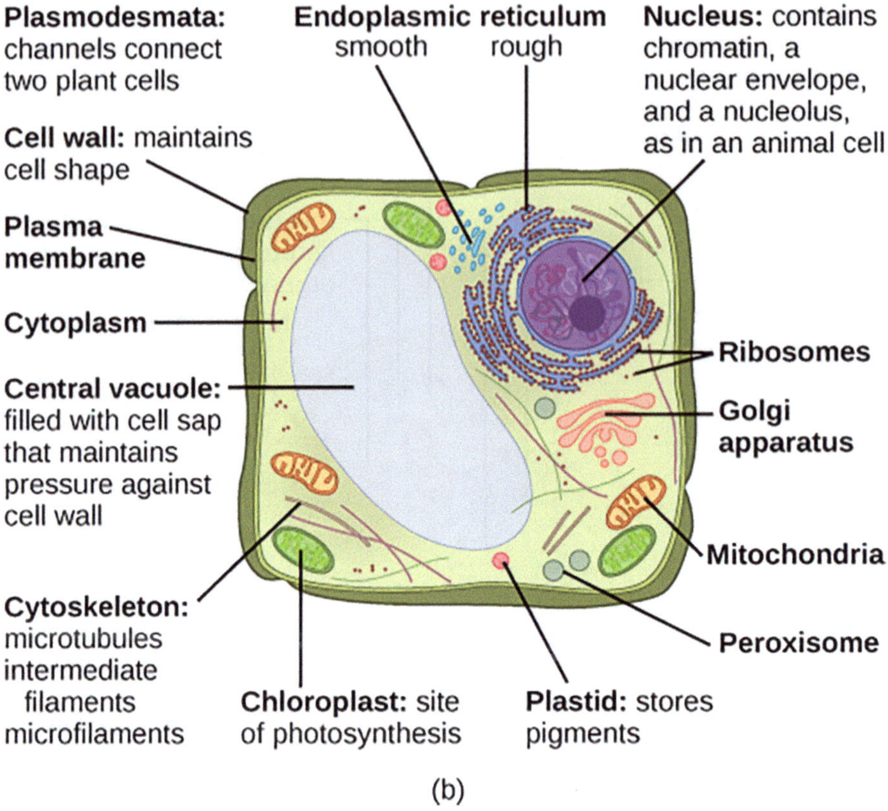

Fig. 30.4 Eukaryotic cell—plant (Molnar and Gair 2015)

30.3 Conclusion

Convinced from her graduate school days that bacteria lived symbiotically in higher cells, Dr. Lynn Margulis advanced the theory of serial endosymbiosis, which has changed humankind's beliefs about evolution. SET demonstrates that symbiotic mergers among bacteria led to plant and animal cells. A prolific author who wrote technical and popular science books, Dr. Margulis received many awards, including the National Medal of Science.

References

Archibald, John M. 2012. Lynn Margulis (1938-2011). Current Biology. Volume 22, Issue 1, PR4-R6, January 10. https://www.cell.com/current-biology/fulltext/S0960-9822(11)01375-3. Accessed October 7, 2025.

Ashworth, Jr., William B. 2024. Scientist of the Day – Lynn Margulis. November 22. https://www.lindahall.org/about/news/scientist-of-the-day/lynn-margulis/. Accessed October 7, 2025.
Douglass Funeral Service. Lynn Margulis. Accessed October 7, 2025.
Goldscheider, Eric. 2009. Teaching & Learning: Evolution Revolution. On Wisconsin. Fall. https://onwisconsin.uwalumni.com/evolution-revolution/. Accessed November 4, 2025.
James, Rachel and Lindsey Fitzsimmons. Mothers in Science – Lynn Margulis: Evolutionary Biologist. 2025. February 3. https://www.mothersinscience.com/trailblazers/lynn-margulis Accessed October 7, 2025.
Library of Congress. 2024. Lynn Margulis Papers. Finding Aid. https://findingaids.loc.gov/exist_collections/ead3pdf/mss/2018/ms018008.pdf. February. Accessed November 4, 2025.
Margulis, Lynn. 2011. Curriculum Vitae. https://hummingbirdfilms.com/wp-content/uploads/2018/02/Margulis-CV-2011.pdf. September. Accessed November 7, 2025.
Margulis, Lynn. Symbiotic Planet: A New Look at Evolution. Basic Books (Science Masters series). 1998.
Molnar, Charles and Jane Gair. 2015. Concepts of Biology – 1st Canadian Edition. https://opentextbc.ca/biology/chapter/3-3-eukaryotic-cells/. Accessed November 11, 2025.
National Academy of Sciences. Lynn Margulis. https://www.nasonline.org/directory-entry/lynn-margulis-o7vmz5/. Accessed November 4, 2025.
National Aeronautics and Space Administration (NASA). Lynn Margulis (1938-2011). https://science.nasa.gov/people/lynn-margulis/. Accessed October 7, 2025.
National Science and Technology Metals Foundation. Lynn Margulis. https://nationalmedals.org/laureate/lynn-margulis/. Accessed October 7, 2025.
Peace Corps. About Colombia. https://www.peacecorps.gov/colombia/about/. Accessed November 18, 2025.
Poundstone, William. 1999. Carl Sagan: a life in the cosmos. Henry Holt and Company, New York.
Sigma Xi. The William Procter Prize for Scientific Achievement. https://www.sigmaxi.org/programs/prizes-awards/william-procter. Accessed November 4, 2025.
Svitil, Kathy A. 2002. The 50 Most Important Women in Science. Discover Magazine. October 31. https://www.discovermagazine.com/the-50-most-important-women-in-science-5574. Accessed November 7, 2025.
Teresi, Dick. 2011. Discover Interview: Lynn Margulis Says She's Not Controversial, She's Right. Discover Magazine. June 16. https://www.discovermagazine.com/discover-interview-lynn-margulis-says-shes-not-controversial-shes-right-1033. Accessed November 7, 2025.
Wikipedia. Lynn Margulis. https://en.wikipedia.org/wiki/Lynn_Margulis. Accessed October 7, 2025a.
Wikipedia. Organelles. https://en.wikipedia.org/wiki/Organelle. Accessed November 11, 2025b.

Chapter 31
Karen K. Uhlenbeck

For her pioneering contributions to global geometry that resulted in advances in mathematical physics and the theory of partial differential equations. Her research accomplishments are matched by her leadership and passionate involvement in mathematics training and education

Keywords Partial differential equations · Gauge theory · Geometric analysis · Minimal surfaces · Harmonic maps · Yang-Mills theory · Integrable systems · Particle physics · Topology · Geometry · Abel Prize · Steele Prize · Geometric partial differential equations · Calculus of variations · Nonlinear waves · Differential geometry

Dr. Karen K. Uhlenbeck, shown in Fig. 31.1, is a mathematician who has used geometrical approaches to the analysis of central problems of modern physics, such as quantum mechanics and particle physics. She has focused throughout her career on using tools developed in one field of mathematics to unleash fundamental knowledge in another field of mathematics. Her rigorous analysis of partial differential equations has been important in enabling computers to "see" by interpreting digital images as well as to our understanding of physics. For her accomplishments, Dr. Uhlenbeck received in 2019, the Abel Prize, widely considered the Nobel Prize in Mathematics, and in 2000, a National Medal of Science.

J. S. Tietjen et al., *Women Laureates 1970 – 2007*, Women in Engineering and Science, https://doi.org/10.1007/978-3-032-23116-1_31

Fig. 31.1 Dr. Karen Uhlenbeck. (Photo by Marsha Miller, courtesy of The University of Texas)

31.1 Life (1942–)

On August 24, 1942, Karen Keskulla was born to parents Arnold and Carolyn (Windeler) Keskulla in Cleveland, Ohio. The family moved to New Jersey in 1945 and ultimately expanded to include four children. Arnold had a degree in mechanical engineering from the Massachusetts Institute of Technology (MIT), but graduating in 1932, during the Great Depression, he had been forced to take jobs at gold mines in California and Utah before landing work with Alcoa in New Jersey. Carolyn, 1 of 12 children in her New Jersey family, had started school as an art student but graduated from City College of New York with an education degree in 1932. She worked for a time teaching art in high school and was a proficient landscape painter (Jackson 2018).

Karen's parents introduced her to art and music and made interesting books available to her. When bored in her classes, Karen would resort to reading a book under her desk. She excelled in school, played the clarinet, and was active in sports.

By high school, Karen had become interested in physics. She had read George Gamow's book *One, Two, Three, Infinity* (Gamow 1947) and had been thrilled to learn about different kinds of infinities (Lipman 2020; Jackson 2018; Ambrose 1997).

Karen applied to three universities to study physics: MIT, Cornell, and the University of Michigan. She was accepted to all three but opted to attend the University of Michigan, as the cost was so much lower than at the other schools. At the University of Michigan, Karen was enrolled in the honors program. She took junior-level mathematics as a freshman and decided to change her major from physics to mathematics. This would also let her avoid physics labs, which she hated and passed by determining the desired answer and fudging data. She did a junior year abroad in Munich and not only loved the quality of the educational offerings but also learned to ski and to love opera. Karen graduated with a BA degree in mathematics in 1964. There were three other women majoring in mathematics in her honors class, all of whom went on to prestigious graduate programs (Ambrose 1997; Wikipedia 2025; Jackson 2018).

In 1964, Karen went to New York University's Courant Institute to continue her studies supported by a National Science Foundation Fellowship. She married Olke C. Uhlenbeck, a biochemist and son of a well-known academic physicist. Because Olke was going to Harvard to continue his graduate studies, Karen transferred to Brandeis University. She chose Brandeis over MIT or Harvard, reasoning that it would be easier to succeed as a woman in mathematics there. There were other women in the graduate program at Brandeis, but Uhlenbeck opted to keep her distance from them. "It was self-evident that you wouldn't get ahead in mathematics if you hung around with women" (Ambrose 1997).

Although Uhlenbeck's parents were the first generation in their families to attend college, they believed that the role of post-secondary education was to prepare you for a well-paying job rather than to satisfy a desire to be a scholar. They were not terribly supportive of their daughter's decision to pursue graduate degrees. Uhlenbeck's in-laws were more intellectual and encouraged her, seeing her on frequent visits she and her husband made to their New York City home. Uhlenbeck's mother-in-law also taught her some of the finer details of good cooking. Uhlenbeck earned her MA from Brandeis University in 1966 and her PhD in 1969 (Ambrose 1997; Jackson 2018; Wikipedia 2025). Her PhD work was on the calculus of variations,

> a technique that involves the study of how small changes in one quantity can help us to find the maximum or minimum value of another quantity – like finding the shortest distance between two points. (Al-Khalili 2019)

After completing her PhD, Dr. Uhlenbeck taught at MIT while her husband finished his PhD at Harvard. The couple then went to the University of California at Berkeley, where they had temporary jobs for 2 years. It was a tumultuous time in the country, with resistance to the war in Vietnam driving protests, and Berkeley was a hotbed of political activism (Ambrose 1997).

As the Uhlenbecks neared the end of their time in Berkeley, Olke was being heavily recruited for academic appointments. Both Stanford and Princeton were interested in having him join their faculty. However, nepotism rules and bias against women meant that these institutions were not interested in providing anything for Dr. Karen Uhlenbeck. As a result, the pair opted to go to the University of Illinois at Urbana-Champaign in 1971, where both could be employed on the faculty. Unfortunately, Dr. Karen Uhlenbeck hated being there. She felt the geography was flat and ugly, and that she didn't fit into the department well. The department seemed to have a real divide along gender lines, and she found it stifling (Ambrose 1997; Jackson 2018).

Dr. Karen Uhlenbeck received a prestigious Sloan Fellowship and used the time that support offered her to change her situation. She separated from her husband, moved to Chicago, and joined the University of Illinois at Chicago in 1976. She also started collaborating with other mathematicians for the first time in her career, having previously always worked alone. Indeed, part of the allure of mathematics for Dr. Uhlenbeck had been its solitary nature (Ambrose 1997).

In 1983, Dr. Uhlenbeck changed her affiliation to the University of Chicago, but she only remained there for 5 years, moving to The University of Texas at Austin in 1988 to hold the Sid Richardson Foundation Regents Chair. By then, she had married fellow mathematician Dr. Robert F. Williams, who studied chaos and loved mountaineering (Wikipedia 2025). She had also by then become established as one of the world's finest mathematicians, generally choosing to use the techniques from one branch of mathematics to examine the problems in another branch of mathematics. Dr. Uhlenbeck's work during this time helped found the field of global analysis, the study of the properties of mathematical objects such as equations by focusing on their geometric characteristics.

By the late 1980s and the 1990s, it was clear to Dr. Uhlenbeck that women in mathematics were not thriving. The number of women on university faculties and attending conferences had plateaued, and there was a decreasing number of women in the ranks behind them. At the University of Texas at Austin, she established the Saturday Morning Math Group in which graduate students in the math department would tutor and encourage high school students (Hunwick 2019). In 1991, she started a program at the Park City Mathematical Institute called the Park City Summer School and encouraged women to apply. In 1994, this program moved to the Institute for Advanced Study at Princeton University as the Women and Mathematics Program. The program is to "provide an immersive education and professional development opportunity for several parallel communities from across the larger umbrella of the mathematics profession" (Wikipedia 2025). It engages mathematicians at all levels of experience. Asked about what motivated her to start this program, Dr. Uhlenbeck said she'd realized that she had been privileged to have been supported in her research, and "I owed the world something" (Lipman 2020). Dr. Uhenbeck is shown with two of the attendees of the summer school from 2006 in Fig. 31.2.

In academic year 1997–1998, Dr. Uhlenbeck held a Visiting Professor position at Princeton University. She returned in 2012, holding Visiting Professor positions until 2019, when her title changed to Distinguished Visiting Professor, an appointment valid through 2027.

Fig. 31.2 Karen Uhlenbeck and two students from the Women and Mathematics Program in 2006 (Dan Komoda, Institute for Advanced Study)

Dr. Uhlenbeck's husband passed away in 2024. He was survived by a daughter, Ellen, from a prior marriage. Dr. Williams and Dr. Uhlenbeck did not have children. Dr. Uhlenbeck, commenting on this perennial question, has said, "I was sad that I was not able to get pregnant, at a time when there were not easy procedures to help solve the problem" (MacLean 2023).

Dr. Uhlenbeck has been the recipient of many honors and awards. In 1983, she received a MacArthur (sometimes called Genius) Award. In 1985, she was made a Fellow of the American Academy of Arts and Sciences, and in 1986, elected as a member of the National Academy of Sciences. In 1988, Dr. Uhlenbeck was named the Noether Lecturer of the Association for Women in Mathematics, and in 1990 became only the second woman to be named a Plenary Speaker at the International Congress of Mathematicians. In 2000, Dr. Uhlenbeck was presented with a National Medal of Science. In 2001, she was made a Guggenheim Fellow, in 2007 received the Leroy P. Steele Prize of the American Mathematical Society, and was made a Fellow of that society in 2012. In 2019, Dr. Uhlenbeck received the Abel Prize from King Harald of Norway. This prize is widely regarded as the equivalent of a Nobel Prize in mathematics. As of 2025, Dr. Uhlenbeck is the only female to have received the Abel Prize, which has been awarded annually since 2003. Dr. Uhlenbeck chose to donate half of the prize money to organizations that promote women and minorities in research in mathematics. In 2019, Dr. Uhlenbeck was also featured in *Time* Magazine's List of History-Making Women. In 2023, she was elected as a foreign member of the Royal Society in England. Dr. Uhlenbeck has received honorary degrees from the University of Illinois at Urbana-Champaign, Ohio State University, University of Michigan, Harvard, and Princeton (Wikipedia 2025; Al-Khalili 2019; Lipman 2020).

31.2 Geometric Analysis and Gauge Theory

Dr. Uhlenbeck describes herself as a messy thinker who works on whatever interests her and what she knows at least a little something about. In her remarks after receiving the Abel Prize, she said she emulated Julia Child: "She knew how to pick a turkey up off the floor and serve it" (Wikipedia 2025). What ties together her many mathematical endeavors has been the use of tools and techniques common in one branch of mathematics on problems in another mathematical branch. Also, the problems she has chosen to investigate have generally been those important in physics.

Dr. Uhlenbeck is best known for her contributions in three areas: geometric analysis, gauge theory and Yang-Mills Equations, and integrable systems and harmonic mappings. In geometric analysis, she has interwoven the tools of geometry and topology with differential equations to understand objects such as curves and surfaces representing important physical quantities, like energy or spatial volume. For instance, she has studied the Dirichlet energy, a measure of how variable a function is. The higher the Dirichlet energy, the more the function wiggles. Minimizing the Dirichlet energy leads to smooth functions, and this can be done by looking at the critical points of the function, the places where the function fundamentally changes its direction. These critical points are defined mathematically by the regions where the function derivatives are zero or undefined. By minimizing the Dirichlet energy of a function, one can find stable functional paths. This has proven very useful, for instance, for computer vision, the ability of computers to "see" by interpreting digital images. It permits the computation of low-distortion images. Dr. Uhlenbeck has extended work in this area to consider what happens when the conditions that guarantee the existence of minima aren't met (The Abel Prize 2019).

Gauge theory is the basis for much of modern physics. Gauge in this context refers to a measurement or mathematical description. The theory is based on the amount of freedom that exists in choosing a measurement, called a gauge quantity, without changing the predictions of the theory. As a non-physics-related example, consider measurements taken at an annual exam by a doctor. Normally, they will choose to measure height and weight, although they could just report the BMI, which is a combined measure of the two. Further, for children and senior citizens, reporting on the trend in height and weight might be more important than the weight or height itself. This illustrates how different gauges can describe the same physical situation. In this example, there is nothing to constrain the choice of metric, but in gauge theory, the issue is the extent to which fundamental physics constrains the choices that can be made.

The fundamental equations of physics are expressed most often in terms of fields. Maxwell's equations, for instance, describe electromagnetic behavior in terms of electric and magnetic fields. Fields are quantities that exist throughout space and time but that are not measured directly. We measure an electric field by moving a charged particle and monitoring the force on it. Physics also defines potentials as quantities that give rise to fields through their variation in time or space. Voltage is an example of electric potential, and it is a gauge quantity, meaning it can be changed in certain ways without affecting the physical electric field. It is also possible to define a different electric potential that still satisfies Maxwell's equations and produces the same observable fields. Gauge theory deals with looking at these possible gauges to understand what they tell us about the underlying physics. This has been particularly useful for quantum mechanical fields, such as general relativity. Maxwell's equations and gravitation are two examples of gauge theories.

The Yang-Mills theory was developed by Chen Ning Yang and Robert Mills to describe particle physics, the study of the fundamental building blocks of the universe (such as quarks and photons) and the forces that govern their interactions. The Yang-Mills theory is another gauge theory. Dr. Uhlenbeck investigated this theory from a rigorous analytical viewpoint, considering four-dimensional Yang-Mills equations (representing three spatial coordinates and one time coordinate). Her work became the basis of all later analytical research on gauge theory and laid the foundation for the Standard Model of particle physics (Al-Khalili 2019; The Abel Prize 2019).

Integrable systems are those that don't become chaotic. They generally involve equations that can be solved exactly. Dr. Uhlenbeck examined such systems using harmonic mappings. Harmonic mappings are the smoothest mapping from one space to another with minimized distortion from stretching and twisting (The Abel Prize 2019).

As of 2025, Dr. Uhlenbeck continues to produce interesting mathematical results and to advocate for women in mathematics.

31.3 Conclusion

Dr. Karen Uhlenbeck is a mathematician who has specialized in application of the tools and techniques from one branch of mathematics to another branch of mathematics. She has focused her efforts on central problems in modern physics, such as gauge theory. She is known, additionally, for her work on geometric analysis of partial differential equations and for the analysis of integrable systems and harmonic maps. Dr. Uhlenbeck has also been a strong advocate for women in mathematics, establishing programs to recruit and support women in mathematics. Dr. Uhlenbeck has received many awards, including the National Medal of Science in 2000 and the Abel Prize (equivalent to the Nobel Prize in mathematics) in 2019.

References

The Abel Prize 2019. https://abelprize.no/sites/default/files/2021-04/citation_English_Karen_Uhlenbeck_2019_abel.pdf. Accessed 27 Sept 2025.

Al-Khalili, Jim. 2019. A biography of Karen Uhlenbeck. https://abelprize.no/sites/default/files/2021-04/biography_English_Karen_Uhlenbeck_2019_abel.pdf. Accessed 27 Sept 2025.

Ambrose, Susan. 1997. *Journeys of Women in Science and Engineering*. Temple University Press. Also available at https://web.ma.utexas.edu/users/uhlen/vita/pers.html. Accessed 27 Sept 2025.

Gamow, George. 1947. *One, Two, Three, Infinity: Facts and Speculations of Science*. Viking.

Hunwick, Cason. 2019. Karen Uhlenbeck Awarded Steele Prize for Lifetime Achievement, College of Natural Sciences, The University of Texas. https://cns.utexas.edu/news/accolades/karen-uhlenbeck-awarded-steele-prize-lifetime-achievement. Accessed 27 Sept 2025.

Jackson, Allyn. 2018. Interview with Karen Uhlenbeck, Celebratio Mathematica. https://celebratio.org/Uhlenbeck_K/article/634/. Accessed 27 Sept 2025.

Lipman, Joanne. 2020. Karen Uhlenbeck's Pioneering Path, Institute for Advanced Study. https://www.ias.edu/ideas/uhlenbeck-pioneering-path. Accessed 27 Sept 2025.

MacLean, Sophie. 2023. A Conversation with Karen Uhlenbeck, Heidelberg Laureate Forum Foundation. https://www.newsroom.hlf-foundation.org/blog/article/a-conversation-with-karen-uhlenbeck. Accessed 27 Sept 2025.

Wikipedia. 9 Sept 2025. Karen Uhlenbeck. https://en.wikipedia.org/wiki/Karen_Uhlenbeck. Accessed 27 Sept 2025.

Chapter 32
Nancy C. Andreasen

For her pivotal contributions to the social and behavioral sciences, through the integrative study of the mind, brain, and behavior, by joining behavioral science with the technologies of neuroscience and neuroimaging in order to understand mental processes such as memory and creativity, and mental illnesses such as schizophrenia

Keywords Creativity · Schizophrenia · Association cortex · Neuroimaging · The brain · Creative minds · Creativity and mental illness · MRI imaging · PET imaging · John Donne · Neurodevelopment · Neuroscience · Mental illness · Mood disorders · Depression · Psychiatry · Renaissance literature · Manic-depression · DSM · Prefrontal lobe

Dr. Nancy C. Andreasen, shown in Fig. 32.1, began her academic career as an expert in Renaissance literature but switched to psychiatry after attending medical school. Her research focuses on two topics: schizophrenia, and creativity and its link to mental illness. She showed that schizophrenia is a developmental disorder rather than the result of progressive brain tissue loss, developed a scale for early assessment of the disease based on objective symptoms, and identified brain abnormalities in schizophrenics using MRI imaging. Dr. Andreasen also studied creativity extensively and learned that creative people are more likely than other people to have histories of mood disorders in near family members. Using imaging techniques, she showed that creative people have more active association cortices than non-creative people. She was awarded a National Medal of Science in 2000.

J. S. Tietjen et al., *Women Laureates 1970 – 2007*, Women in Engineering and Science, https://doi.org/10.1007/978-3-032-23116-1_32

Fig. 32.1 Dr. Nancy Andreasen. (Courtesy of the University of Iowa, image by Jude Gustaphson)

32.1 Life (1938–)

Nancy Coover was born on November 11, 1938, in Lincoln, Nebraska. From early on, she was seen as a rebel and has quoted her school report cards as having comments such as "Nancy is an excellent leader, but she tends to lead children in the wrong direction" (YouTube 2010). Although she found that she loved science subjects in school, her parents steered her toward the arts. She attended the University of Nebraska, studying English, history, and philosophy, earning a bachelor's degree in English.

After graduating, Nancy studied literature at Radcliffe as a Woodrow Wilson Fellow and then went to Oxford University on a Fulbright scholarship. She earned her PhD from the University of Iowa in 1963, having met and married George F. Andreasen, an academic orthodontist, along the way. Both Drs. Andreasen joined the faculty of the University of Iowa in 1963—George in the orthodontics department and Nancy as the first woman hired into a tenure-track position in the English department (Wikipedia (b) 2025; Andreasen 2014).

From 1963 to 1968, Dr. Andreasen worked on a scholarly book: *John Donne: Conservative Revolutionary* (Andreasen 1967). She also gave birth to two daughters: Suz in 1964 and Robin in 1968. Unfortunately, she acquired a very serious post-partum infection after Robin's birth, leaving her hospitalized and on intravenous antibiotics for several days. That experience led her to reconsider her life direction. Dr. Andreasen wondered if she shouldn't do something more impactful than scholarship on Renaissance literature. "I was lying there, having my life saved by penicillin, and thinking 'what if I dedicated my life to medicine instead?'" (Mohammadi 2017; Wikipedia (a) 2025).

After recovering, Dr. Andreasen quit her faculty appointment and entered medical school at the University of Iowa, graduating with her medical degree in 1970. She was one of only six women in her class. She knew immediately that she wanted to do research as well as patient care. During her second year in medical school, Dr. Andreasen took a psychiatry elective, doing lab work on the effect of lithium on cortisol production. It was the first time she worked with people with schizophrenia or bipolar disease, and it convinced her that psychiatry was the direction she should go.

> Because I relish complexity, I chose psychiatry. It's more complicated than neurology. And I chose brain research because the brain is the most complicated organ in the body. (Dreifus 2008)

Psychiatry requires learning the symptoms an individual exhibits and relating them back to the brain in the context of the family history and environment (Moran 2007; Mohammadi 2017). Dr. Andreasen completed her psychiatric residency in 1973.

The first job held by Dr. Andreasen was as a staff psychiatrist in the burn unit of the University of Iowa Hospital. The director of the unit had seen that medicine was increasingly capable of saving burn victims, but that the trauma inflicted upon them and the residual disfigurement led to serious mental problems. Dr. Andreasen worked with a four-person team to wrestle with tough questions, such as when a burn victim should be permitted to see their disfigured face (Moran 2007).

After roughly 2 years, Dr. Andreasen joined the faculty of the psychiatry department of the University of Iowa. Shortly after arriving, she ran into the chair of her department, a man well known for his male chauvinism. He told her, "Andreasen, you may be an MD/PhD, but that PhD of yours isn't worth sh--, and it won't count favorably toward your promotion" (Andreasen 2014). Dr. Andreasen was determined to prove him wrong and decided to use her PhD as an entry point into research on creativity. This led to a study of writers participating in the Iowa Writer's Workshop, one of the top recognized gatherings of writers in the USA. She followed 30 well-known writers, including Kurt Vonnegut and John Cheever, for 15 years. Her findings from this study showed extraordinary links between creativity and psychiatric mood disorders, such as panic disorder and depression. This was the first modern observational study linking creativity and mental illness.

Next, Dr. Andreasen worked on schizophrenia, a disease in which two patients can exhibit very different symptoms. She recognized early on that there are two types of symptoms in schizophrenics: positive symptoms, such as hallucinating or being delusional, and negative cognitive impairment symptoms, such as an inability to think or speak fluently or an inability to experience or express emotions. Through her patients, she learned that the cognitive impairment symptoms were generally found to be more debilitating than the positive symptoms. By using emerging imaging techniques, Dr. Andreasen was able to make several important findings about the brains of schizophrenic people (Mohammadi 2017; Andreasen 2014). She has, more recently, suggested that the next studies should incorporate genetic testing for specific symptoms to further refine our understanding of schizophrenia and how to detect it early.

In 2010, Dr. Andreasen began her second study of creativity, using modern imaging techniques and a broader group of subjects. The results of this study have reinforced the findings of her first study of creativity and added significant new information, both at an anatomical and at a psychological level.

Dr. Andreasen has written three books for the general public based on her studies of creativity and mental illness: *The Broken Brain: The Biological Revolution in Psychiatry* (Andreasen 1985), *Brave New Brain: Conquering Mental Illness* (Andreasen 2001), and *The Creating Brain: The Neuroscience of Genius* (Andreasen 2006). She has also authored or co-authored a dozen scholarly books and over 600 articles (Wikipedia (a) 2025).

As her work progressed, despite the department chair's sentiments, Dr. Andreasen was promoted. She now holds the Andrew H. Woods Chair of Psychiatry at the University of Iowa medical school and has received numerous awards and recognitions. She received the National Medal of Science in 2000. She was elected to the American Academy of Arts and Sciences in 2002. She was also elected a member of the National Academy of Medicine and has received numerous awards from the American Psychiatric Association. Dr. Andreasen was awarded the Latour Prize of the Belgian Academy of Science and the inaugural Vanderbilt Prize in Biomedical Science in 2006. The University of Iowa Psychiatry Department established an annual Andreasen lecture in 2018 in her honor (Wikipedia (a) 2025; Marino 2006).

Dr. Andreasen has also played important roles in the broader profession of psychiatry. She served as the Editor-in-Chief of the *American Journal of Psychiatry* for 13 years, as President of the American Psychiatric Association, and as the Founding Chair of the neuroscience section of the American Association for the Advancement of Science. She served on task forces for the Diagnostic and Statistical Manual of Mental Disorders (DSM), chairing a working group on schizophrenia for DSM V (Wikipedia (a) 2025).

32.2 Schizophrenia and Creativity

While it might seem that creative people and schizophrenic people occupy opposite ends of the spectrum—one type struggling to function and the other operating above the normal level—Dr. Andreasen saw links between these two groups and used the same interviewing and imaging techniques to examine them both. Additionally, there had been reports nearly forever that creative people also suffered from mental illness, and she initially thought that the link might be to schizophrenia.

Prior to Dr. Andreasen's work on schizophrenia, it was believed that schizophrenia was caused by brain tissue loss, much like dementia. Thanks largely to her research, schizophrenia is understood instead to be a disease of brain development (Harrell 2019).

Early on in her work with schizophrenic patients, Dr. Andreasen noted that a crucial trait they shared was emotional blunting, the inability to relate to others in a meaningful way. Her department chair pushed her to get measurable data to support this claim, and this led her to develop an instrument to gather that data. She also created a set of widely used scales to rate the positive and negative symptoms of schizophrenic patients (Moran 2007; Mohammadi 2017).

CT scans became a medical tool in the 1970s. Dr. Andreasen immediately saw the potential they had to scan the brains of healthy and mentally ill patients to discover anatomical differences. Although the University of Iowa was one of the early adopters of CT scans, she was initially denied their use to study the brains of subjects on the grounds that it would expose them unnecessarily to ionizing radiation. It was also assumed that comparing brain scans of normal and schizophrenic people would produce no meaningful results. Nonetheless, Dr. Andreasen persisted and, in 1982, published a study of schizophrenic brains using CT scans. She was able to show that there were anatomical differences from normal brains—notably ventricular enlargement (Dreifus 2008; Moran 2007).

The 1980s introduced another important imaging technique, MRI, which had the advantage of not using ionizing radiation. She published the first quantitative study of psychiatric patients using MRI in 1986 and began a long-term study of schizophrenic patients in 1989 to answer the question, is schizophrenia a neurodegenerative disease? Her study produced a series of important conclusions. She showed that people with schizophrenia lose brain tissue faster than normal people of the same age. Further, the more drugs a patient has taken to treat his illness, the more brain mass they are likely to lose. Nonetheless, she showed that the negative symptoms of schizophrenia are caused by decreased brain prefrontal lobe activity and not an effect of long-term treatment (Dreifus 2008; Moran 2007).

Dr. Andreasen's study of creativity began even earlier than her key work on schizophrenia. Work that preceded hers had shown that an individual's IQ was not a good measure of their creativity. One of the key questions, then, was what is a good indicator of creativity? Additionally, Dr. Andreasen sought to understand why so many creative people suffer from mental illnesses.

Studies of creativity immediately run into the problem of how to define creativity. There have been two approaches to this, which Dr. Andreasen refers to as "little c" and "big C." The little c approach is to develop a quantitative assessment of creativity based on the assumption that it is divergent thinking, or the ability to come up with many responses to a question. The big C approach takes advantage of the adage "if it walks like a duck and quacks like a duck, it is a duck," and simply uses people widely acknowledged to be creative as subjects. It is the big C approach that Dr. Andreasen used in her studies (Andreasen 2014).

In Dr. Andreasen's first creativity study, she used 30 well-known writers as her target subjects, and "non-creative people" by her judgment (accountants, administrators, and social workers) as the control group. She conducted extensive interviews with all the subjects, including evaluation of the creativity and mental struggles of near relatives. Her results showed that 80% of creative subjects had experienced some kind of mood disturbance in their life, while only 30% of the control group had. Additionally, she found that creativity ran in families. This study raised a number of questions, among them: Why does creativity run in families? How much of creativity is attributable to nature versus nurture? Why are writers so prone to mood disorders? Would creative scientists be similar to writers or different from them? (Andreasen 2014)

In her second long-term creativity study, beginning in 2010, Dr. Andreasen used functional MRI to capture the images of brain activity while subjects performed tasks. She had previously done a PET imaging study of brain activity with two different kinds of memory: episodic (event and time-linked) and semantic (general). She chose to focus on tasks that would ask subjects to have random episodic silent thought (REST), essentially free association. Dr. Andreasen thought REST might show activity in the association cortices. Association cortices are the portions of the cortex that help us interpret sensory information. For instance, they convert an optical pattern into words with meaning. She hypothesized that this part of the brain would be more active in creative people: "I decided that people who are creative probably have an enhanced ability to make connections or associations, and there is a part of the brain that does nothing but this called the association cortex" (Mohammadi 2017). A view of the association cortices is provided in Fig. 32.2.

For this study, Dr. Andreasen broadened the creative subjects to include scientists as well as creative artists. She aimed to consider again the link between familial

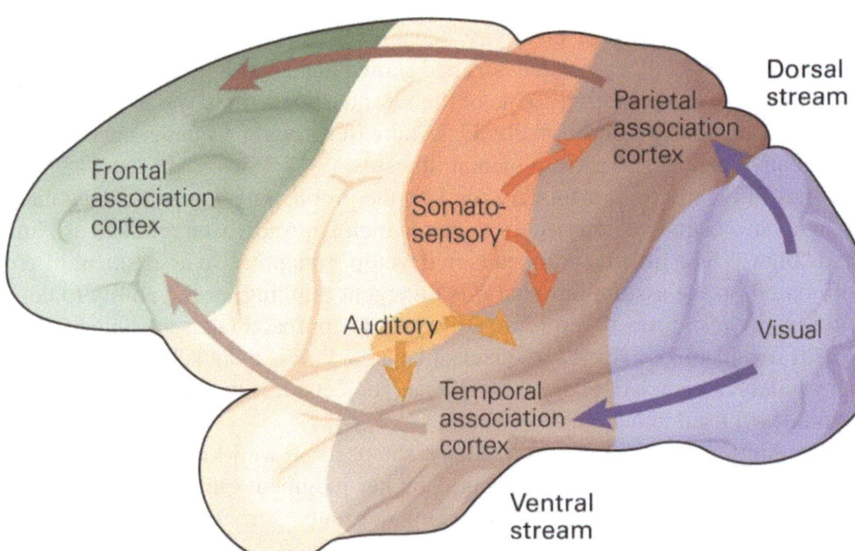

Fig. 32.2 The locations of the association cortices of the brain. (Image courtesy of Neupsy Key)

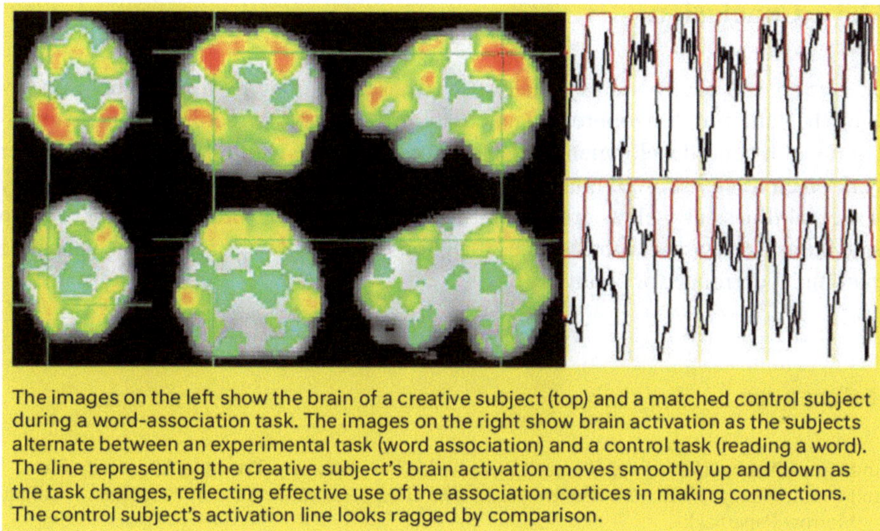

Fig. 32.3 The brain scans of a creative subject on top, compared to a normal subject on bottom. The colors indicate level of activity, with red showing more activity than yellow (Andreasen 2011)

creativity and mental illness, as well as to compare the brains of scientists and artists. In addition to extensive interviews, each subject was put through one and a half days of brain scans while doing various activities: REST, word association, picture association, and pattern recognition (Andreasen 2014).

Results of the study supported her earlier creativity study, finding creativity and mental illness running at higher rates in the families of creative people. Additionally, brain scans showed that creative people had greater activity in their association cortices than did normal subjects. This is shown, for instance, by the image in Fig. 32.3, which compares the functional MRI of a creative subject to a normal subject.

Through interviews, Dr. Andreasen also drew many conclusions about creative people: Many creative people associate their gifts with strong feelings of joy and excitement. Eureka moments often come during REST, after long periods of incubation. Many creative people are autodidacts—they are self-taught. This suggests that we should find ways in schools to encourage creativity by letting students find their own ways of learning. Additionally, many creative people are polymaths. They have broad interests and abilities. There is no distinction between creativity in science and in the arts. Finally, creative people tend to be persistent, even in the face of rejection (Andreasen 2014).

Having found imaging an invaluable tool in her research, Dr. Andreasen has been instrumental in the development of three-dimensional brain imaging techniques. She has served as the director of the Iowa Neuroimaging Consortium and has promoted consideration of the latest techniques and tools in medicine, such as genomics, to further elucidate the causes of mental illness and creativity.

32.3 Conclusion

Dr. Nancy Andreasen is a psychiatrist who has used imaging tools to help understand the nature of schizophrenia and creativity. Her work on schizophrenia showed that it was a neurodevelopmental disease rather than the result of degeneration. Her work on creativity showed that the families of creative people tend to have a history of mental illness as well as creativity. She also showed that creative people tend to have more active association cortices than non-creative people, indicating a possible anatomical correlate for creativity.

References

Andreasen, Nancy. 1967. *John Donne: Conservative Revolutionary.* Princeton University Press.

Andreasen, Nancy. 1985. *The Broken Brain: The Biological Revolution in Psychiatry*. William Morrow Paperbacks.

Andreasen, Nancy. 2001. *Brave New Brain: Conquering Mental Illness*. Oxford University Press.

Andreasen, Nancy. 2006. *The Creating Brain: The Neuroscience of Genius*. Penguin Publishing.

Andreasen, Nancy. 2011. A Journey into Chaos: Creativity and the Unconscious. *Mens Sana Monogr* 9(1);42-53.

Andreasen, Nancy. July/Aug 2014. Secrets of the Creative Brain. *The Atlantic*.

Dreifus, Claudia. 15 Sept 2008. Using Imaging to Look at Changes in the Brain. *NY Times*.

Harrell, Kelsey. 27 March 2019. UI Professor pioneer in schizophrenia and cerebellum research. *The Daily Iowan*. https://dailyiowan.com/2019/03/27/ui-professor-pioneer-in-schizophrenia-and-cerebellum-research/. Accessed 31 Oct. 2025.

Marino, Melissa. 2006. Vanderbilt Prize Winner Probes Nature of Genius. *VUMC News*. https://news.vumc.org/reporter-archive/vanderbilt-prize-winner-probes-nature-of-genius/. Accessed 31 Oct 2025.

Mohammadi, Dara. 2017. Nancy C. Andreasen: creativity and mental illness. *Lancet Psych*. 4: 192.

Moran, Mark. 1 June 2007. Imaging Breakthrough Provides Key to Clinical Symptoms, Psychiatry Online. https://psychiatryonline.org/doi/10.1176/pn.42.11.0012. Accessed 31 Oct 2025.

Wikipedia (a). 6 Oct 2025. Nancy Coover Andreasen. https://en.wikipedia.org/wiki/Nancy_Coover_Andreasen. Accessed 31 Oct 2025.

Wikipedia (b). 14 Oct 2025. George Andreasen. https://en.wikipedia.org/wiki/George_Andreasen. Accessed 31 Oct 2025.

YouTube. 2010. 2010 UCSF Medal Recipient: Nancy C. Andresen, MD, PhD. https://www.youtube.com/watch?v=hMx2bBhyRBA. Accessed 31 Oct 2025.

Chapter 33
Ann M. Graybiel

For her pioneering contributions to the understanding of the anatomy and physiology of the brain, including the structure, chemistry, and function of the pathways subserving thought and movement

Keywords Brain · Basal ganglia · Striatum · Striosomes · Striosomal matrix · Parkinson's disease · Huntington's disease · Neurotransmitters · Matrisomes · Memory · Habit · Habitual behavior · Decision-making · Learning · Mindfulness · OCD · Anxiety · Depression · Schizophrenia · Acetylcholine · Dopamine · Chunking · Neocortex · Inhibition · Compulsion · Compulsive behavior · Stress · Addiction · Neuroscience · Cognitive science

Dr. Ann M. Graybiel, shown in Fig. 33.1, is one of the first scientists to link cellular chemistry in the brain to human behavior. She has spent most of her career studying the basal ganglia, a large organ deep in the brain. Her research has revealed the structure of the basal ganglia and shown that it is linked to the formation of habits, learning, and decision-making as well as to movement. Her research has implications for the treatment of many ailments, including Parkinson's disease, Huntington's disease, schizophrenia, and mood disorders, such as anxiety and depression. Dr. Graybiel received the National Medal of Science in 2001.

J. S. Tietjen et al., *Women Laureates 1970 – 2007*, Women in Engineering and Science, https://doi.org/10.1007/978-3-032-23116-1_33

Fig. 33.1 Dr. Ann M. Graybiel. (Photo by Justin Knight, McGovern Institute, MIT)

33.1 Life (1942–)

Ann M. Graybiel was born in 1942 in Boston, Massachusetts, but she and her brother grew up in Pensacola, Florida. Her father, Ashton, was a cardiologist and researcher who studied the effects of weightlessness and high accelerations on airplane pilots and astronauts. The family move to Pensacola had been prompted by the observation toward the end of World War II that pilots were becoming disoriented—a problem the government hoped Dr. Graybiel could help them unravel (Graybiel 2012). Ann's mother came from a family of doctors in Northern Ireland and would have pursued that profession herself but found it nearly impossible: "My mother would have been a physician. It was tough in those days for women" (Infinite MIT 2025).

Ann's family instilled in her a love of nature and science. From an early age, when her father would work on research papers, they would discuss them together. Unfortunately, the public school district in Pensacola did not permit women to take science classes, steering them into home economics courses instead. In response, the Graybiels sent Ann to attend and board at the National Cathedral School in Washington, DC, starting when she was 14 years old. Ann loved the school, especially the music, arts, and history offerings, and decided she would become a historian (Graybiel 2012).

Upon graduating from high school, Ann entered Harvard University. After taking a small-group freshman seminar in chemistry, she changed her mind about history and decided instead she would be a chemist. Ultimately, Ann graduated with an AB in biology in 1964 (Graybiel 2012).

Ann's experience at Harvard was common for the time, with women being treated quite differently from men. Ann was not allowed, for example, to use the main library at Harvard. Also, when she started as a student, women had to sign in and out of the dormitories and be in their rooms by 10 pm (with a limited number of exceptions). Ann said, "I never saw a female professor … I knew one woman, turned out she was an instructor" (Infinite MIT 2025). Ann also tells a story about being accosted by another student while she was using a microscope:

> And sitting at my microscope at night, and a guy coming in and saying, 'How dare you work? How dare you do this?' (Infinite MIT 2025)

Following her time at Harvard, Ann was a Woodrow Wilson Fellow at Tufts University and earned an MA in biology in 1966. At this point, she knew that she was very interested in how the brain operates and how its basic biochemistry links to human behavior. At the time, the Massachusetts Institute of Technology (MIT) was one of the only universities studying the link between cellular level activity and behavior, so Ann enrolled in the PhD program there and earned her PhD in neuroanatomy in 1971 (Nichols 2006; Graybiel 2012; Graybiel and Smith 2014). While a graduate student, she was asked to teach a course with another student, James Lackner, in what has been speculated to have been an intentional attempt at matchmaking by the department chair. The two were married as graduate students. James, a 1966 graduate from MIT at the undergraduate level, earned his MIT PhD in 1970. He has been a Professor and worked at the laboratory named for Ann's father at Brandeis, studying spatial orientation and movement control (Brain and Cognitive Sciences Newsletter 2002).

From 1971 to 1973, Dr. Graybiel worked as a research associate at MIT. At the time, the focus of brain studies was on how people sense. With no brain imaging tools yet available, Dr. Graybiel decided to see if she could study pathways in the brain by using stains of the chemicals known to be related to neuron communication, i.e., by staining neurotransmitters. Using this approach, she made a surprising discovery—namely, that the basal ganglia are not, as thought, simply a mass of cells, but have a sophisticated architecture with chemically distinct compartments (Graybiel 2012).

Since discovering the structure of the basal ganglia, Dr. Graybiel has focused her research on that brain organ, gathering information on basal ganglia anatomy and the roles it plays in behavior. She has been able to make important discoveries about the formation of habits, the impact of stress on decision-making, and how humans learn. Her work potentially holds important implications for the treatment of diseases, such as Parkinson's disease, Huntington's disease, schizophrenia, anxiety, depression, Tourette's syndrome, and obsessive compulsive disorder.

Dr. Graybiel joined the faculty of MIT in 1973. She was given no startup funding and difficult space:

> They gave me a room in the small place where they kept the monkeys. I used to clean the floor myself. … For a long time, I was the only female faculty member in a six-floor building. (Gruber Foundation 2018)

In 1994, Dr. Graybiel was named the Walter A. Rosenbluth Professor of Neuroscience (Graybiel 2012). She has earned many awards and honors. In 2000, Dr. Graybiel received the Teaching Prize for Excellence in Graduate Education of the MIT School of Science. In 2001, she was presented with the National Medal of Science and the Outstanding Women in Neuroscience Award from Brown University. In 2002, Dr. Graybiel received the Robert S. Dow Neuroscience Award and MIT's Jaime R. Killian Faculty Achievement Award. In 2004, she received the Woman Leader of Parkinson's Science Award, in 2005, the Prix Plasticité Neuronale of the IPSEN Foundation, and in 2007, the NARSAD Distinguished Investigator Award. In 2008, Dr. Graybiel received the Vanderbilt Prize in Biological Science, and in 2012, the Kavli Prize. In 2018, she received the Gruber Neuroscience Prize (McGovern Institute n.d.).

Dr. Graybiel has accepted honorary doctorates from Hebrew University, Queens University (Belfast), Tufts University, and Mount Sinai School of Medicine. She has been elected to membership of the National Academy of Sciences, the National Academy of Medicine, the American Academy of Arts and Sciences, the Royal Academy of Medicine in Spain, the Royal Irish Academy, and the American Philosophical Society (Wikipedia 2026; McGovern Institute n.d.).

Given her educational experiences, Dr. Graybiel has been active in encouraging women to consider careers in science and seeking to remove obstacles. In 1994, she was 1 of the 16 women faculty who signed a letter to the President of MIT documenting bias against women (Wikipedia 2026). Her lab is known for having lots of female undergraduates as well as some high school students working in it. It is also known for the number of children born to members of the lab, a fact that has prompted some discussion of the nearby water fountain as providing fertility water. Although the Drs. Graybiel and Lackner wanted children, they did not have any (Infinite MIT 2025).

33.2 The Basal Ganglia

At the point that Dr. Graybiel was in graduate school, the emphasis of the scientific community studying the brain was understanding how people sense. The brain is not a homogeneous organ but rather is comprised of many distinct, interconnected parts. In studying eye movement, Dr. Graybiel found indications that the basal ganglia, organs deep in the brain, played a role (Infinite MIT 2025). The left and right basal ganglia, shown in Fig. 33.2, are the largest of the organs deep in the brain.

It was known at the time that the basal ganglia played a role in Parkinson's disease and Huntington's disease, so researchers thought the basal ganglia might be linked to motor skills (Nichols 2006). However, with none of the modern imaging techniques available to them, it was difficult to gather more information about the structure and function of such deeply embedded organs as the basal ganglia.

Considering this challenge, Dr. Graybiel, who worked with animal brains and human brain specimens from the Massachusetts General Hospital, noted:

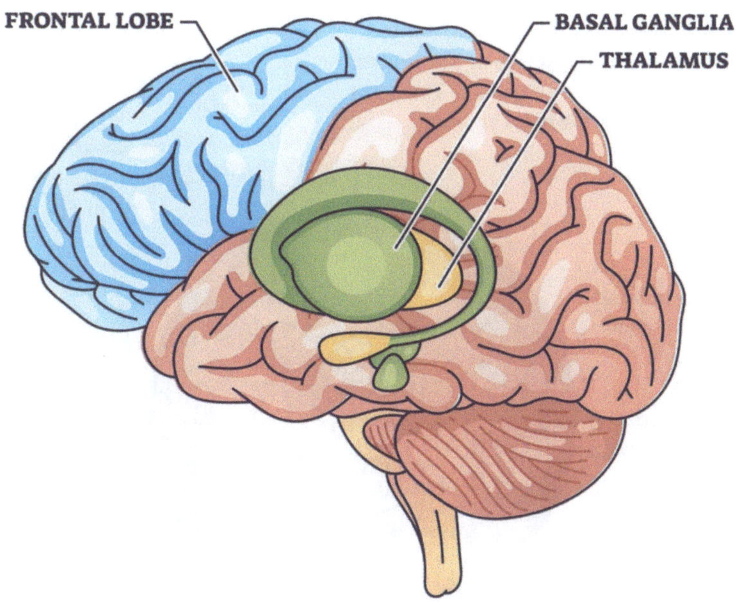

Fig. 33.2 The basal ganglia in the human brain (Guy-Evans 2025)

> And it dawned on me that if you could see the distribution of the chemicals in the brain, you know what you could do, you could go from an experimental animal to a human brain. (Infinite MIT 2025)

Based on this approach, Dr. Graybiel looked at chemicals that might mark connections in an animal brain (neurotransmitters) and then found agents to stain those chemicals so she could look at the phenomenon in a human brain (Brans 2017).

Dr. Graybiel started her studies of chemicals in the brain by staining cholinergic markers. Of particular interest were dopamine and acetylcholine, two neurotransmitter chemicals normally in balance that were known to play key roles in Parkinson's disease (too much acetylcholine) and schizophrenia (too much dopamine). Surprisingly, her studies showed strong involvement of the basal ganglia, and that they were not simply masses of cells, but organs with chemical architecture. Specifically, she showed that the striatum, the largest nucleus in the basal ganglia, had chemically distinct compartments that had not been seen previously. She named these striosomes. They are shown in Fig. 33.3. The striosomes and their surrounding striosomal matrix were found to connect to unique areas of the neocortex, the part of the brain used for high-level thinking. Additionally, Dr. Graybiel found that the matrix had modules in it as well, and she named these matrisomes. Her work was able to show that the striosomes and matrisomes connect to different nearby brain structures (Gruber Foundation 2018; Graybiel 2012).

By studying the striosomes, Dr. Graybiel was able to show that they mainly work on dopamine pathways and play a role in reward reaction and in movement. By focusing on the dopamine pathways, Dr. Graybiel has been able to reveal much

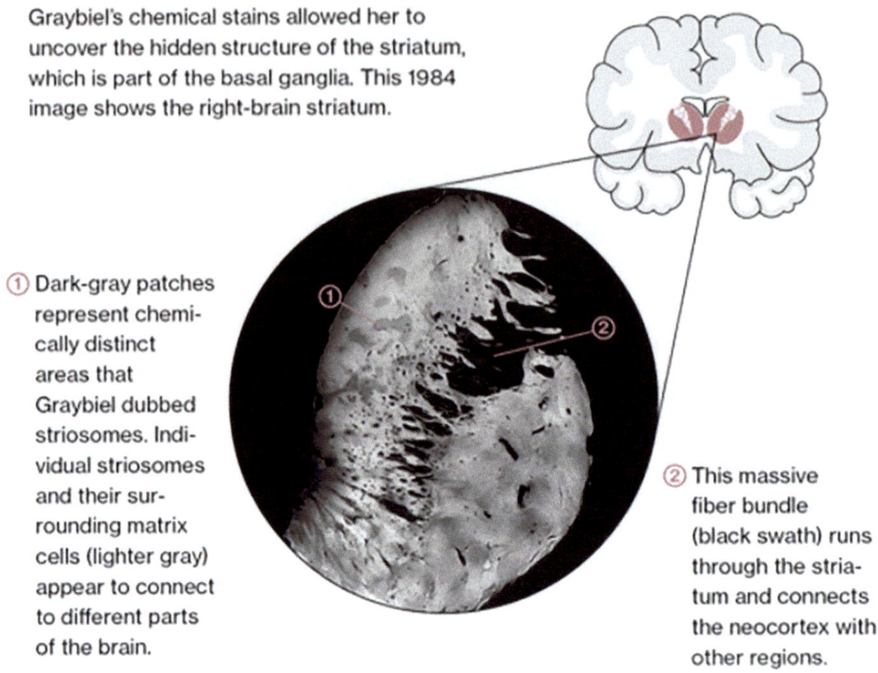

Fig. 33.3 Image of the striatum showing striosomes and striosomal matrix. (Courtesy of *MIT Technology Review*)

about the neuroscience of habit forming. Her work has been aided by a growing number of imaging techniques, by increasing knowledge of molecular genetics, and the development, in her lab, of microelectrode methods permitting examination of the deep brain (Gruber Foundation 2018; Infinite MIT 2025).

In a key animal experiment, Dr. Graybiel first trained rats to run in a T-shaped maze and to respond to an auditory cue to turn left or right for a treat. The habit was deemed established when the rats continued to follow the pattern even when the tasty reward was replaced by one distasteful to them. By studying the brain activity while learning the habit, Dr. Graybiel found that in early learning, the striatum was very active throughout the test, but as the habit was established, the activity was more limited to the task start and finish. This establishment of a shorthand for a habitual event is referred to as chunking and describes how we manage to do some tasks on automatic mode without being terribly aware we are doing them. Dr. Graybiel also learned that she could prevent the habit from forming using light to turn off the neurons in the prefrontal cortex, the part of the brain in which the chunking signal was embedded (Zeliadt 2013; Graybiel 2014; Brans 2017).

As Dr. Graybiel and her colleagues have continued to study habits and the basal ganglia, they have shown that compulsive behaviors in mice could be inhibited. They have also found and controlled the brain circuit that causes animals under stress to opt for high-risk, high-reward approaches rather than safe, low-reward options (Trafton 2018; Humphries 2013).

Dr. Graybiel's research offers insights into many psychiatric conditions and might ultimately provide treatment options, particularly for involuntary repetitive actions such as seen in autism, Parkinson's disease, Huntington's disease, Tourette's syndrome, addiction, and OCD.

33.3 Conclusions

Dr. Ann M. Graybiel is a neuroscientist who has studied the basal ganglia embedded deep in the brain. She has shown that these organs are responsible for the formation of habits and affect decision-making as well as some movements. She has also shown means of inhibiting the formation of habits through control of neurons in specific brain sites, leading to the possibility that psychiatric disorders with involuntary repetitive behaviors might be treatable.

References

Brain and Cognitive Sciences Newsletter. Fall 2002. Ann Graybiel vol V(1). https://bcs.mit.edu/sites/default/files/newsletters/bcsnewsletter_fall_2002_0.pdf. Accessed 17 Nov 2025.

Brans, Pat. 02 Jan 2017. What Warren Buffett and Ann Graybiel Advise on Habits. *Forbes*.

Graybiel, Ann. (2012). Discovering New Features of the Brain's Organization and Function, The Kavli Prize. https://www.kavliprize.org/ann-martin-graybiel-autobiography. Accessed 17 Nov 2025.

Graybiel, Ann and Smith, Kyle. (Jun 2014). Good Habits, Bad Habits. *Sci Am* 310(6); 38–43.

Gruber Foundation. 2018. Ann Graybiel. https://gruber.yale.edu/recipient/ann-m-graybiel. Accessed 17 Nov. 2025.

Guy-Evans, Olivia. 31 July 2025. Basal Ganglia, Simply Psychology. https://www.simplypsychology.org/basal-ganglia.html. Accessed 17 Nov 2025.

Humphries, Courtney. 17 Dec 2013. Why We Do What We Do. *MIT Technology Review*. https://www.technologyreview.com/2013/12/17/175086/why-we-do-what-we-do/. Accessed 17 Nov 2025.

Infinite MIT. 2025. Ann M. Graybiel PhD '71. https://infinite.mit.edu/video/ann-m-graybiel-phd-71. Accessed 17 Nov 2025.

McGovern Institute, n.d. Ann M. Graybiel. https://mcgovern.mit.edu/profile/ann-graybiel/. Accessed 17 Nov 2025.

Nichols, Catherine. 11 July 2006. Ann Graybiel, PhD '71. *MIT Technology Review*. https://www.technologyreview.com/2006/07/11/228627/ann-graybiel-phd-71/. Accessed 17 Nov 2025.

Trafton, Anne. 21 Feb 2018. The Risky Business of Stress. *MIT Technology Review*. https://www.technologyreview.com/2018/02/21/145223/the-risky-business-of-stress/. Accessed 17 Nov. 2025.

Wikipedia (14 April 2026) Ann Graybiel. https://en.wikipedia.org/wiki/Ann_Graybiel, accessed 26 April 2026.

Zeliadt, Nicolette. 9 Sept 2013. QnAs with Ann M. Graybiel, PNAS. https://www.pnas.org/doi/10.1073/pnas.1315012110. Accessed 17 Nov 2025.

Chapter 34
Evelyn M. Witkin

For her insightful and pioneering investigations on the genetics of DNA mutagenesis and DNA repair that have increased our understanding of processes as varied as evolution and the development of cancer

Keywords DNA · DNA repair · SOS repair · Metagenesis · SOS response · *Escherichia coli* · *E. coli* · Cold Spring Harbor Lab · Rutgers University · Genome repair · Bates Seven · Ultraviolet radiation · Ultraviolet radiation damage · UV-induced mutations · RecA · LexA · Mutation frequency decline · MFD · Lasker Award · Wiley Prize · Joseph Witkin · Hyman Witkin · Miroslav Radman

Dr. Evelyn Witkin, shown in Fig. 34.1, was a biochemist who discovered that DNA repairs itself when damaged. Through her studies of *Escherichia coli*, a common bacterium, she elucidated a mechanism for DNA damage from ultraviolet radiation and an avenue for repair of the damage in an error-prone fashion, leading to mutations in the DNA. Her work has implications for many cell-related processes, including aging and the development of cancer. Dr. Witkin was awarded the National Medal of Science in 2002.

J. S. Tietjen et al., *Women Laureates 1970 – 2007*, Women in Engineering and Science, https://doi.org/10.1007/978-3-032-23116-1_34

Fig. 34.1 Dr. Evelyn Witkin. (Photo by Nick Romanenko, Rutgers University)

34.1 Life (1921–2023)

Evelyn Ruth Maisel, born March 9, 1921, in Manhattan, was the second daughter born to parents Joseph and Manya Maisel. Joseph was a pharmacist who died when Evelyn was only 3 years old. Manya remarried when Evelyn was about 9 years old and moved the family to Queens. Evelyn's stepfather, Joseph Bersin, was also a pharmacist. In Queens, the pharmacy was in the building in which they lived and included a food service that Manya supervised (Gitshier 2012; Risen 2023).

Evelyn and her sister were quite accomplished students and routinely skipped every other half year of class to jump forward. Both attended Washington Irving High School in Manhattan, despite the long commute, as it was an all-girls high school that was held in high regard (Gitshier 2012; Wikipedia 2025). Dr. Witkin credits the school and her parents for encouraging her to pursue a scientific career: "My parents had never made me feel that there was any direction I couldn't take because I was a girl and I was grateful for that" (Albert and Mary Lasker Foundation 2016; Lasker Foundation 2016).

Evelyn left for New York University (NYU) when she was only 15 years old. She struggled with choosing a major, as she was interested in poetry and literature as well as science. Ultimately, Evelyn chose to pursue science because, unlike literature or poetry, it could not be done in isolation, and training would provide access to the tools required, such as microscopes and temperature-controlled incubators (Albert and Mary Lasker Foundation 2016).

In her senior year at NYU, Evelyn served as the head of the student council. It was a contentious time on campus as students objected to the gentleman's agreement that Black athletes would not travel with their team to play in venues where segregation was still the norm. Evelyn led a campus wide protest when it was learned that Leonard Bates, a star football player, would not accompany the team to play at the University of Missouri. The protests lasted for months, and a group of seven students (including Evelyn), who came to be known as the Bates Seven, were suspended for 3 months after 2000 students picketed the administration building, chanting "No Missouri Compromise," and over 4000 signed a petition decrying the blatant racism. As a result, Evelyn did not graduate with her class in May 1941 but was awarded her BA in Zoology after her suspension ended (Lally 2023; Volkert 2023).

The suspension she faced had a significant impact on Evelyn's thinking about her future:

> I had planned to stay at NYU for graduate work in genetics, but I decided to go to Columbia. My having gone to Columbia was the greatest blessing that ever happened to me professionally. I'm not sure I would be a National Science Laureate if New York University hadn't decided that I was a bad girl in 1941. (Lally 2023)

Evelyn started her graduate career at Columbia University without having lined up an advisor. She knocked on the door of Theodosius Dobzhansky and asked if he would advise her. His reply was "A woman in the lab. This should be interesting" (Volkert 2023). She began her work in the lab, focusing on fruit flies.

When Evelyn was 16 years old, she met Herman Witkin, who was 21 years old, through his sister, who was part of the graduate student group Evelyn's sister associated with. They met at a fair to raise money for loyalists in Spain who were busy fighting the Spanish War. In 1943, Herman and Evelyn married. Herman had earned a doctorate in psychology and had been working as an instructor at Brooklyn College since 1941.

In the summer of 1944, after 1 year of marriage, Herman joined the Army and traveled to Texas to work with the School of Aviation Medicine to continue his research on perception. Dr. Dobzhansky arranged for Evelyn to spend the summer at Cold Spring Harbor Laboratory (CSHL) on Long Island, NY, which was a hotbed of work on genetics at the time. She arrived and was assigned the task of generating mutations in cultures of *Escherichia coli* (*E. coli*), a common bacterium. In describing this period, Evelyn has said, "Oh, it was heaven. So, never having had a course in microbiology, not knowing a thing about sterile techniques, nothing about bacteria, I went to CSHL for the summer" (Gitshier 2012).

At the time Evelyn started at CSHL, DNA had not been identified as the home of genetic material. It was known that ultraviolet (UV) light caused changes in the characteristics of organisms, but it was not known how that process worked. Evelyn put several cultures of *E. coli* under a UV lamp, exposing them to mistakenly high levels of radiation. All the cultures died except four. The result left Evelyn wondering: "At this point I asked, why did they survive? Maybe a mutation made them resistant" (Risen 2023). This experiment turned Evelyn's focus from generating mutations to understanding mutation resistance and set her on the path for her PhD.

Evelyn remained at CSHL to conduct the research necessary for her PhD. Herman and she moved to Long Island to be close to that lab once he returned from Texas, and Herman commuted to Brooklyn to teach, a situation that continued for 10 years. Evelyn was awarded her PhD in 1947 (Cold Spring Harbor Lab). Son Joseph was born in 1949 and son Andrew in 1952. Dr. Vannevar Bush, head of CSHL at the time, arranged for Dr. Witkin to work part-time at the lab as her children were born. She was permitted to work on planning, data analysis, and writing at home rather than at CSHL, a very unusual arrangement for the time (Wikipedia 2025). Starting in 1950 and lasting until 1964, Dr. Witkin organized and edited the *Microbial Genetics Bulletin*, one of the few publications aimed at the emerging field of microbial genetics.

In 1955, Dr. Witkin moved to the SUNY Downstate Medical Center in Brooklyn to minimize the commuting to work in the family. In 1971, with both the Witkin sons off to college or beyond, she moved to Douglas College, the women's college of Rutgers University in New Brunswick, New Jersey. She remained at Rutgers until her retirement in 1992, which was thrust upon her by law rather than circumstances. She had been named the Barbara McClintock Professor of Genetics in 1979, the same year her husband died, and been named the director of the Waksman Institute, a research institute in microbiology at Rutgers, in 1983 (Gitshier 2012; Cold Spring Harbor Laboratory).

Throughout her career, Dr. Witkin continued to conduct research on the process of mutation generation and repair in *E. coli*, ultimately producing evidence that the genetic material had the ability to repair itself but tended to use an error-prone process, which she called the SOS response, that would yield many mutations (Volkert 2023).

In retirement, Dr. Witkin returned to her early interest in literature and poetry. She conducted historical research on Charles Darwin and Robert Browning and joined the Princeton Research Forum science book club (Gitshier 2012).

Dr. Witkin received many awards for her work. She was elected a member of the National Academy of Sciences in 1977 and a member of the American Academy of Arts and Sciences in 1980. In 1980, she received the American Women of Science Award for Outstanding Research and was named a Fellow of the American Academy of Microbiology. Dr. Witkin received the Thomas Hunt Morgan Medal of the Genetics Society of America in 2000. In 2002, she received the National Medal of Science, and in 2015, she received the Albert Lasker Award with Stephen Elledge, another geneticist. That same year, she received the Wiley Prize in Biomedical Sciences, which recognizes research leadership in the life sciences.

In honor of Dr. Witkin's 100th birthday, a symposium was held involving over 300 scientists, and a lab at the Waksman Institute was renamed for her. A seminar was established in her name at Rutgers University and was first given in 2023.

In 2010, her son Andrew, then a computer animator, died in a scuba diving accident. In 2023, at the age of 102, Dr. Witkin died of the complications from a fall. She is survived by her son Joseph, once a founder of the band ShaNaNa and now an ER doctor, and four grandchildren.

34.2 DNA Repair

When Dr. Witkin first began her research work at CSHL, DNA was not yet known as the location of hereditary information, although it was certainly known that organisms had traits that were inherited. While it was also understood that radiation could damage an organism, it was thought to do so by directly damaging the genetic information. It also was then, and remains now, difficult to study cells of organisms high on the evolutionary ladder, so the preferred biological research organisms were plants, Drosophila (fruit flies), and *E. coli*. Dr. Witkin started with fruit flies but moved to *E. coli* as she began her work with CSHL. The advantages for geneticists of working with *E. coli* are that it multiplies every 20 min and is so small that a test tube can hold millions of them.

E. coli is a common bacterium found in the intestines of warm-blooded animals, including humans. As shown in Fig. 34.2, it is rod-like in structure and has a tail that helps propel it.

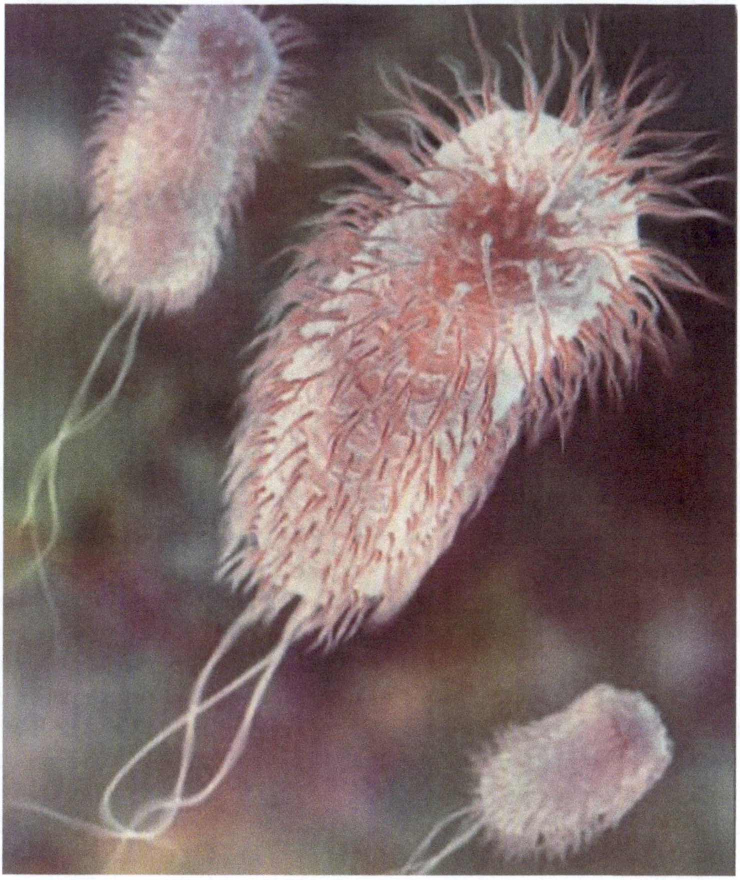

Fig. 34.2 Image of *E. coli*. (Public domain)

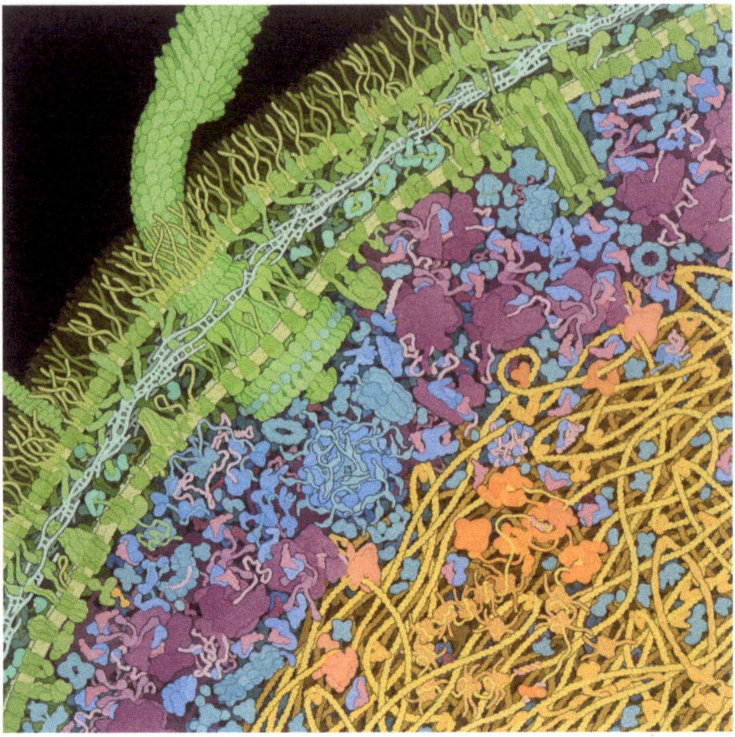

Fig. 34.3 The interior of *E. coli*, showing the central nucleoid (yellow), ribosomes (purple) and cell membrane features (green). (Image by David S. Goodsell at RCSB Protein Data Bank)

The interior of *E.coli*, as shown in Fig. 34.3, has a central region containing DNA forming a nucleoid (a nucleus without a membrane defining its border) and other organelles in the cytoplasm.

Dr. Witkin is best known for two important discoveries: radiation-resistant *E. coli* and a mechanism of DNA repair named the SOS response. When her first experiment at CSHL produced four cultures of *E. coli* that survived UV radiation, it was the first time that mutants conferring radiation resistance were isolated (Wikipedia 2025). Dr. Witkin studied them under a microscope and found that they grew into long filaments. She realized that the radiation had inhibited cell division, the standard means of replication of bacteria, and instead the cells were growing in size. She found that after roughly 3 h, the filaments were 50 times longer than normally seen. After this extensive growth, the *E. coli* organism would die (Gitshier 2012). Figure 34.4 shows an image of normal *E. coli* and the mutated *E. coli*, clearly demonstrating the filamentary elongation.

Continuing her studies of the mutation process in *E. coli*, in the early 1950s, Dr. Witkin found that the number of mutations in the cultures depended upon the nutrients available. If she added nutrients to the cultures, the number of mutations increased, while if only a single amino acid (tryptophan) was available, the number

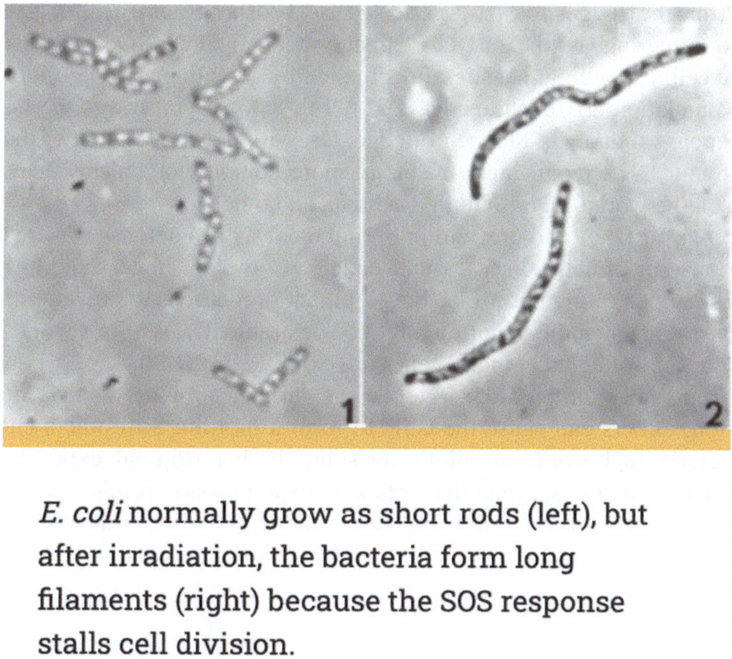

Fig. 34.4 Normal (left) and mutated *E. coli* (right). (Image courtesy of Lasker Foundation)

of mutations would be low. She referred to this as mutation frequency decline, Mfd, and it provided a clue to the biochemical processes ongoing when the *E. coli* DNA was damaged (Gitshier 2012).

During her career, Dr. Witkin was able to develop greater understanding of the biochemical process going on in the radiation-resistant *E. coli*. She discovered that two proteins were involved: LexA and RecA. In normal, healthy cells, LexA bonds to a part of the DNA called the SOS box. This area of the DNA codes for about 50 genes. The LexA prevents the genes in the SOS box from being expressed.

RecA floats around in the cell looking for DNA to repair. It identifies damaged DNA as that which is not in its double helix shape and bonds to it. The bonding of RecA to the DNA stimulates breaking of the LexA protein, which allows the SOS box genes to be expressed, making proteins that deal with damage (Gould 2023).

The initial cellular response is to try simple repair, appropriate for DNA damage that might involve just one nucleotide (base element) of the DNA needing to be added or replaced. If this works, more LexA is made, and the repair search stops. If the simple repair is insufficient, then the cell goes into a process of pulling together whatever nucleotides it can gather in a slapdash fashion to try to repair the DNA. This process is quite error-prone and likely to produce mutations. Because the process works by grabbing whatever nucleotides are available, it makes sense that cultures fed more nutrients are more mutation rich, i.e., have more mutagenesis.

Based on the suggestion of geneticist Dr. Miroslav Radman, this error-prone process of repairing significant DNA damage was named the SOS response, alluding to the cell responding to a stressful situation by signaling an SOS (Gould 2023; Lasker Foundation 2016; Wikipedia 2025). The SOS response was the first DNA repair response found. Dr. Stephen Elledge, Dr. Witkin's co-recipient of the Lasker Award, found DNA repair mechanisms in more complex organisms than *E. coli*, building on the work of Drs. Witkin, Radman, and others.

The repair of DNA and prevention of mutagenesis is a defense system that has implications for genetic diseases and cancer. Complete understanding of how mutations happen might permit the development of methods to prevent them or to encourage them when the mutation might be desirable. Dr. Witkin's work with the SOS response revealed that many genes are expressed when the LexA bond to the DNA is removed. Some of these genes can turn on proteins that promote cell hibernation. Others can cause the cells to form a biofilm. Still others encourage some bacteria to become antibiotic-resistant. In her original experiment, she learned that it was the sfi gene that was key. Its expression produced the sfi protein, which inhibited cell division and caused the long filaments to form (Gitshier 2012).

34.3 Conclusion

Dr. Evelyn Witkin was a microbiologist and geneticist who studied *E. coli*, a common bacterium found in the intestines of warm-blooded organisms. She isolated the first mutants that were radiation resistant. By studying these mutants, she was able to show that DNA works to repair itself, using an error-prone method if necessary. She called this response the SOS response and uncovered the biochemistry behind its operation. Dr. Witkin received the National Medal of Science in 2002.

References

Gitshier, Jane. 2012. It was Heaven: An Interview with Evelyn Witkin. *PLoS Genet* 8(10):e1003009.

Albert and Mary Lasker Foundation. 2016. *An Interview with Evelyn Witkin*. https://www.youtube.com/watch?v=ERFXATwiswQ. Accessed 6 Dec 2025.

Cold Spring Harbor Laboratory. *Evelyn Witkin*. https://www.cshl.edu/personal-collections/evelyn-witkin/. Accessed 6 Dec 2025.

Lally, Robin. 9 July 2023. Rutgers University, *Evelyn M. Witkin, World Renowned Geneticist, Dies at 102*. https://www.rutgers.edu/news/evelyn-m-witkin-world-renowned-geneticist-dies-102. Accessed 6 Dec 2025.

Risen, Clay. 13 July 2023. Evelyn M. Witkin, Who Discovered How DNA Repairs Itself, Dies at 102. *NY Times*.

Wikipedia. 9 Oct 2025. *Evelyn M. Witkin*. https://en.wikipedia.org/wiki/Evelyn_M._Witkin. Accessed 6 Dec 2025.

Volkert, Michael. 18 Sept 2023. Remembering Evelyn M. Witkin (1921 – 2023) and the SOS Response. *Proc Natl Acad Sci USA* 120(39): e2314904120.

Gould, S. 22 Dec 2023. The SOS Response: How bacteria deal with damaged DNA. *Sci Am.* https://www.scientificamerican.com/blog/lab-rat/the-sos-response-how-bacteria-deal-with-damaged-dna/#:~:text=The%20bacterial%20response%20to%20DNA,on%20when%20DNA%20damage%20occurs. Accessed 6 Dec 2025.

Lasker Foundation. 2016. *Discoveries concerning the DNA-damage response*. https://laskerfoundation.org/winners/discoveries-concerning-the-dna-damage-response/. Accessed 6 Dec 2025.

Chapter 35
Dace Viceps Madore and Maya Koster

For their work in the discovery, development, and commercialization of Prevnar, the first-ever vaccine to prevent the deadly and disabling consequences of Streptococcus pneumoniae infection in children

Keywords *Streptococcus pneumoniae* · Meningitis · Sepsis · Pneumonia · Prevnar® · Ronald Eby · Velupillai Puvanesarajah · Praxis Biologics · Lederle · Wyeth · Pfizer · Vaccines · Conjugate vaccines · Serotypes · Pneumococcal capsule · Pneumococcus · Capsular polysaccharide · PPSV · PCV · CRM197 · T cells · B cells · Hib vaccine · Haemophilus influenza type b

Dace Viceps Madore and Maya Koster (Fig. 35.1) were half of Wyeth Pharmaceutical's team presented with a National Medal of Technology and Innovation in 2005 for their invention of Prevnar®, a conjugate vaccine targeting the deadly *Streptococcus pneumoniae* bacterium, also known as pneumococcus. Pneumococcus is prevalent among children less than a year old and senior citizens over 65 years old and causes meningitis, sepsis, pneumonia, and ear infections. By bonding (conjugating) the immunization agent with a protein, Prevnar® provokes a strong immune response in youngsters with immature immune systems. Earlier pneumococcal vaccines were unable to produce a strong immune reaction in youngsters. Additionally, Prevnar® vaccines have been designed to work on multiple serotypes, or different types of pneumococcal bacteria.

J. S. Tietjen et al., *Women Laureates 1970 – 2007*, Women in Engineering and Science, https://doi.org/10.1007/978-3-032-23116-1_35

Fig. 35.1 Dr. Dace Viceps Madore (courtesy of Ryan K. Morris and the National Science and Technology Medals Foundation) and Maya Koster (courtesy of Wyeth)

35.1 Dace Viceps Madore Life (1948–)

At the end of World War II, many people had been displaced from their homes and were living in resettlement camps. Dace Viceps was born in 1948 in such a camp in Germany to Latvian parents Karlis and Velta Viceps. When she was 2 years old, the family emigrated from the camp in Bremerhaven, Germany, arriving in New York City on the USS *General Taylor*. From New York City, her family traveled to Baldwinsville, New York, a suburb of Syracuse, where they settled. There Karlis worked at the AH Pond Company, a Syracuse jewelry firm, until his retirement in 1972. In 1956, Dace's brother, also named Karlis Viceps, was born (National Science and Technology Medals Foundation; Ancestry.com; Karlis Viceps 1975; Velta Viceps Obituary 2013).

After graduating from Baker High School in 1965, Dace attended the University of Rochester, earning a bachelor's degree in biology in 1969. From there, she attended Temple University in Philadelphia, earning an MA degree in 1971 and a PhD in 1974, both in biology (Hauser 2007; Dace Viceps Madore 2014).

After completing her PhD, Dace went to work at the Wistar Institute of Anatomy and Biology in Philadelphia. The Wistar Institute was founded in 1892 as the first independent biomedical research lab in the USA. Originally focused on anatomy, it had shifted its attention to cancer, immunology, and infectious diseases. While working at the Wistar Institute, Dace met and married H. Paul Madore from Waterbury, Connecticut. Paul had earned a BS in biology from Fairfield University in Connecticut and a PhD in microbiology from the University of Pennsylvania (Madore–Viceps Vows are Heard 1975; Wistar Institute). In 1980, the union produced daughter Elizabeth, and in 1983, daughter Laura.

In 1982, Drs. Viceps Madore and Madore moved to Rochester, NY. Dr. Viceps Madore joined Praxis Biologics, a small pharmaceutical startup company that grew out of research at the University of Rochester Medical Center. The focus of the work at Praxis Biologics was the creation of a vaccine for *Streptococcus pneumoniae* (also known as pneumococcus) that would be effective on young children as well as adults. Dr. Paul Madore joined the University of Rochester School of Medicine and Dentistry in the Department of Pediatric and Adult Infectious Diseases (National Science and Technology Medals Foundation; Hauser 2007). Dr. Viceps Madore remained at work for Praxis Biologics (under its various company names) until her retirement in 2004.

After roughly 15 years of research work by a dedicated team, the first-ever vaccine focused on pneumococcus that would be highly effective on young children was released in 2000 as Prevnar®. This initial release was designed to be effective on seven different types of pneumococci. Later versions were expanded to 13 and then 20 types. Dr. Viceps Madore led the team that focused on tests to evaluate the effectiveness of the vaccine. She has commented,

> Upon reflection, working on this project was a unique opportunity. Very few people have the opportunity to be involved in a project like this from its beginning, through its evolution, and to its end. (Hauser 2007)

In 2003, Dr. Viceps Madore was part of the team that received the Heroes of Chemistry Award from the American Chemical Society for the development of Prevnar®. In 2005, she, Maya Koster, Ronald Eby, and Veluplillai Puvanesarajah received the National Medal of Technology and Innovation from President George W. Bush for the work on Prevnar® (see Fig. 35.2). Commenting on Prevnar® work, Dr. Viceps Madore has said:

> All of us loved working in the vaccine business, because we knew that we were preventing disease and saving lives. We always had this feel-good sensation because we knew we were making a difference. (Dace Viceps Madore 2014)

35.2 Maya Koster Life (1951–)

Maya Koster was born in Kishinev, Moldova, a small town in the Russian area of the country. Upon graduating from high school, she hoped to become a medical doctor, but decided that her family's connections were not adequate to help her achieve that goal. She instead pursued an undergraduate degree in the Chemistry Department of the Agricultural Institute in Kishinev and then a master's degree equivalent in the Soil Science and Agronomy Department (Koster 2025).

While at school, Maya met and married Yakov Koster. Her husband's family were Russian Jews (as were her own), and they applied for and were granted emigration papers to go to Israel. The group approved included Yakov, his parents, his two sisters, and Maya. The stated plan to travel to Israel was a ruse, however, and the family moved instead to the USA. Approval to emigrate to the USA would not have been granted by the Moldovan authorities. Maya was 3 months pregnant when the Koster family arrived in New York City, settling in Brooklyn in 1976. Daughter

Fig. 35.2 The Wyeth team with President George W. Bush (center) after the award of the National Medal of Technology and Innovation in 2005. (Photo courtesy of Ryan K. Morris and the National Science and Technology Medals Foundation). From left to right, the team members are Veluplillai Puvanesarajah, Ronald Eby, Dace Viceps Madore, and Maya Koster

Terry was born 6 months later. Two years later, in 1978, a second daughter, Adrianne, was born (Koster 2025).

The Koster family members found it very difficult to find jobs suitable to their talents given their foreign degrees. One of Yakov's sisters moved with her family to Rochester, New York, and suggested Yakov and Maya and the girls join them there. It was noted that the job opportunities would be greater in Rochester than Brooklyn, although the weather would be worse. They moved to Rochester, and Maya decided to pursue a US degree to help her find work. "Like any immigrant, I wanted to find success, and I thought a degree would help" (Hauser 2007). The University of Rochester gave Maya enough credits toward a BS in Chemistry degree that she was able to complete the degree in one and a half years, finishing in 1983 (Koster 2025).

Upon finishing her US undergraduate degree, Maya volunteered at the Biomedical Department of the University of Rochester for a couple years. At the end of that time, one of the professors in the department recommended her for a position in a new startup company—Praxis Biologics. She interviewed and got the job starting in 1986. She remained with the company and successor companies for the rest of her career (Koster 2025).

At Praxis Biologics, Koster worked with manager/mentor Ronald Eby on a pneumococcus vaccine that would be effective in babies. She was instrumental in developing the chemistry as part of the team effort that generated Prevnar7 and Prevnar13. In 1989, Praxis Biologics was merged with Lederle into Lederle Praxis. The company was next acquired by American Cyanamid, and in 1994, the company was purchased by American Home Products and merged into Wyeth Pharmaceuticals. Still later, the company was purchased by Pfizer. Regardless of the company name or owner, the long project to create an effective pneumococcal vaccine for infants was supported, and the team was permitted to remain independent (Hauser 2007).

In 2003, the Rochester facility was closed, and operations moved to the New York City area (Hauser 2007). The Koster family moved with the company.

Koster retired in 2014. She moved to Long Island, New York, to be closer to her daughter, Adrianne, and Adrianne's husband and two children. By the time of her retirement, Koster's mother and sister had immigrated to the USA.

Maya Koster received the American Cyanamid Scientific Achievement Award for her work in the development of conjugate vaccines for nine different pneumococcus types. In 2003, she was one of the team awarded the Heroes of Chemistry Award of the American Chemical Society for her work on the development of Prevnar®. In 2005, she was part of the four-person team awarded the National Medal of Technology and Innovation for Prevnar®, along with her manager/mentor Ronald Eby. Commenting on her career she has said: "It was a favor to young kids. That's why I was so happy to be working on the vaccines. I truly loved my work" (Koster 2025).

35.3 Pneumococcal Vaccination History and Prevnar®

Streptococcus pneumoniae is a common and dangerous pathogen. In the lungs, it can cause pneumonia; in the blood, bacteremia and sepsis; in the lining of the brain or spinal cord, meningitis. It is also a common pathogen causing ear and sinus infections.

Meningitis is inflammation of the lining of the brain and spinal cord. It can develop very quickly and can be caused by bacteria, fungi, or viruses. The bacterial form of meningitis is estimated to be fatal in about 15% of cases and to cause severe complications in 20% of cases (World Health Organization (a)). There are four main bacteria that can cause meningitis: meningococcus, pneumococcus, *Haemophilus influenzae*, and *Streptococcus agalactiae*. Newborns are most at risk from pneumococcus. In 2019, an estimated 236,000 people died of bacterial meningitis (World Health Organization (b)).

Bacteremia is a blood bacterial infection. If it causes systemic infection and organ dysfunction, it is classified as sepsis. Pneumococcus is a leading cause of sepsis and is reported to have a roughly 20% fatality rate (World Health Organization (c)). Globally, sepsis strikes 1.3–3.9 million patients each year, causing 400,000—700,000 deaths, accounting for nearly 15% of young infant deaths (Gan et al. 2022).

Pneumococcus is also the leading cause of community-acquired pneumonia, accounting for 15% of the cases in the USA and a larger fraction in Europe and developing countries (Dion and Ashurst 2023). Close to 800,000 children under the age of 5 die of pneumonia worldwide each year (World Health Organization (b)).

In 1891, three different groups of investigators experimented with using killed pneumococci to inoculate animals. Although they showed that this method worked, it was not until the 1910s that a similar approach was used to vaccinate humans working in gold mines in South Africa. There, a vaccine based on introduction of killed pneumococci led to a 50% drop in deaths of workers (Musher et al. 2022).

In the early 1910s, a series of researchers were able to show that the ability to vaccinate against pneumococcal disease depended on which form of the bacterium was present. These different forms were called serotypes. Figure 35.3 shows a typical pneumococcal infection. Work from 1914 to 1934 discovered that the different serotypes were determined by the pneumococci capsule, the gelatinous coating of the bacterium by a polysaccharide (a carbohydrate). Each serotype corresponded to a different polysaccharide coating, and each required a specific vaccine. That meant that rather than using pieces of the pneumococci bacterium for vaccination, it would be sufficient to vaccinate using the capsular polysaccharide (CPS) (Musher et al. 2022).

The introduction of antibiotics caused interest in pneumococcal vaccines to wane from 1940 to 1970, but Merck introduced a CPS vaccine (known as PPSV) capable of providing protection from 14 serotypes in 1977 and 23 in 1983. While these

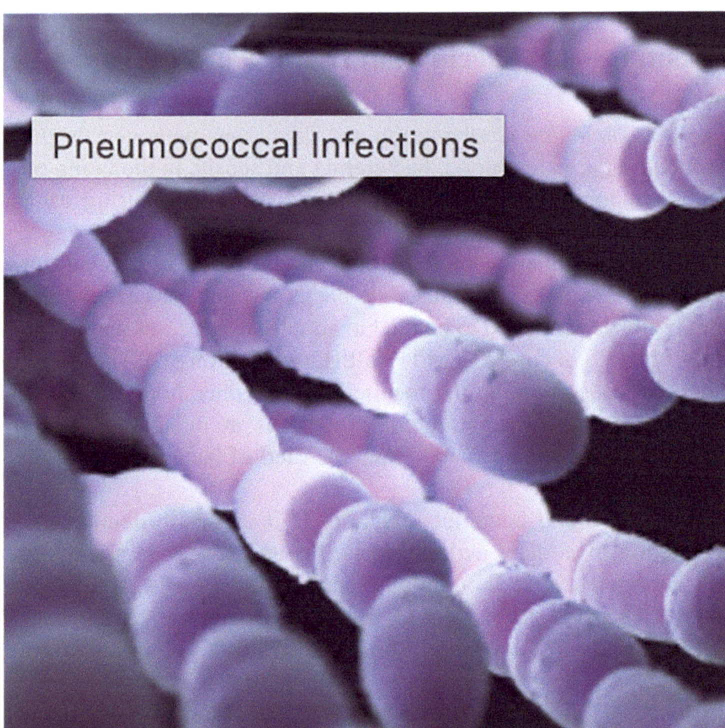

Fig. 35.3 Pneumococcus infection. (From MedlinePlus)

vaccines were very effective on adults, they were shown to be far less effective on infants and young children, a group in which pneumococcal infections were common (Musher et al. 2022).

In the early 1980s, David Smith, Richard Insel, and Porter Anderson developed the first conjugate vaccine. This work conjugated (bonded) a serotype polysaccharide to a protein. They showed that such an approach could produce a much stronger immune reaction in youngsters to *Haemophilus influenza* type b. Nonetheless, the pharmaceutical industry was skeptical, so the three researchers formed Praxis Biologics to develop and commercialize what became known as the Hib vaccine (Hauser 2007). It was this research that inspired the work on pneumococcus to see whether a similar approach could increase immune response in infants and young children.

Generally, the aim of conjugate vaccines is to pair a weak antigen (immunization agent) to a strong antigen carrier, thus prompting a stronger immune response. Antigens prompt production of two types of white blood cells in the body: T cells and B cells. T cells kill infected cells and regulate the immune response, while B cells make antibodies and remember the antigen so that they can respond if it is reintroduced. In immature immune systems, the CPS vaccines produce B cells but not T cells, so the immune response is muted. By adding the conjugate protein, T cell generation is prompted, and the response is made more vigorous and longer lasting (Wikipedia 2025).

The Hib vaccine, which was released in 1987, used CRM197 as the strong antigen carrier. CRM197 is the diphtheria toxin protein with a single amino acid substitution to render it nontoxic (Malito et al. 2012). From 1987 to 1991, the rate of Hib infection dropped 90.7% (Wikipedia 2025). Prevnar® is a pneumococcal vaccine that also uses CRM197 as the conjugating protein. It is known as a PCV vaccine and has been shown to be as effective as PPSV vaccines in adults but more effective in infants and young children (Musher et al. 2022).

Prevnar7, aimed at the seven most common serotypes of pneumococcal disease in children, was released in 2000. Prevnar13, targeting 13 serotypes, was released in 2007. Prevnar13 is estimated to have prevented over 175 million cases of pneumococcal diseases and to have prevented more than 624,000 deaths between 2010 and 2019 (Chapman et al. 2020). The impact has been especially important in sub-Saharan Africa, where epidemics of meningitis are common. As noted by Maya Koster, “We knew that would work, but we didn’t know that it would be as successful as a product as it has turned out to be” (Hauser 2007). Prevnar20 was released in 2021.

35.4 Conclusion

Dace Viceps Madore and Maya Koster were part of the team that developed the first vaccine to target pneumococcal infections in infants and young children. Their work led to the release of Prevnar®, a conjugated vaccine targeting multiple types of pneumococci. Prevnar® uses a strong antigen carrier conjugated (bonded) to the weak pneumococcal antigen to induce a strong immune reaction. Prevnar® is credited with preventing over 175 million cases of pneumococcal infection and saving more than 600,000 lives. For their work, the team developing Prevnar® was awarded the National Medal of Technology and Innovation in 2005.

References

Ancestry.com. *Africa, Asia and Europe, Passenger Lists of Displaced Persons, 1946-1971.* Database. Accessed October 5, 2025.

Chapman, Ruth, Sutton, Kelly, et al. 2020. Ten year public health impact of 13-valent pneumococcal conjugate vaccination in infants: A modelling analysis. *Vaccine* 38(45):7138-45.

Dace Viceps Madore, Class Notes, *Temple University Magazine.* Winter 2014, p42. Available at https://news.temple.edu/sites/news/files/Publications/Temple-Magazine/2014/14win.pdf. Accessed 16 Oct 2025.

Dion, Christopher and Ashurst, John. 2023. Streptococcus pneumoniae, StatPearls. https://www.ncbi.nlm.nih.gov/books/NBK470537/. Accessed 16 Oct 2025.

Gan, Ming, Lee, Wen, et al. 2022. Contemporary Trends in Global Mortality of Sepsis Among Young Infants Less Than 90 Days: A Systematic Review and Meta Analysis. *Front Pediatr.* 10:890767.

Hauser, Scott. Sept-Oct, 2007. Childhood Vaccine with Rochester Roots Recognized. *Rochester Review* Vol 70(1). Available at https://www.rochester.edu/pr/Review/V70N1/gazette2.html. Accessed 16 Oct 2025.

Karlis Viceps. 6 Dec 1975. *Syracuse Herald Journal.*

Koster, Maya. Oct 2025. Private interview.

Madore-Viceps Vows are Heard. July 14, 1975. *Syracuse Post-Standard.*

Malito, Enrico, Bursulaya, Badri, Chen, Connie, et al. 2012. Structural basis for lack of toxicity of the diphtheria toxin mutant CRM197. *Proc Natl Acad Sci* 109;14:5229-34.

Musher, Daniel, Anderson, Ronald and Feldman, Charles. 2022. The remarkable history of pneumococcal vaccination: an ongoing challenge. *Pneumonia* 14:5.

National Science and Technology Medals Foundation, Dace Viceps Madore, https://nationalmedals.org/laureate/dace-Viceps Madore/. Accessed 16 Oct 2025.

Velta Viceps Obituary. 2013. *Syracuse Post Standard.*

Wikipedia. 5 Sept 2025. Conjugate vaccine. https://en.wikipedia.org/wiki/Conjugate_vaccine, Accessed 16 Oct 2025.

Wistar Institute. https://www.wistar.org.

World Health Organization (a). Meningitis, https://www.who.int/news-room/fact-sheets/detail/meningitis. Accessed 16 Oct 2025.

World Health Organization (b). Pneumonia, https://www.who.int/health-topics/pneumonia#tab=tab_1. Accessed 16 Oct 2025.

World Health Organization (c). Sepsis, https://www.who.int/news-room/fact-sheets/detail/sepsis#:~:text=Sepsis%20is%20one%20of%20the,complication%20of%20receiving%20health%20care. Accessed 16 Oct 2025.

Chapter 36
Nina V. Fedoroff

For her pioneering work on plant molecular biology, and for her being the first to clone and characterize maize transposons. She has contributed to education and public policy pertaining to recombinant DNA and genetic modification of plants

Keywords Plant molecular biology · Transposons · Maize · Recombinant DNA · rDNA · Genetic modification of plants · DNA sequencing · Maize genetics · Barbara McClintock · Genetically engineered organisms · Genetically modified food · Maize cloning · Maize transposable elements

The daughter of Russian immigrants, Nina V. Fedoroff (Fig. 36.1), took an unconventional path to her stellar career in plant molecular biology. While a postdoc, she met future Nobel Laureate Barbara McClintock at the Cold Spring Harbor Laboratory. That meeting, the follow-up research that Dr. Fedoroff undertook and her hiring at the Carnegie Institute, allowed her to follow her conviction that she could clone maize. Dr. Fedoroff is credited with pioneering the field of plant molecular biology and is an outspoken advocate of genetically modified organisms. The recipient of numerous honors, Fedoroff received the National Medal of Science in 2006.

J. S. Tietjen et al., *Women Laureates 1970 – 2007*, Women in Engineering and Science, https://doi.org/10.1007/978-3-032-23116-1_36

Fig. 36.1 Nina Fedoroff and her maize plants. (Courtesy of Carnegie Science)

36.1 Life (1942–)

Born in Cleveland, Ohio, to a family of Russian immigrants who had been noblemen, scientists, clerics, and diplomats under the Russian czars and who spoke Russian at home, Nina Vsevolod Fedoroff was indoctrinated with the message that girls were not worth very much. She says she was stubborn, independent, and ornery—something that her mother told anybody who would listen. "I knew that I wanted to be something long before I knew what it was that I wanted to be." She took up the flute when she was a pre-teen and was determined to be a musician (Bennett 2025). By 16, she was taking college-level courses and teaching music. By 17, she was pregnant, had quit school, and had gotten married to an Air Force serviceman. Determined to finish high school despite her pregnancy and marriage, she took the New York Regents Exams on her own—graduating first in her class (Bennett 2025; Fedoroff 1996; Library of Congress 2025).

She was terrified of being stuck in low-paying jobs, a fate that befell many teenage mothers—and she says was "completely motivated by fear," and thus was convinced that she needed to continue her education. While her Air Force husband was stationed out of the country, she managed to get accepted at Syracuse University, found someone to take care of the baby, and went to college. When her husband returned and demanded that she quit college—and she refused—he beat her until she was unconscious, took all the money out of the joint account, and left. At the time, Syracuse University had a rule that offspring scholarships could be awarded—Federoff's mother was teaching Russian at the Air Force School associated with the university. But Federoff didn't technically qualify as she was married and no longer a dependent and her "family wasn't giving me a penny: I'd broken its rules for girls, and they were ashamed of me (Fedoroff 1996)." Through the good-heartedness of the financial aid officer at Syracuse University who broke this rule, Fedoroff was awarded a scholarship. She supplemented her income as needed by translating Russian science abstracts into English and teaching music (Bennett 2025; Fedoroff 1996).

Before graduating, Fedoroff left for Philadelphia with her daughter Natasha in tow to study flute with the principal flutist in the Philadelphia Orchestra. "I had as keen a love of music as I had of books, so one day I took off again (Fedoroff 1996)." She got a clerical job at an organization called Biological Abstracts—she believes she was hired because she knew Russian. There she translated, wrote abstracts, and became a key person in their Translation Bureau. Biological Abstracts, which was a life sciences company that published journals, and provided indexing and abstracting services, later became BIOSIS—and Fedoroff served on its board of trustees later in her life (Bennett 2025; Wikipedia 2025; Fedoroff 1996).

During the 2 years, Federoff was in Philadelphia, she was able to get a divorce, had a second child (a son—whom she gave up for adoption), and earned money by translating and teaching music (Bennett 2025). Fedoroff got her scholarship back at Syracuse University and returned with the intention of becoming a medical doctor. Her zoology professor, Roger Milkman, invited her into his lab, and "Once I started doing experiments, I never thought about medical school again (Bennett 2025)." With her National Science Foundation summer undergraduate research award, she spent the summer at Woods Hole Oceanographic Institution, the world's premier institution dedicated to ocean sciences, in Falmouth, Massachusetts, under Milkman's tutelage (Fedoroff 1996; Woods Hole 2025). In 1996, she graduated summa cum laude from Syracuse University with a dual major in biology and chemistry (Library of Congress 2025; Santa Fe Institute 2025).

Fedoroff got accepted to graduate school at Rockefeller University. There, the climate was not very hospitable to women. Her colleagues in Norton Zinder's lab said, "You'd better be glad you're not a girl, or we'd run you out of here!" She called it a "daft compliment of a sort." Perhaps they didn't see her as a threat because she had a daughter (Fedoroff 1996). Her advisor's wife told her, "Nina, I think that you're going to change Norton's mind about women in the lab." Nevertheless, she finished her PhD in molecular biology in 1972. In the interim, Fedoroff had gotten married again and had another child (a son, Kyr). That marriage also ended in divorce (Fedoroff 1996; Library of Congress 2025).

With a Damon Runyan Fellowship in hand, Dr. Fedoroff began her postdoc, first at the University of California-Los Angeles (1972–1974) and then, starting in 1975, at the Carnegie Institution of Washington's Embryology Department (Fedoroff 1996; Library of Congress 2025; Santa Fe Institute 2025). The field of molecular biology was very new (Bennett 2025). There, in Donald Brown's lab, she got to sequence some of the first genes ever to be sequenced (Fedoroff 1996). "I became something of a sensation overnight (Bennett 2025)."

Fig. 36.2 Nina Fedoroff (center) with Dr. Barbara McClintock (left). (Courtesy of Carnegie Science)

Invited to present her work on DNA sequencing at the Cold Spring Harbor Laboratory, she ran into a "tiny elderly woman"—whom Dr. Fedoroff immediately surmised was Dr. Barbara McClintock—a future Nobel Laureate and the first female recipient of the National Medal of Science (1970) (see Chap. 2). McClintock apologized for missing Dr. Fedoroff's lecture and invited Dr. Fedoroff into her office (Fig. 36.2) (Fedoroff 1996). "Although I had attended meetings at Cold Spring Harbor before, I had never met Barbara. That day, she invited me down to her lab, and we talked the afternoon away. We didn't talk just science, we talked about all kinds of things, and I was intrigued with her as a person. She had such an utterly unique way of thinking and talking about everything (Bennett 2025)."

Dr. McClintock's work had not been accepted in the scientific community, but she had still written up her results. Those results were published in the *Year Books* of the Carnegie Institution of Washington. Dr. Fedoroff was intrigued and read Dr. McClintock's reports. Dr. Fedoroff then decided that it would be fascinating to clone the maize gene. This would require Dr. Fedoroff to learn maize genetics, develop molecular cloning techniques for plants, and clone and characterize transposable elements, on top of needing to teach and get tenure at a university. It seemed a heavy lift given that she was a single parent (Fedoroff).

But opportunity did present itself. In 1978, Dr. Fedoroff was offered a staff position at Carnegie and a faculty position at Johns Hopkins University in the Biology Department (Library of Congress 2025; Santa Fe Institute 2025). Dr. Fedoroff made what she said was "an impulsive commitment to work out the molecular biology of maize transposable elements. Yet doing it turned out to be much harder than I'd imagined at worst." She worked—and fought—with Dr. McClintock in the early years. Dr. Fedoroff says, "McClintock's treasure box of mutants made it possible to make beautiful sense out of extraordinary molecular complexity." In the end, Dr. McClintock became a huge supporter of Dr. Fedoroff's work (Fedoroff 1996). Dr. Fedoroff became one of the first scientists to clone plant DNA at a time when "Everyone kept saying that cloning plant DNA wasn't possible" (Bennett 2025). At that time, there wasn't even a scientific discipline called plant molecular biology. Today, it is a thriving field (Bennett 2025; Fedoroff 1996). Funding from the National Science Foundation made it possible for Dr. Fedoroff to move forward (Bennett 2025). The work that she did laid the foundation for genetic engineering of plants to improve crops and create organisms with advantageous characteristics, such as pest control and drought tolerance (Bennett 2025).

In 1995, Dr. Fedoroff was named the Verne M. Willaman Professor of Life Sciences on the faculty at the Pennsylvania State University (Santa Fe Institute 2025). Two days after she joined Penn State (there had been a big spread about her joining the university in the print media), she got a call from a man who said, "My name is Jim Rathmell, and I think you're my mother." In her 1996 article, she calls this "a wonderful personal postscript" (Fedoroff 1996).

She served as Director of the Biotechnology Institute from 1995 to 2002. She organized and served as Founding Director of the Life Sciences Consortium (today the Huck Institutes of the Life Sciences). Named an Evan Pugh Professor of the Pennsylvania State University in 2002, Dr. Fedoroff became a member of the External Faculty of the Santa Fe Institute in 2003 (Santa Fe Institute 2025). From 2007 to 2010, she served as the Science and Technology Advisor to the Secretary of State and to the Administrator of the U.S. Agency for International Development. From 2011 to 2014, Dr. Fedoroff was the Distinguished Professor of Biosciences at the King Abdullah University of Science and Technology in Saudi Arabia. She retired in 2015 (Library of Congress).

Dr. Fedoroff served as editor of *Gene* from 1981 to 1984 and for *Perspectives in Biology and Medicine* from 1990 to 2001 (Library of Congress 2025). She also served on the editorial boards of *Proceedings of the National Academy of Sciences*, *Science*, and *Plant Journal*. She chaired the National Academy of Sciences (NAS) Publications Committee and served on the Council of the NAS. In addition to her board service for BIOSIS, she served on boards for the International Science Foundation, the Genetics Society of America, the American Association for the Advancement of Science (of which she was President in 2012), the National Science Board (which oversees the National Science Foundation), and Sigma-Aldrich Corporation. Fedoroff served on the International Scientific Advisory Board of the Engelhardt Institute of Molecular Biology in Moscow and the Science Steering Committee of the Santa Fe Institute (Santa Fe Institute 2025).

Dr. Fedoroff's many honors include election to the National Academy of Sciences in 1990 (National Academy of Sciences 2025). She received the University of Chicago's Howard Taylor Ricketts Award in 1990, the New York Academy of Sciences' Outstanding Contemporary Woman Scientist Award in 1992, and Sigma Xi's McGovern Science and Society Medal in 1997. She is a member of the American Academy of Sciences, the European Academy of Sciences, the American

Academy of Microbiology, and the International Academy of Food Science (Santa Fe Institute 2025). She received the National Medal of Science in 2006 (National Science & Technology Medals Foundation 2025).

Dr. Fedoroff has published 3 books and more than 170 papers in scientific journals (Santa Fe Institute 2025). Her 1992 book, published by the Cold Spring Harbor Press, is titled *The Dynamic Genome: Barbara McClintock's Ideas in the Century of Genetics* (Library of Congress). Her papers are at the Library of Congress (Library of Congress 2025; National Academy of Sciences 2025).

36.2 Recombinant DNA

In the early 1980s, Dr. Fedoroff was appointed as a member of the National Institutes of Health (NIH) Recombinant DNA Advisory Committee (RAC). This committee was convened by the NIH Director to provide advice on how recombinant DNA research should be conducted (Fedoroff 2017). Recombinant DNA is a technique whereby DNA is manipulated and analyzed such that specific DNA sequences can be isolated. Then, fragments can be cut to be spliced and placed into different cells. Recombinant DNA (rDNA) has applications in agriculture, including genetically modified plants and edible vaccines. rDNA has led to advances in medicine and environmental science in addition to agriculture (Karcher 2023).

Dr. Fedoroff says, by the time she had joined the committee, it had already published a set of guidelines for the conduct of rDNA research. And that although the RAC was purely advisory, everybody played by the rules either because to not do so would result in the loss of NIH funding but, more importantly, the loss of their colleagues' respect. The committee also was able to consider new information and alter the guidelines based on the new data presented (Fedoroff 2017).

The first requests to field test transgenic (genetically modified) plants came before the RAC in about 1983. Although Dr. Fedoroff's term on the RAC expired by 1984, she remained involved in the regulation and issues around rDNA. As a member of the Council of the National Academy of Sciences, she was involved in the production of a 1987 white paper titled "Introduction of Recombinant DNA-Engineered Organisms into the Environment—Key Issues." The conclusions of that white paper were (Fedoroff 2017):

> There was no evidence that unique hazards attend the use of rDNA techniques or the movement of genes between unrelated organisms.
>
> The risks associated with the introduction of genetically engineered organisms into the environment were the same as those associated with the introduction of unmodified organisms or organisms modified by more traditional techniques (including chemical and radiation mutagenesis).
>
> An assessment of risks of introducing genetically engineered organisms into the environment should be based on the nature of the organisms and the environment into which it is being introduced and not the method by which it was produced.

Dr. Fedoroff took every opportunity to espouse the benefits of genetically engineered organisms, including her 2004 book, *Mendel in the Kitchen: A Scientist's View of Genetically Modified Foods*, written with science writer Nancy Marie Brown and published by the Joseph Henry Press of the National Academy of Sciences (Fedoroff 2017).

The primary transgenic crops are those with the greatest acreage in production around the world—corn, soybeans, cotton, and canola, with 90% of the world's 18 million farmers being smallholder, resource poor farmers. Through 2014, genetically modified crops had reduced pesticide use by 37%, increased crop yields by 22%, and increased farmers' profits by 68%. Farmers around the world use the genetically modified crops because their yields increase and their costs decrease (Fedoroff 2017).

Dr. Fedoroff, a significant advocate for genetically modified food, told the *New York Times*,

> "There's almost no food that isn't genetically modified. Genetic modification is the basis of all evolution. Things change because our planet is subjected to a lot of radiation, which causes DNA damage, which gets repaired, but results in mutations, which create a ready mixture of plants that people can choose from to improve agriculture. In the last century, as we learned more about genes, we were able to devise ways of accelerating evolution" (National Science & Technology Medals Foundation 2025).

36.3 Conclusion

Although she did not follow a conventional path into science, Dr. Nina Fedoroff, a pioneer in the field of plant molecular biology, decided she would clone maize after meeting and getting to know future Nobel Laureate Dr. Barbara McClintock at Cold Spring Harbor Laboratory. Dr. Fedoroff laid the foundation for plant biotechnology, is an advocate for genetically modified foods, and received the National Medal of Science in 2006.

References

Bennett, Allison. 2025. https://carnegiescience.edu/news/how-nina-fedoroffs-groundbreaking-gene-research-transformed-global-agriculture. Biosphere Sciences & Engineering, Carnegie Institution for Science. March 28. Accessed December 9, 2025.

Fedoroff, Nina V. 1996. Two Women Geneticists. The American Scholar. Volume 65, Number 4, Autumn. Pp. 587-592.

Fedoroff, Nina. 2017. From Mazie Transposons to the GMO Wars. In Privalle, Laura, editor. Women in Sustainable Agriculture and Food Biotechnology: Key Advances and Perspectives on Emerging Topics. Springer.

Karcher, Susan J. 2023. DNA: recombinant technology. EBSCO Knowledge Advantage. https://www.ebsco.com/research-starters/biotechnology/dna-recombinant-technology. Accessed December 13, 2025.

Library of Congress. Nina V. Fedoroff papers, 1914-2012 (Finding Aid). https://findingaids.loc.gov/db/search/xq/searchMfer02.xq?_id=loc.mss.eadmss.ms018036&_faSection=overview&_faSubsection=bioghist&_dmdid=d19038e20. Accessed December 11, 2025.

National Academy of Sciences. Nina Fedoroff. https://www.nasonline.org/directory-entry/nina-fedoroff-mr56ru/. Accessed December 9, 2025.

National Science & Technology Medals Foundation. Nina V. Fedoroff. https://nationalmedals.org/laureate/nina-v-fedoroff/. Accessed December 9, 2025.

Santa Fe Institute. Nina Fedoroff: Science Board Fellow. https://www.santafe.edu/people/profile/nina-fedoroff. Accessed December 9, 2025.

Wikipedia. BIOSIS Previews. https://en.wikipedia.org/wiki/BIOSIS_Previews. Accessed December 16, 2025.

Woods Hole Oceanographic Institution. https://www.whoi.edu/. Accessed December 16, 2025.

Chapter 37
Rita R. Colwell

For her in-depth research that has contributed to a greater understanding of the ecology, physiology, and evolution of marine microbes, most notably Vibrio cholerae, the causative agent of pandemic cholera, which has elucidated critical links between environmental and human health

Keywords *Vibrios* · *Vibrio cholerae* · Cholera · Marine microbes · National Science Foundation · University of Maryland · Chesapeake Bay · Women in science · Women in STEM · Science education · Math education · Infectious diseases · Weather and disease · Water temperature and disease · Water salinity and disease · Plankton · Zooplankton · Copepods

Dr. Rita R. Colwell, shown in Fig. 37.1, is known for her scientific prowess and leadership ability. Her research on cholera led to conclusions that its causative bacterium occurs naturally in brackish waters and that it can go into dormancy when environmental conditions are poor but revive when the conditions improve. Dr. Colwell served as the Director of the National Science Foundation during a period in which it was able to grow and add focus programs on emerging science areas, including nanotechnology, biocomplexity, information technology, and social sciences. She is also a staunch supporter of science and math education and an advocate for women and underrepresented groups in STEM fields. Dr. Colwell received the National Medal of Science in 2006.

J. S. Tietjen et al., *Women Laureates 1970 – 2007*, Women in Engineering and Science, https://doi.org/10.1007/978-3-032-23116-1_37

Fig. 37.1 Dr. Rita R. Colwell, Director Emeritus of the National Science Foundation and distinguished professor at the University of Maryland and Johns Hopkins University. (Courtesy of Rita Colwell)

37.1 Life (1934–)

Rita Rossi was born on November 23, 1934, in Beverly, Massachusetts, to parents Louis and Louisa Rossi. She was the seventh of the eight children the Rossis had, five of whom would survive to adulthood. Louis Rossi was an Italian immigrant to the USA who worked as a stonemason and ultimately founded his own construction company. Louisa Rossi also emigrated from Italy to the USA. She had been educated only through elementary school before starting to work in the candle shop owned by her family and, once in the USA, worked in a factory in Beverly. She passed away from a heart attack when she was only 51 years old, and Rita was 16 (Colwell and McGrayne 2020).

Rita attended a four-room schoolhouse and enjoyed walking on the beach and studying rocks and shells. She also read voraciously and enjoyed basketball. In sixth grade, when IQ tests were administered to the students, Rita tested higher than anyone had in the school's history. The principal called her into the office and told her she had a responsibility to go to college (Steedman 2011; Colwell and McGrayne 2020; Morrison and Colwell 2025).

By the time Rita was considering what school to attend after high school graduation, she had spent time hearing her sister Yolanda's husband talking about his studies of physics. He and his friends seemed happy in their chosen science field and were apparently leading interesting lives, which inspired Rita to consider science despite having been told while in high school that science wasn't for women. She applied to some highly selective universities and was accepted, but their offers of financial support were inadequate. Yolanda's husband suggested Rita apply to Purdue University, where he was on leave from MIT while on a Fulbright Fellowship, and she did, landing a full scholarship inclusive of room and board as well as tuition (Schulman 2024; Morrison and Colwell 2025; Colwell and McGrayne 2020).

At Purdue, Rita started in chemistry but got discouraged. The classes were huge, and the teaching assistants had a creepy habit of trying to date her. She also noted, "It felt very peculiar wanting to be a scientist when everyone around me majored in home economics" (Steedman 2011). She persevered, joined a sorority and a debate society, and graduated with a BS in bacteriology in 1956, the same year she married Jack Colwell, a physical chemistry graduate student at Purdue. As reported by Rita, she and Jack had decided to get married after a single date (Steedman 2011; Colwell and McGrayne 2020).

Because Jack needed to finish his graduate work, Rita decided to remain at Purdue for an MS degree. She applied for fellowship support but was told by the department head that those weren't wasted on women. She asked her advisor, a genetics professor, for advice, and he offered her a fellowship in his lab. Rita accepted his offer and switched her work to genetics, graduating in 1957 with an MS degree (Fincham 2016; Wikipedia 2025).

In 1958, Rita and Jack Colwell moved to Seattle and joined the University of Washington—Jack as a faculty member and Rita as a PhD student. In addition to their studies, the couple enjoyed racing dinghies on Lake Union and skiing. Rita's first supervisor turned out to be a man who resented female academics and would therefore not waste his time advising them. After she understood that this pairing wouldn't work, Rita again shifted her research focus and joined the oceanography department, where she found a great advisor. In 1961, she became the first woman to earn a PhD from the College of Oceanography (Wikipedia 2025; Colwell and McGrayne 2020; Marmor 2000).

In 1961, the Canadian National Research Council (CNRC) initially awarded both Drs. Colwells fellowships, but a short time later, cancelled the offer to Rita as the policy at CNRC was not to fund spouses. Fortunately, Rita enlisted help and was able to obtain NSF funding to support her work and lab space at CNRC in Ottawa, so the Colwells left for Canada. Daughter Allison was born during the second year of that Canadian sojourn (Colwell and McGrayne 2020; Steedman 2011).

In 1964, Rita and Jack moved to the Washington, DC area, where Rita joined the Biology Department of Georgetown University, and Jack went to work at the National Bureau of Standards. At this point, Rita's research focused on the microbes associated with cholera. She was the first to determine that *Vibrio cholerae*, the causative agent for cholera, was found in waters that were not contaminated by fecal matter, leading her to believe that the microbe was naturally occurring. This result was in opposition to the beliefs at the time about cholera, and her results were met with great skepticism. Her response to the skeptics was to continue to collect data until the conclusion was overwhelmingly obvious. Daughter Stacie was born in 1965 (Wikipedia 2025; Colwell and McGrayne 2020).

Dr. Colwell's lab grew at Georgetown as support for her research increased, but in 1971, she was passed over for a promotion to Full Professor despite being the most productive faculty member in the department. She considered where she might move while staying in the Washington, DC, area and opted to relocate to the University of Maryland at College Park in 1972 (Wikipedia 2025).

In 1977, with her research thriving and research funding pouring in, Dr. Colwell was tapped to be the founding director of the National Sea Grant College Program at the University of Maryland. The National Sea Grant Program was established in 1966 by Congress to partner the federal government and universities to create and maintain a healthy coastal environment. This appointment made Dr. Colwell the first woman appointed director of one of the National Sea Grant programs. She served in this position for 7 years (Fincham 2016).

In 1983, Dr. Colwell became the Chief Academic Officer of the university's five-campus system, and in 1984, she was appointed by President Reagan to a 6-year term on the National Science Board, the entity that establishes policies for the National Science Foundation and advises Congress and the President. In 1985, she founded and directed the University of Maryland Biotechnology Institute, aimed at promoting basic science but connecting with private companies interested in making use of the developed inventions. Although her leadership skills were being increasingly called upon, Dr. Colwell continued to devote time to her research as well (Steedman 2011).

In 1998, Dr. Colwell was appointed the eleventh Director of the National Science Foundation (NSF), the first female among the directors of this agency and the first biologist. In her 4-year term in this position, Dr. Colwell managed to push major increases in NSF funding through Congress. She emphasized improving K-12 science and math education and increasing the participation of women and underrepresented minorities in science. She also led the establishment of major initiatives in emerging science areas, including nanotechnology; biocomplexity; information technology; and social, behavioral, and economic sciences. As part of her responsibility as Director, she managed the US Antarctic Research Program. *Colwell Massif*, a geologic formation in Antarctica, named in her honor in 1994, recognizes the work she did in polar regions (Gulf of Mexico Research Initiative n.d.).

After completing her term as NSF Director, Dr. Colwell joined Canon US Life Sciences, Inc., as a senior scientist, ultimately becoming Chairman of their board. She subsequently founded CosmosID Inc. in 2008, a company designed to leverage new tools and computational methods to provide services to analyze microorganisms in samples. Not temperamentally suited to retirement, she has most recently used her vast connections and skills to organize a consortium in support of safe drinking water throughout the world (Colwell and McGrayne 2020). As of 2025, she also continues to give public talks and to publish scientific research. Her husband, Jack, passed away in 2018.

Dr. Colwell has been a prodigious author. She has produced over 800 publications, 20 books, and an award-winning film about how microbiologists study the ocean called *Invisible Seas* (Wikipedia 2025). Her memoir is entitled *A Lab of One's Own: One Woman's Personal Journey Through Sexism in Science* and documents much of her life history (Colwell and McGrayne 2020). A passage from this book summarizes the message that Dr. Colwell regularly delivers to female students considering the obstacles to a career in a STEM field:

> …I think what's really important is for the individual girl or woman to understand that these obstacles are systemic. They're not your problem; they're not your fault. Too often as women, we are conditioned to take blame for whatever happens or goes wrong or to take responsibility when in fact it's really the system against which you are working. (Colwell and McGrayne 2020)

This advice is given along with the suggestion to persevere, find a way through obstacles, and to gather a "posse" for support and advice.

Dr. Colwell is the recipient of too many awards to list fully. Among them, she was elected to the National Academy of Sciences in 2000 and received the Central Intelligence Agency Civilian Recognition Award in 2004. In 2005, Dr. Colwell was honored with the "Order of Rising Sun, Gold and Silver Star" by the Emperor of Japan. She received the National Medal of Science in 2006 and clean water-related prizes from Sweden, Saudi Arabia, and Singapore. In 2011, she received the CRDF Global George Brown Award for International Scientific Collaboration, and in 2017, received the Vannevar Bush Award of the National Science Foundation. In 2024, Dr. Colwell received the Lifetime Achievement Award of the American Society of Microbiology. Dr. Colwell has received honorary doctorates from 63 colleges and universities. She was inducted into the National Women's Hall of Fame in 2005. She has also served as President of five professional societies (University of Maryland; Wikipedia 2025).

37.2 Cholera

Cholera is a serious disease in which bacteria colonize the small intestines of humans and cause diarrhea and vomiting that can be fatal due to the amount of body fluid lost. It is found in more than 50 countries and causes from 21,000 to 145,000 deaths per year among the 1,300,000–4,000,000 annual cases (CDC 2025). Cholera tends to appear as an epidemic and then largely disappear until the next epidemic in that area.

Bacteria are generally identifiable by their shape. The common shapes are shown in Fig. 37.2. *Vibrio cholerae* is a *Vibrio*, shown on the right of that figure, which has a comma shape. It is the causative agent of cholera. Figure 37.3 shows a typical *Vibrio cholerae* infection. When *Vibrio cholerae* enters the small intestine of a person, it attaches to the wall and releases a toxin that makes the intestine release water and minerals quickly. This lowers the blood volume of the host and makes it difficult to oxygenate the body, potentially leading to death (Colwell and McGrayne 2020).

Dr. Colwell's research on cholera produced four important breakthroughs: she showed that *Vibrio cholerae* is naturally occurring, she proved that it can go dormant when conditions are poor, she helped develop an inexpensive and easy method for filtering water where cholera is prevalent, and she developed a model of environmental conditions to predict where cholera is likely to break out.

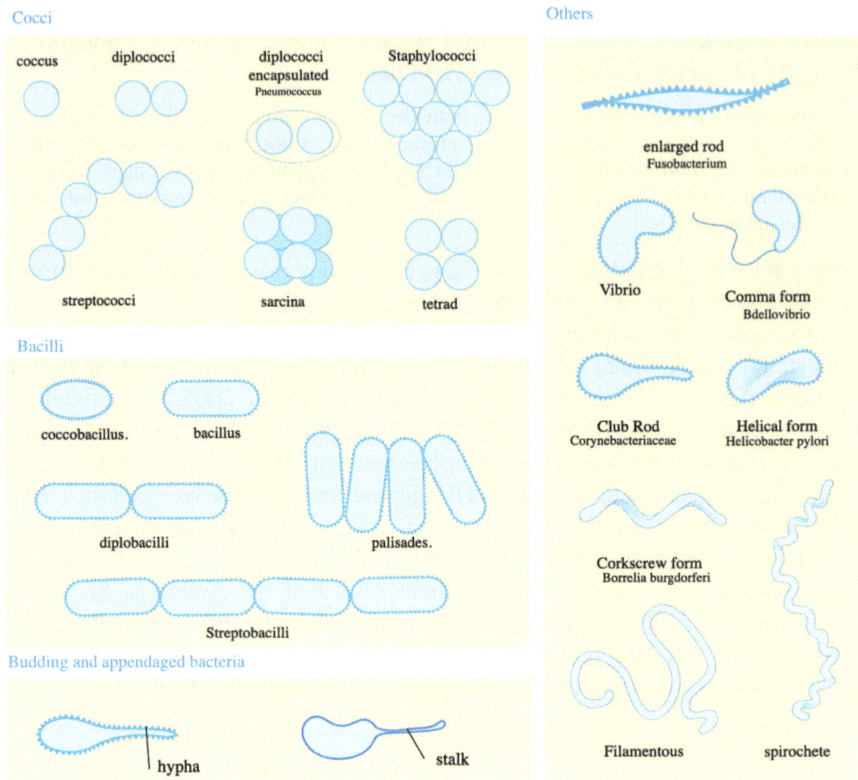

Fig. 37.2 The shapes of common bacteria that affect humans. (Ladyofhats, public domain)

Prior to the work of Dr. Colwell, it was believed that cholera was spread by exposure to infected feces in contaminated water, but this didn't explain where the bacteria went between pandemics. Dr. Colwell considered three clues she found in her lab as evidence that *Vibrio cholerae* was a naturally occurring marine bacterium. She found that *Vibrio cholerae* needed salt water to survive. She also learned that the bacteria could break down chitin, a common material found in sea animal shells but not in humans. Dr. Colwell also learned that *Vibrio cholerae* is bioluminescent, as are many marine bacteria. Taken together, these facts led Dr. Colwell to hypothesize that *Vibrio cholerae* is naturally occurring in salt water (Colwell and McGrayne 2020). Fortunately, she was living near the Chesapeake Bay, and her team showed that samples of the bay water contained *Vibrio cholerae* even when fecal matter was not present. This was met with skepticism from cholera scientists, but Dr. Colwell simply kept collecting data from various locations around the world until her conclusion was found to be overwhelmingly, obviously correct.

In her work on the Chesapeake Bay, Dr. Colwell showed that *Vibrio cholerae* can attach to copepods. She said, "We were able to show that cholera is a vector-borne disease, and the vector happens to be the copepod" (Fincham 2016). Copepods are

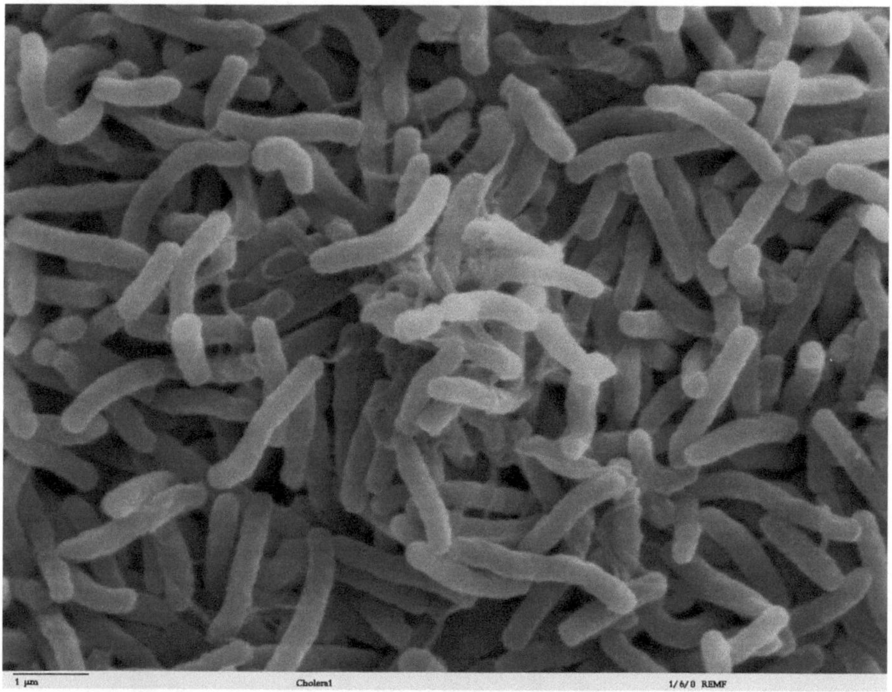

Fig. 37.3 Image of a *Vibrio cholerae* infection. (Public domain)

small crustaceans about the size of a grain of rice. They come in a variety of shapes and are illustrated in Fig. 37.4. Copepods are the largest biomass of animals on the planet and are found in fresh and salt water. They are a component of zooplankton, a community of tiny animals living in water (Colwell and McGrayne 2020).

When Dr. Colwell first published her results with copepods as a cholera vector in *Science*, it created concern that cholera might affect oysters and blue crabs in the Chesapeake. A follow-up article appeared in the *Washington Post*, as the oyster and crabbing industries were large in the area. This resulted in the immediate loss of some research funds, as they came from a federal agency aligned with the fishing industry, and they feared her work would harm the commercial interests (Fincham 2016).

Understanding that *Vibrio cholerae* attaches to copepods led Dr. Colwell to look for the bacteria in sediment in the winter, since copepods tend to burrow into sediment when it is cold. She found *Vibrio cholerae* present, but it was round rather than comma-shaped, small, and shriveled. She thought perhaps it was dead but found instead that these bacteria could be revived. She had found that they entered a dormant state. Dr. Colwell referred to these specimens as viable but not culturable (VBNC), a phrase that became common as a descriptor of such a state of being. Again, this result originally produced skepticism, as bacteria were not known to have an ability to become dormant except in a spore stage, which *Vibrio cholerae*

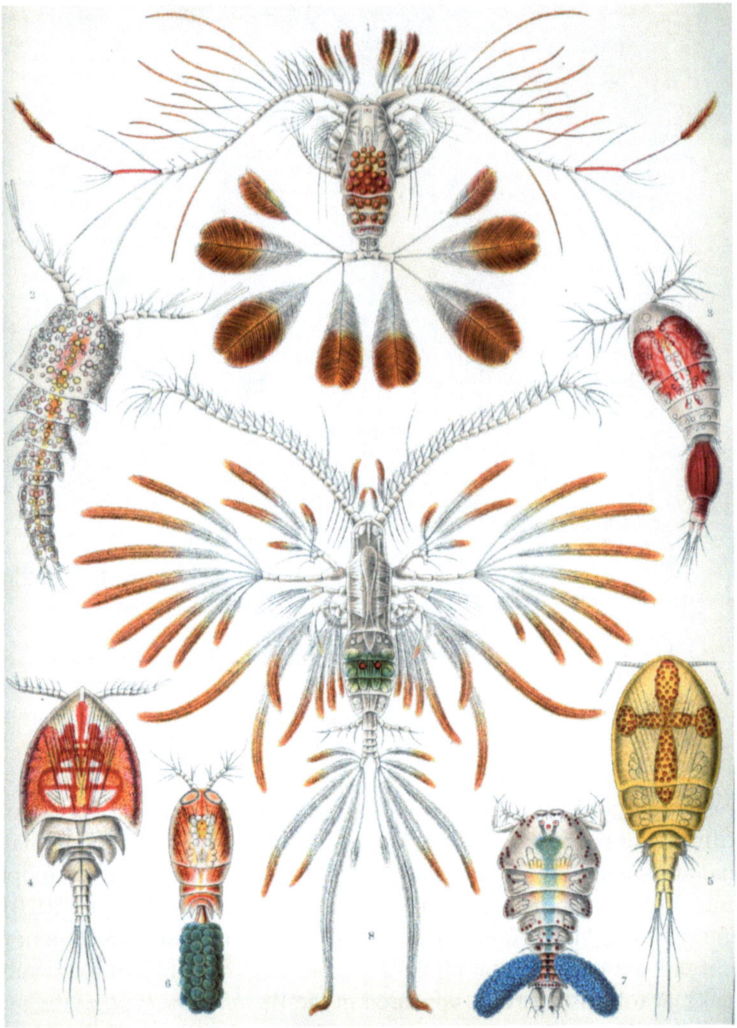

Fig. 37.4 Copepod shapes. (Ernst Haeckel, Kunst former der Natur, public domain)

does not have. This result explained where the bacteria went between outbreaks (Colwell and McGrayne 2020).

Dr. Colwell has also contemplated the impact of climate change on cholera. Warmer water and more extreme storms will result in more stirring of sediment and more nutrients being provided to the *Vibrio cholerae*. Thus, global warming could explain why the incidence of cholera has been increasing.

Knowing that copepods were the disease vector for *Vibrio cholerae*, Dr. Colwell also ran a study to find a low-cost means of straining water so that communities where cholera outbreaks were common could have safer drinking water. In one study conducted over 3 years in 65 villages in Bangladesh housing over 133,000 people, sari cloth folded over mesh screens and held over water pots as a filter reduced the incidence of cholera by nearly 50% (Wikipedia 2025). The Gulf Research Initiative referred to this work as Dr. Colwell using "very high-tech analysis to produce extremely low-tech solutions" (Gulf of Mexico Research Initiative n.d.).

Over the years, Dr. Colwell studied a variety of measures of bodies of water and searched for patterns relating them to cholera outbreaks. Aided by satellite data, she considered weather patterns, surface water temperatures, chlorophyll concentrations, sea surface heights, and rainfall patterns. She found that the cholera infection rate correlates well with water surface temperature. As temperatures increase, algae blooms and it hosts the bacteria, whose population increases. Rainfall and extreme weather aid in spreading the cholera. Using these studies, her team has developed a computer model capable of predicting cholera outbreaks before they happen. This permits public health efforts to be mobilized effectively and efficiently (Colwell and McGrayne 2020; Wikipedia 2025).

37.3 Conclusion

Dr. Rita Colwell is a talented microbiologist, leader, and entrepreneur. Her work on cholera has shown that its causative agent is naturally occurring and that it is capable of going dormant and reviving when conditions improve. She has devised a simple and inexpensive way to reduce cholera by straining drinking water with a mesh wrapped in sari cloth and has developed a computer model capable of predicting where cholera is likely to break out so that public health resources can be mobilized. Dr. Colwell has held many leadership positions, including as the director of the National Science Foundation. She has also been a strong advocate for women in science, engineering, and math.

References

CDC. 29 May 2025. *Cholera*. https://www.cdc.gov/cholera/about/index.html#:~:text=What%20it%20is,ill%20patients%20can%20survive%20cholera. Accessed 12 Dec 2025.

Colwell, Rita and McGrayne, Sharon. 2020. *A Lab of One's Own: One Woman's Personal Journey Through Sexism in Science*. Simon and Schuster.

Fincham, Michael. 12 April 2016. *The Woman Who Launched Maryland Sea Grant,* Sea Grant Maryland. https://www.mdsg.umd.edu/onthebay-blog/woman-who-launched-maryland-sea-grant. Accessed 12 Dec 2025.

Gulf of Mexico Research Initiative, n.d. *Rita R. Colwell, PhD*. https://gulfresearchinitiative.org/gri-research-board/board-roster/rita-r-colwell-phd/. Accessed 12 Dec 2025.

Marmor, John. Jun 2000. *Rita Colwell, '61, honored for a career spent breaking barriers,* University of Washington Magazine. https://magazine.washington.edu/feature/rita-colwell-61-honored-for-a-career-spent-breaking-barriers/#gsc.tab=0. Accessed 12 Dec 2025.

Morrison, Lydia and Colwell, Rita. 2025. *Interview with Dr. Rita Colwell: A lifetime in science*, New England Biolabs. https://www.neb.com/en-us/podcasts/podcast-67-interview-with-rita-colwell. Accessed 12 Dec 2025.

Schulman, Madison. 26 Sept 2024. *Scientific Trailblazer Dr. Rita Colwell Shares Renowned Research*, Endicott College. https://www.endicott.edu/news-events/news/news-articles/2024/09/scientific-trailblazer-dr-rita-colwell-shares-renowned-research. Accessed 12 Dec 2025.

Steedman, Emily. 2011. *Rita R. Colwell, Ph.D. (1934 -)*, Archives of Maryland. https://msa.maryland.gov/megafile/msa/speccol/sc3500/sc3520/011500/011592/html/11592bio.html. Accessed 12 Dec 2025.

University of Maryland. *Rita Colwell*, Department of Cell Biology and Molecular Genetics. https://cbmg.umd.edu/people/rita-colwell. Accessed 12 Dec 2025.

Wikipedia. 9 Dec 2025. *Rita R. Colwell*. https://en.wikipedia.org/wiki/Rita_R._Colwell. Accessed 12 Dec 2025.

Chapter 38
Fay Ajzenberg-Selove

For her pioneering contributions in nuclear physics that have advanced research into many applications, including energy generation from fusion, dating of artifacts, and nuclear medicine; her passion for teaching; and her outstanding service to her profession

Keywords Light elements · Nuclei · Nuclear spectroscopy · Nuclear physics · Cosmic rays · Nuclear structure · Nuclear energy levels · Nuclear decay · Isotopes · Excitation states

Dr. Fay Ajzenberg-Selove (Fig. 38.1) became characterized as the nerve center for the entire world's understanding and quantification of the behavior of light elements (those lighter than magnesium). Her work is considered one of the most referenced and most useful data available in the nuclear physics community. Her research in nuclear spectroscopy and the energy levels of light nuclei—both in absorbing and emitting energy—is considered as providing the groundwork needed for fusion, carbon dating, and the use of light nuclei in nuclear medicine. She received the National Medal of Science in 2007.

J. S. Tietjen et al., *Women Laureates 1970 – 2007*, Women in Engineering and Science, https://doi.org/10.1007/978-3-032-23116-1_38

Fig. 38.1 Fay Ajzenberg-Selove. (Courtesy of AIP Emilio Segrè Visual Archives, Physics Today Collection)

38.1 Life (1926–2012)

Born in Berlin in 1926 to Polish/Russian parents—an engineer father and stay-at-home mother—Fay Ajzenberg's family moved to Paris when she was 4 years old (Lubkin 2013). She was an avid reader and assumed she would grow up to be an engineer, like her father. The Nazi invasion of France in 1940 caused her family to flee France. After a dangerous journey, they arrived in Lisbon, Portugal. A stop in the Dominican Republic was followed by internment at Ellis Island in New York City. Needing visas, they went to Havana, Cuba, before being allowed entry into the USA. The family settled in New York City in 1941 when Fay was 15 (Shalvi 2025).

After high school in New York City, Fay matriculated at the University of Michigan as the only woman in the engineering entering class—with 100 men. There, she discovered engineering was not to her liking. She preferred physics. Although she had a difficult time with the coursework, Fay graduated with a BS in engineering physics in 1946 and returned to New York City (Physics Today 2018; Shalvi 2025). There, despite an abysmal score on the physics GRE, she took physics classes at Columbia University, many of which she failed, and took a part-time job teaching physics at Hunter College, where she was mentored by future Nobel Laureate and National Medal of Science awardee Rosalyn Yalow (Howes and Herzenberg 2015). Fay became a US citizen in December 1946 (Howes and Herzenberg 2015). After returning from a 1947 summer visit to Paris, she spent the 1947–1948 academic year at the Navy Pier in Chicago, teaching introductory physics and mathematics courses at a community college to 150 returning veterans (Ajzenberg-Selove 1994; Waggoner 2007).

Fay matriculated at the University of Wisconsin in 1948 after summer travel with her mother to Europe. Her mission was to get her master's in physics in 1 year—and she succeeded, she earned her MS in 1949. A visit to the cosmic ray observatory in the Swiss Alps confirmed her interest in nuclear physics, and she went back to Wisconsin to pursue her doctorate, earning her PhD in 1952. Her dissertation was titled "Energy Levels of Some Light Nuclei and Their Classification" (Howes and Herzenberg 2015; Lubkin 2013; Physics Today 2018; Shalvi 2025).

Finding an academic job in physics was not easy for a woman, and Dr. Ajzenberg would deal with discrimination throughout much of her career. She applied to Bryn Mawr and Smith College and accepted a 1-year position at Smith College for the fall of 1952. In the interim, she had written to Tom Lauritsen at Caltech and asked if he would hire her for the summer to rewrite an article he had written 2 years earlier. That article summarized the work done in the field of nuclear spectroscopy on the so-called light nuclei (those lighter than magnesium). Lauritsen hired her. She characterizes that summer as determining her future. One small problem did have to be overcome at Caltech. Lauritsen forgot to tell the dean that F. Ajzenberg was a woman; she became the first woman appointed to the physics department there (Ajzenberg-Selove 1994; Lubkin 2013).

Dr. Ajzenberg-Selove worked together with Lauritsen until his death in 1973. She became the specialist in producing a nuclear bible each year summarizing the work in nuclear spectroscopy of light nuclei and providing data on nuclear structure and decay of the nuclei lighter than magnesium. She published in this field for 38 years, critically evaluating the experimental results on the structure of light nuclei. She and Lauritsen collaborated on this work until he died. Thereafter, she continued until 1990. During that time, she published 26 reviews—first in *Review of Modern Physics* and, starting in 1959, in *Nuclear Physics*. In a typical year, 1400 papers were published. Her 1979 and 1992 editions were called citation classics by the Institute for Scientific Information (Lubkin 2013; Urban-Klaehn 2025).

During her time at Smith, Dr. Ajzenberg-Selove also obtained a visiting research fellow position at the Massachusetts Institute of Technology (MIT), performing research at the MIT Van de Graff laboratory, commuting the 90 miles between the locations with an apartment at both locations (Howes and Herzenberg 2015; Shalvi 2025). After her year at Smith was concluded, she was named an assistant professor at Boston University, where she stayed until 1956. That position came with an initial salary offer that was 15% lower than the salary offered to men. She refused to accept the offer until the original salary was restored (Howes and Herzenberg 2015; Urban-Klaehn 2025).

In 1954, friends decided it was time for her to get married and declared they had found a mate for her. Skeptical, she went to hear him lecture, fell in love, pursued Walter (Wally) Selove, and they were married in 1955 (Howe 2015; Shalvi 2025). At the time, Selove was an experimental physicist at Harvard University (Shalvi 2025). Unable to have children, the marriage became mutually supportive and dual career. Wally Selove discovered a particle he named for her—the f-zero or "faon" (Shalvi 2025).

In 1956, the couple both accepted positions in Pennsylvania. Dr. Ajzenberg-Selove became an associate professor at Haverford College, where she was the first tenured woman faculty member. She loved the students and her job, where she conducted research on documenting the properties of nuclei with bright undergraduate students, became involved with the Federation of American Scientists, was promoted to full professor (1962), and served as acting chair of the physics department (Howes and Herzenberg 2015; Lubkin 2013; Shalvi 2025). She began traveling to international physics meetings in 1957. Awarded a Guggenheim Fellowship in 1964, she used the money for a year of research at the Lawrence Berkeley National Laboratory (Shalvi 2025).

In 1969, a new President took the helm at Haverford. She and he had a disagreement about his approach to a tenure decision in the physics department, where she was acting chair. As a result of this disagreement, Dr. Ajzenberg-Selove took a sabbatical, which she was due, and found a different position. She accepted an untenured position as a research professor at the University of Pennsylvania (where her husband was employed) as of the beginning of 1970. In 1972, when the university posted three openings for tenured physicists, she applied but was denied. After the successful resolution of the complaints she filed with the Federal Equal Employment Opportunity Commission and the Pennsylvania Human Relations Commission, she was made a tenured professor in physics in 1973. She was the second female professor in the School of Arts and Sciences at the University of Pennsylvania (Shalvi 2025).

Dr. Ajzenberg-Selove's first bout of cancer occurred in 1970—breast cancer; the second bout was in 1982. By 1988, the side effects of the spread of the cancer and the chemotherapy caused her to stop teaching, but she continued to do research. In 1990, the cancer, a collision with a motorcyclist, and concerns about Wally's health, led her to cease research but return to teaching. At her death, she was a professor of physics emerita and had been advocating for women in science for more than 20 years (DeBakcsy 2023; Lubkin 2013; Shalvi 2025).

Dr. Ajzenberg-Selove organized the first meeting on "Women in Physics" during the American Physical Society's (APS) 1971 annual meeting. Six hundred people showed up for the meeting, almost all of them men. Due to the success of this effort, a Committee on the Status of Women in Physics was formed by the APS (Lubkin 2013; Shalvi 2025).

Fluent in French, Russian, and German in addition to English, Dr. Ajzenberg-Selove was the first female member of an Atomic Energy Commission delegation to the Soviet Union. She helped found the APS nuclear division and served as chair from 1973 to 1974. Instrumental in the founding of the Department of Energy/NSF Nuclear Science Advisory Committee, she was also active in the International Union of Pure and Applied Physics. Dr. Ajzenberg-Selove served as chair of the Union's commission on nuclear physics from 1978 to 1981 (Ayala 2025; Lubkin 2013).

In 1999, she received the APS Nicholson Medal for Humanitarian Service (Lubkin 2013). She was a Fellow of the APS and the American Association for the Advancement of Science (CWP 1997). Dr. Ajzenberg-Selove received the National Medal of Science in 2007 (Ayala 2025).

38.2 Research Areas

Spectroscopy is a scientific method that examines objects and materials based on the understanding that light and matter interact in specific ways. In addition, it takes advantage of the fact that objects and materials emit unique patterns of colors (wavelengths). Spectroscopy is used in many areas, including astronomy, materials science, Earth science, medicine, forensics, national security, and food safety (Webb a 2022). Light, or electromagnetic radiation, includes gamma rays, X-rays, ultraviolet light, visible light, infrared light, microwaves, and radio waves. Light behaves like waves, and wavelength can be measured. Light interacts with matter, and measuring the wavelength provides significant information about that matter (Webb b 2022).

Dr. Ajzenberg-Selove specialized in an area of nuclear physics called nuclear spectroscopy. This area of science uses the nucleus of elements to determine the properties of materials. Radiation is either absorbed or emitted from the nucleus, and then the structure of the material can be obtained. Specific techniques used for these purposes include, among others, gamma spectroscopy, hypernuclear spectroscopy, and muon spin spectroscopy (Wikipedia 2025).

Dr. Ajzenberg-Selove found her scientific niche when she was at Caltech—the study of the light elements (those lighter than magnesium). Writing a periodic review of the work done on light nuclei for the next 38 years required reading at first hundreds, later thousands, of papers that announced new measurements for different excitation states of various isotopes of the light nuclei. She used her knowledge of the results—and the reputation and experience of the labs themselves—to ascertain the validity of the studies (DeBakcsy 2023).

She recorded the known energy levels and updated the information as new data came in. "She was the nerve center for the entire world's efforts in quantifying and understanding the behavior of the light elements, and she would remain so for the next thirty-eight years." The first Lauritsen-Ajzenberg review was 80 pages. Later reviews would be hundreds of pages. She later estimated that her work on the light elements alone was 5000 pages "of the most referenced and useful information available to the nuclear physics community" (DeBakcsy 2023).

As she did her review work, she often found holes in the data and devoted her research to designing and running experiments to fill in those holes. Often, she had to use the equipment of others as the institutions where she was housed didn't have the facilities needed. She partnered with people she had met at institutions with accelerators, and together they would work on the problem; she would arrive with the element to be examined (the target). This included facilities at MIT, Princeton University, Indiana University, and Brookhaven, among others (DeBakcsy 2023; Lubkin 2013).

Her research in nuclear spectroscopy and the energy levels of light nuclei—both in absorbing and emitting energy—is considered as having provided the groundwork needed for fusion, carbon dating, and the use of light nuclei in nuclear medicine.

38.3 Conclusion

Nuclear physicist Dr. Fay Ajzenberg-Selove always found a way—just as she and her family had in making their way out of Germany and just as she did in navigating through her three degrees and battling discrimination in her career. Her work was seminal to the nuclear physics community, leading to many applications today across a broad range of fields.

References

Ajzenberg-Selove, Fay. 1994. A Matter of Choices: Memoirs of a Female Physicist. New Brunswick, New Jersey: Rutgers University Press.

Ayala, Melissa. National Science & Technology Medals Foundation. Fay Ajzenberg-Selove. https://nationalmedals.org/laureate/fay-ajzenberg-selove/. Accessed June 8, 2025.

CWP at UCLA. Fay Ajzenberg-Selove. 1997. http://cwp.library.ucla.edu/Phase2/Ajzenberg_Selove,Fay@88123456.html. April 30.

DeBakcsy, Dale. 2023. Corralling the Light Elements: The Nuclear Spectroscopy of Fay Ajzenberg-Selove. https://www.wisarchive.com/post/corralling-the-light-elements-the-nuclear-spectroscopy-of-fay-ajzenberg-seloveThe Women in Science Archive. June 1. Accessed June 18, 2025.

Howes, Ruth H. and Caroline L. Herzenberg. After the War: Women in Physics in the United States. 2015. San Rafael, California: Morgan & Claypool Publishers.

Lubkin, Gloria B. Fay Ajzenberg-Selove. 2013. https://pubs.aip.org/physicstoday/article/66/6/62/414485/Fay-Ajzenberg-Selove. Accessed June 18, 2025.

Physics Today. 2018/ https://pubs.aip.org/physicstoday/online/5915/Fay-Ajzenberg-Selove. February 13. Accessed June 8, 2025.

Shalvi, Alice. Fay Ajzenberg-Selove. Jewish Women Archives: Sharing Stories Inspiring Change. https://jwa.org/encyclopedia/article/ajzenberg-selove-fay. Accessed June 8, 2025.

Urban-Klaehn, Jagoda. 2025. "Women in Nuclear History" series #10 Dr. Fay Ajzenberg-Selove – author of "nuclear bible" for nuclear spectroscopy. U.S. Women in Nuclear. https://www.winus.org/the-winning-edge/women-in-nuclear-history-series-10-dr-fay-ajzenberg-selove-author-of-nuclear-bible-for-nuclear-spectroscopy/. January 3.

Waggoner, Kate. 2007. Women's History in Michigan Science and Engineering Oral History Project – Recordings – Ajzenberg-Selove, Fay. https://deepblue.lib.umich.edu/handle/2027.42/83521. And https://public.websites.umich.edu/~whimse/selove.html. January 29.

Webb a. Webb telescope. 2022. Spectroscopy 101 – Introduction. https://webbtelescope.org/contents/articles/spectroscopy-101%2D%2Dintroduction. July 7. Accessed June 21, 2025.

Webb b. Webb telescope. 2022. Spectroscopy 101 – Light and Matter. https://webbtelescope.org/contents/articles/spectroscopy-101%2D%2Dlight-and-matter. July 7. Accessed June 21, 2025.

Wikipedia. Nuclear spectroscopy. https://en.wikipedia.org/wiki/Nuclear_spectroscopy. Accessed June 19, 2025.

Index

J. S. Tietjen et al., *Women Laureates 1970 – 2007*, Women in Engineering and Science, https://doi.org/10.1007/978-3-032-23116-1

MIX
Papier aus verantwortungsvollen Quellen
Paper from responsible sources
FSC® C105338

FSC
www.fsc.org

If you have any concerns about our products,
you can contact us on
ProductSafety@springernature.com

In case Publisher is established outside the EU,
the EU authorized representative is:
Springer Nature Customer Service Center GmbH
Europaplatz 3, 69115 Heidelberg, Germany

Printed by Libri Plureos GmbH
in Hamburg, Germany